AF531498

Fisheries Biology, Aquaculture and Post-Harvest Management

NIPA® GENX ELECTRONIC RESOURCES & SOLUTIONS P. LTD.
New Delhi-110 034

About the Editors

Dr. Sudhan Chandran is working at Fisheries College and Research Institute, Tamil Nadu. Dr. J. Jayalalithaa Fisheries University, as Assistant Professor in Department of Fisheries Biology and Resource Management. He is teaching Fisheries Biology and Resource Management related course to under graduate students. He is a recipient of ICAR-CIFE fellowship for Doctor of Philosophy in Fisheries Resource Management at ICAR-Central Institute of Fisheries Education, Mumbai. To his credit, he published several research papers in national and international journals including 5 years of experience in research and 3 years in teaching programmes.

Th. Ranjithkumar Kolanjinathan is working at State Fisheries Department, Cuddalore District, Tamil Nadu as Inspector of Fisheries. He is a receipt of ICAR-CIFE fellowship for Master of Fisheries Science in Fish Biotechnology at ICAR-Central Institute Fisheries Education, Mumbai. To his credit, he published several research papers in national and international journals including 3 years of experience in research and 3 years in fish marketing and management programmes.

Fisheries Biology, Aquaculture and Post-Harvest Management
Volume 1

Sudhan C
Fisheries College and Research Institute
Tamil Nadu Dr. J. Jayalalithaa Fisheries University
Thoothukudi, Tamil Nadu, India

Ranjithkumar K
State Fisheries Department
Cuddalore, Tamil Nadu, India

NIPA® GENX ELECTRONIC RESOURCES & SOLUTIONS P. LTD.
New Delhi-110 034

NIPA, GENX ELECTRONIC RESOURCES & SOLUTIONS P. LTD.

101,103, Vikas Surya Plaza, CU Block
L.S.C.Market, Pitam Pura, New Delhi-110 034
Ph : +91 11 27341616, 27341717, 27341718
E-mail:newindiapublishingagency@gmail.com
www: www.nipabooks.com

For customer assistance, please contact
Phone: + 91-11-27 34 17 17
Fax: + 91-11-27 34 16 16
E-Mail: feedbacks@nipabooks.com

ISBN: 978-81-19072-07-1

Composed and Designed by NIPA.

A. Antony Xavier

मात्स्यिकी विकास आयुक्त
भारत सरकार
मत्स्यपालन, पशु एवं डेरी मंत्रालय
मत्स्यपालन विभाग
कृषि भवन, नई दिल्ली - 110001

Fisheries Development Commissioner
Government of India
Ministry of Fisheries, Animal Husbandry & Dairying
Department of Fisheries
Krishi Bhawan, New Delhi-110001

Foreword

Fisheries and Aquaculture have become an inevitable source of nutritional security for the human population with high quality of protein at cheaper cost. India also has a thriving seafood export market comprising of both captured and cultured aquatic food, valued at US$ 7 billion per year pegged at about 5 percent of the global sea food trade. India has shown continuous and sustained increments in shrimp production from the level of 75,000 MT during 2008-09 to a record 11.84 lakh tons (provisional figures) in 2022-23. India's seafood exports doubled from Rs. 30,213 crore in 2013-14 to Rs. 63,969 crore in 2022-23 with shrimp contributing the lion's share of exports i.e. Rs. 43,135 crore.

It is with great pleasure and honour, I introduce this edited volume on "**Fisheries Biology, Aquaculture and Post-Harvest Management**." This comprehensive work represents a significant milestone in the field of fisheries and aquaculture, providing a valuable resource for researchers, students, and other stakeholders alike. It is a testament to the collaborative spirit and dedication of the contributors, who have generously shared their knowledge and insights. The contributions of a diverse group of experts from around the world offer a broad and balanced perspective on the subject matter, encompassing a range of topics.

This book addresses these pressing concerns by diving deep into various aspects of fisheries biology, aquaculture, and post-harvest management that include the latest advancements in fish biology, innovative aquaculture techniques, and effective post-harvest management strategies with in-depth explorations of the biology and ecology of various fish species, their life cycles, behaviour, and habitats. Cutting-edge research and practices in aquaculture, from breeding and husbandry to disease management and sustainability.

Room No. 242-C, Krishi Bhawan, New Delhi-110001, Tel.: (O): +91-11-23386379, Mob.: +91 9894735385
E-mail: fdc-india@dof.gov.in, antonyxavier@caa.gov.in

Comprehensive guidance on post-harvest management, including fish handling, processing, and marketing, to reduce waste and improve the overall quality of seafood products.

I express my sincere gratitude and appreciation to the contributors and the editor of this book for their dedication in advancing our understanding of fisheries biology, aquaculture and post-harvest management.

A. Antony Xavier

Room No. 242-C, Krishi Bhawan, New Delhi-110001, Tel.: (O): +91-11-23386379, Mob.: +91 9894735385
E-mail: fdc-india@dof.gov.in, antonyxavier@caa gov.in

Preface

In a world where sustainability, food security, and environmental conservation have emerged as foremost concerns, aquaculture and fisheries are essential to providing worldwide demand for aquatic foods. The diverse and dynamic field of aquaculture and fisheries science has witnessed significant advancements and innovations in recent years, making it imperative to compile and disseminate the latest knowledge in this domain. The book you are holding is a result of the teamwork of specialists and researchers who have committed their careers to expanding the boundaries of aquaculture and fisheries. Readers will gain a full grasp of the many sides of aquaculture and fisheries from biological processes to economics, from technology to nutrition, by reading this extensive text, which is divided into five separate sections. Section 1: Aquaculture and Fisheries - This section passes into the principles and practices of aquaculture, examining the latest techniques and strategies employed to ensure sustainable production of aquatic species. The chapters encompass a wide range of topics, from breeding and rearing to water quality management and disease control. Section 2: Fisheries Biology, Resources, and Post-Harvest Management - In this theme, the focus shifts to the biology of fish species, the management of fisheries resources, and the crucial post-harvest processes that determine the quality and safety of seafood products. Section 3: Fish Processing, Fisheries Engineering, and Fishing Technology - Cutting-edge technologies, engineering innovations, and the processing methods that transform raw fish into market-ready products are the subjects of exploration in this theme. The chapters in this section encompass an array of state-of-the-art techniques in fish processing and preservation, as well as the technologies employed in fisheries and their engineering aspects. Section 4: Fisheries Economics, Extension, and Statistics - Economics is an integral part of any industry, including aquaculture and fisheries. This theme delves into the economic aspects of the field, along with extension services that facilitate knowledge dissemination and statistical methods used for informed decision-making. Section 5: Fish Pathology, Genetics, Biotechnology, and Nutrition - The final theme focuses on the health of aquatic organisms, exploring fish pathology, genetics, biotechnology, and the essential role of nutrition in aquaculture. It underscores the critical interplay between genetics, health, and nutritional management.

This book is the outcome of a various collaboration that draws on the knowledge of scholars as well as professionals from various backgrounds. We are thankful to the contributors who volunteered their expertise and experiences so as to provide this book a comprehensive and up-to-date resource for students, researchers, and professionals working in aquaculture and fisheries. As editors, it has been our privilege to compile this invaluable collection of knowledge, and we hope that readers find it both informative and inspiring. It is our belief that the insights within these pages will contribute to the sustainable growth of the aquaculture and fisheries sectors, and ultimately, to the well-being of our country. We express our heartfelt gratitude to all the authors, reviewers, and supporters who have made this publication possible. We invite readers to embark on a journey through the wealth of knowledge contained within these pages and to join us in our commitment to advancing the fields of aquaculture and fisheries.

Editors

Contents

Foreword v
Preface vii

Section 1: Aquaculture and Fisheries

1. Epigenetics in Aquaculture and Fisheries 1
   Kolanjinathan Ranjithkumar, Rajendiran Rajeshkannan, Chandran Sudhan, D. Kaviarasu and R. Bavithra
2. Integrated Multi-Trophic Aquaculture: A Balanced System for Sustainable Aquaculture 19
   Kalidoss Radhakrishnan, P. Dheeran, Rajendiran Rajeshkannan, P. Seenivasan, N. Raghuvaran, Radhakrishnan Naveenkumar, Abhilash Thapa and Kanagaraja Anantharaja
3. COVID-19 Pandemic - Impact on Fisheries and Aquaculture 55
   Hari Prasad Mohale and Sanjay Chandravanshi
4. Nutraceuticals as Functional Feed Additives: Roles and Scope in Aquaculture 63
   Raghuvaran N.

Section 2: Fisheries Biology, Resources and Post-Harvest Management

5. Jellyfish Apocalypse - An Opportunity? 89
   Meenatchi S., Abuthagir Iburahim S, Vidhya V., Kamalii A. and Ulaganathan Arisekar
6. Cybertaxonomy: A New Biodiversity Science Tool 105
   Sanjay Chandravanshi, Sudhan Chandran, Hari Prasad Mohale, Sahil Rai and H.S. Mogalekar
7. Delineation of Stocks through Integrated Approaches for Management and Conservation of Fishes 111
   Dhanlakshmi M., Shivkumar, Rinkesh N. Wanjari and Harshavarthini M.

8. Current Biodiversity Status of the Chilika Lake and Their Sustainable Management 129
Sanjay Chandravanshi, Rishikesh Venkatrao Kadam, Koshalya Rohidas, Sudhan Chandran and Abinaya R.

9. Integrative Taxonomical Approach for Portunid Crab Species Identification 137
Preetysh Nanda Patnaik, Adyasha Sahu, Sudhan C., Venkataramani V.K., Jayakumar N., and R. Durairaja

10. Need of the Hour: To Conserve Portunid Crab Along The Tamil Nadu Coast 157
Adyasha Sahu, V.K. Venkataramani, N. Jayakumar, R. Durairaja Sudhan C., and Preetysh Nanda Patnaik

Section-3: Fish Processing, Fisheries Engineering and Fishing Technology

11. Pathogenicity of Salmonella and Their Impact in Seafood Rejection.. 181
Thamizhselvan Surya, Balasubramanian Sivaraman, Geevaretnam Jeyasekaran, Robinson Jeyashakila and Shanmugam Sundhar

12. Overview of Surimi Technology and Its Challenges and Opportunities 197
Mohammed Akram Javith S.

13. Nucleic Acid Amplification: Alternative Methods of Polymerase Chain Reaction 211
Nahida Quyoom, Rajendiran Rajeshkannan, A. Pradeep and Darshan Pawaskar

14. Recent Trends in Packaging 229
Angela Brighty R.J

Section 4: Fisheries Economics, Extension and Statistics

15. Climate Change Vulnerability in Fisheries: Impacts and Adaptation Strategies 245
Suvetha V.

16. Fisheries Technologies and Its Adoption 253
Ankush Kamble, Jayapratha T., Rujan J. and Shivaji Argade

17. Management Implications for Hokersar Wetland: A Ramsar Site in Jammu and Kashmir 263
Regu Atufa, Nadeem Qadri, Saba N. Reshi and Sadaf Gul

18. Management Options for Sustainable Exploitation of Fishes on the Coast of Gulf of Mannar of Tamil Nadu, India 271
N. Neethiselvan, S. Archana, J. Amala Shajeeva and D. Sundareswari

19. Role of NGOs in Fisheries Development 285
J. Rujan, Ankush L. Kamble & P. Seenivasan

Section 5: Fish Pathology, Genetics Biotechnology and Nutrition

20. Organ on A Chip 299
Rajendiran Rajeshkannan

21. Bacterial Diseases of Shrimp 343
Manimozhi E. and Porkodi M.

22. Types of Markers and Their Application in Fisheries 353
Sanjay Chandravanshi, Narshing Kashyap, Sudhan Chandran and Hari Prasad Mohale

23. Importance of Essential Fatty Acid Nutrition in Fish 365
Shivkumar, Dhanalakshmi M., Mukkeri Krantirekha and Akshaya Vinod Mayekar

24. Role of Macroalgae in Reducing the Oxidative Stress 377
P. Chellamanimegalai, Amom Mahendrajit and Geetanjali Deshmukhe

25. Herbs in Fish Nutrition: As a Boon for Present Fish Feed Consequences 385
Shivkumar, Dhanalakshmi M., Rinkesh N. Wanjari and Harshavarthini M.

26. Impacts of Xenobiotics in Fish Reproduction 401
Sanjay Chandravanshi, Sudhan Chandran, Sahil Rai Narsale Swapnil Ananda and H.S. Mogalekar

Section 1
Aquaculture and Fisheries

1

Epigenetics in Aquaculture and Fisheries

[1]Kolanjinathan Ranjithkumar, [2]Rajendiran Rajeshkannan [3]Chandran Sudhan, [4]D. Kaviarasu and [5]R. Bavithra

[1]*Senior Technician, CPF Pvt. Ltd., Bhimavaram, Andhra Pradesh*

[2]*Ph.D. Research Scholar, Department of Fish Biotechnology, Fish Genetics & Biotechnology Division, ICAR - Central Institute of Fisheries Education (ICAR-CIFE), Mumbai, Maharashtra*

[3]*Ph.D. Research Scholar, Department of Fisheries Resource Management Fisheries Resources, Harvest & Post-Harvest Management Division ICAR-CIFE, Mumbai Maharashtra*

[4]*Assistant Professor, Department of Aquatic Animal Health Management Dr. M.G.R. Fisheries College and Research Institute, Ponneri, Tamil Nadu*

[5]*Scientist, ICAR-Central Marine Fisheries Research Institute, Mandapam Regional Centre of ICAR-CMFRI, Marine Fisheries, Mandapam Ramanathapuram, Tamil Nadu*

Abstract

Epigenetics plays an energetic role in various fields of aquaculture production with its realizable value, especially under conduciveenvironments that can be influenced, or normal distinction occurs. At this juncture, we discuss important features and mechanisms of epigenetic changes, including DNA methylation, histone modifications, chromatin remodeling, and non-coding RNAs. Assess the present consideration of the impacts of epigenetics in cultured aquatic animals, and suggest prime areas of aquaculture such as feeds, breeding, disease management, and others where epigenetics could be applied. Because culture fisheries are enlarging worldwide, we must innovate new technologies to overcome the surging production demand and species diversity problems. Environmental stimuli can change the organism's phenotype by inducing epigenetic alterations. Therefore, sympathetic to the ecofriendly-influenced epigenetic markers associated with disease resistance will give a new route to establish favorable breeding conditions with affordable economic revenue. Several researchers suggest that epigenetic effects in diverse species, stimulated by various cultivable conditions, are essential for the aquatic animals and pieces of evidence for heritability of the gained adaptive phenotypic traits over generations, making them even more appropriate in a production background. In the case of nutritive epigenetics, the chances of initial dietary programming to gain the performance of broodstocks or even the long-term performance of their progeny has been suggested. This chapter presented the primary epigenetic mechanisms, various analysis methods, and essential applications in aquaculture and fisheries.

Keywords: Epigenetic alterations, Environmental changes, Impacts on aquatic animals and Xenobiotics

Introduction

Aquaculture production is promptly increasing all over the world. To increase the demand and for best management practices, we required additional technological innovations in various fisheries and aquaculture fields such as breeding, feeds, disease management, *etc.* The study of epigenetics provides a new way to comprehend ecofriendly-induced epigenetic markers linked to economically relevant features, disease resistance, and other qualities, allowing for the establishment of more suitable breeding conditions in aquaculture. CH Waddington invented the term "epigenetics" in 1942 to define the "study of the interactions between genes and their products that bring the phenotype into being," which establishes the relationship between DNA segment and phenotype (Waddingto, 1942). Epigenetics is defined as processes or events that influence gene activity inheritance without relying on variations in DNA base sequences. The actions of genes are mostly determined by whether or not they are existing to transcription factors, which is heavily influenced by chromatin remodelling dynamics (Li et al., 2007). Environmental stressors can impact phenotypic modifications in organisms via epigenetic mechanisms, it has been widely established over the last decade. These strategies can control gene expression at the transcription, translation, or post-translational levels without changing the DNA. DNA methylation, histone modifications and variations, chromatin remodeling, and noncoding RNA activity are all epigenetic processes. These changes are thought to have happened mostly during embryonic development, are thought to promote cell type differentiation, and are mitotically and/or meiotically heritable.

Even though some epigenetic markings may be reversible, epigenetic marks can be acquired throughout one's life and are quite durable (Bishop & Ferguson, 2015). As a result, epigenetic mechanisms are also responsible for conveying diverse gene expression profiles in a variety of cells and tissues (Moore et al., 2013). Transgenerational epigenetic inheritance is a characteristic property of epigenetic effects and phenotypic traits that can be acquired by later generations but cannot be explained by changes in the DNA sequence or Mendelian genetics (Shao et al., 2014). Both mitotic and transgenerational cellular inheritance are included by epigenetic inheritance. The first happens in mitotically dividing cells, while the second is linked to intergenerational epigenetic inheritance (Jablonka and Raz, 2009). Epigenetic programming erases and resets most epigenetic marks in animals, according to rare reports. Except for fish, it does not appeal to all animals (Shao et al., 2014). The effect of epigenetic alterations induced by the environment on an organism phenotypic can be either harmful or beneficial to the organism (Babu et al., 2012). According to several studies, unfavorable environmental stimuli such as climate stressors or sublethal chemical concentrations can cause epigenetic changes over time, increasing an individual's susceptibility to diseases and disorders (Mirbahai et al., 2011). Some of them believe that cancer is caused by epigenetic changes in organisms, such as specific alterations in DNA methylation patterns that can occur before to mutational changes, detecting interactions, and complementarity between epigenetic and genetic changes (Sharma et al., 2010). In the case of polyploid species, epigenetic changes can provide effective and adaptable means for neopolyploids to rapidly respond to significant genetic material modifications, resulting in the survival and potential reproduction of those organisms (Li et al., 2011). This chapter gives an overview of the most investigated epigenetic mechanisms and their accompanying analysis methodologies, as well as the state of the art in epigenetic studies in aquaculture, as well as possible applications and problems in this sector.

Mechanisms of Epigenetic Modifications

(i) DNA methylation of certain sections of the genome, such as controlling sections and genetic factors (ii) histone alterations and histone disparities, and (iii) the activity of non-coding RNAs are all examples of epigenetic alterations. Environmental factors can cause epigenetic changes such as DNA methylation, histone alterations and chromatin structural fluctuations (remodeling), which can last a lifetime or numerous generations.

DNA Methylation

Yan et al., 2011 explained that is a process in which a methyl group (-CH_3) is added to the carbon 5 position of a cytosine residue, generally within a CpG dinucleotide, and also reported outside of the CpG context. DNA methylation plays a vital role in epigenetic alteration and it is inducing autoimmunity when the experimental (disease) animal is treated with the DNA methylation inhibitor 5-azacytidine (Hewagama and Richardson, 2009). In mammals, the DNA methylation typically encompasses the part of a methyl group to cytosine moieties in CpG dinucleotide, whereas the genome 70%–80% of CpG sites occupied. (Ehrlich et al., 1982). Interestingly, the spreading of CpG dinucleotides seems to be diminished entire the length of the genome, probably due to the high variability of the methylated cytosine bases and much higher occurrence at which the deamination of a methyl6ated cytosine can create thymine (Bird, 1986). Consequently, this mechanism has been proposed as the most important evolutionary driver shaping the vertebrate genome. CpG islands, alternatively, are compact recurrences of CpG dinucleotides found transversely in the genome. The CpG islands are a common structure of the vertebrate genome, it was located at the 5' ends of almost all housekeeping genes along with a number of tissue-specific genes (Deaton and Bird, 2011). They are primarily associated with the promoter region (Antequera and Bird, 1993), therefore less prone to impulsive deamination (Bird, 1986). DNA methyltransferases (DNMTs), a comprehensive enzyme family that has advanced to be evolutionarily well-maintained in animals and plants (Law and Jacobsen, 2010), govern methylation methods by whichever de novo marking of a cytosine nucleotide or maintaining existing methylation patterns. The stage of methylation at the promoter region is generally contrariwise linked through a gene's transcriptional action due to conformational variations to DNA and histone assemblies (Robertson and Wolffe, 2000). Even if CpG islands, mainly the ones that are in closeness to promoters, are hardly methylated, methylation of these places can pointedly decrease or entirely suppress the expression of a gene by obstructive the binding of transcription features to DNA and conscripting methyl binding proteins (Jones, 2012). Still, the particular mechanisms of such developments are immobile, they are unstated (Moore et al., 2013).

Even in well-studied model species, we don't have adequate knowledge of the Spatio-temporal variations in methylation position of genes, despite the fact that such actions have been proposed as a main component of disturbing progress, genomic marking, sexual activity chromosome inactivation, and compatible element silencing (Zentner and Henikoff, 2014). In spite of the fact that DNA methylation is generally thought to be a stable genomic modification, particularly for imprinted genes, a number of studies have discovered that genome-wide DNA demethylation, whichever way *i.e.* active or passive processes, is a common structure of some cells at the time the growth developmental stages (Wu and Zhang, 2010).

Histone Modification

Histones play a vital role in epigenetics. Nucleosomes are made up of H2B, H2A H3 and H4. Histone joined with DNA when they entered the nucleosomes. After the completion of translation process, it will be updated, with the most of the changes occurring on the histone end region, including stilation, methylation, phosphorylation, obi coientination, and ADP rebozilation (Portela and Esteller, 2010). These adjustments play a big role in transcription control and chromosomal compacting. Based on the transcription Process genomes are split into two types they are Eu-chromatin active form and de active hetero chromatin form (Portela and Esteller, 2010).

Acetylation and deacetylation, as well as methylation and demethylation, seem to be the furthermost important factors in gene regulation. Acetylation of particular lysine deposits in the tail portion of nucleosomal histones eliminates the positive charge on the lysine amino group that is acetylated, slackening the interaction between DNA, histone and preventing chromatin from the folding into the 30 nm fibre (Fischle et al., 2005), which generally helps the gene expression. On the other hand, Deacetylation is commonly connected to gene repression. Histone H1 phosphorylation looks to play a role in chromatin folding into advanced-order assemblies (Neumann et al., 2009). The properties of methylation of lysine or arginine remains in the histone tails on gene regulation are reliant on on both the exact location of the altered residues in the histone tail and the number of methyl groups supplementary to these scums (Li et al., 2010). The same amino acid in a histone tail can do this in occasional environments (Chen and Li, 2010).

Acetylation of lysine and methylation of lysine or arginine plays a vital role in coordinating multiprotein complex growth in the histone tails. Some of these mixture' components can "read" histone alterations through chemical interactions between the methyl/acetyl chains and architecturally protect protein modular fields found in many chromatin regulators and transcription factors, which work to alter chromatin components at mark gene loci (Kouzarides et al., 2007).

Noncoding RNA's

Approximately 98 percent of the transcribed human genome does not code for proteins 55, and advanced research has revealed a variety of non-coding RNAs that function post-transcriptionally to regulate a variety of biological processes through connections with messenger RNA (mRNA) or DNA (Meda et al., 2011). MicroRNA and noncoding RNAs are now regarded to be the most relevant to autoimmunity, while it is projected that additional research will reveal that as a minimum few of the other noncoding RNA types are also significant. MiRNAs are tiny molecules but it has a length of around 22 nucleotides that control the translation process of more than 60% of protein-coding genes (Friedman et al., 2009).

Chromatin Remodeling

In the interior part of nucleus has eukaryotic cells, chromatin is a DNA and protein complex that creates chromosomes. It controls who has access to genetic information. As a result, the nucleosome is a fundamental component of chromatin. Euchromatin and heterochromatin are two different types of chromatin. Euchromatin has an open structure that contains active gene transcription sites that are less electron-dense and gene-rich. Gene-poor, considerably denser, and silenced genomic locations persist in heterochromatin (Labbe et al., 2017). Constitutive heterochromatin and facultative heterochromatin are the two types of heterochromatin that exist. Throughout the cell cycle, constitutive heterochromatin has tightly packed and terminally locked genomic loci. Facultative heterochromatin, on the other hand, resembles the locations where gene activity can be restored in response to cell signaling. The existence of the transcriptionally lenient alteration H3K4me3 was found in euchromatin. H3K9ac, H4K9ac, and H3K27ac are three more significant permissive changes that relate with histone acetylation. Trimethylation, the most famous of which is H3K9me3, but also H3K64me3 and H4k20me3, is used to inscribe repression. H3K27me3 indicates a transitory silent state in facultative heterochromatin, and this spot can be originating alongside tolerant ones like H3K4me3 (H3,4 – histone; K – lysine; ac – acetyl; me3 – trimethyl). The polysome repressive complexes (PRCs), which are tangled in H3K27 trimethylation and chromatin compaction, and the trithorax complexes (TRX), which are involved in H3K4 trimethylation and gene expression conservation, are two of the most investigated (Noordermeer and Duboule, 2013).

Chromatin remodelling is defined as the dynamic transformation of chromatin architecture from a condensed to a transcriptionally accessible state. Because a greater connection between DNA and histone proteins keeps chromatin in a summarized form, chromatin compaction is inversely proportional to transcriptional activity. The chromatin will be less condensed when this linkage decreases due to other epigenetic mechanisms, permitting gene expression initiation (Mahmoud & Poizat, 2013; Moore et al., 2013). These pathways are linked to chromatin remodelling, and chromatin remodelers can target them. The chromatin structure can be changed by enzymatic post-translational histone alterations and ATP-dependent chromatin remodeling enzymes. Histone chaperones, DNA methylation, and the growth initiated of chromatin-binding proteins can then alter chromatin architecture (Mahmoud & Poizat, 2013).

Chromatin Remodelling Assessment

Several technologies have been developed to regulate the 3D structure of chromatin and book of maps chemical changes. For genome-wide plotting of histones and alterations of interest, chromatin immunoprecipitation (ChIP) is a regularly used approach in chromatin study. Upgraded ChIP approaches include ChIP-chip, ChIP-seq, ChIP-exo, and ChIA-PET. Using corresponding-end tag sequencing, ChIA-PET is utilized to investigate chromatin interaction. It was initially defined by Fullwood et al. (2009) and consists of ChIP with Chromatin Conformation Detention (3C) technology, in which the efficient chromatin structure can be transformed into lots of tall tag sequences, allowing for a global, high-quantity, and impartial mapping of transcription factor necessary sites and estimation of connections between them. However, it necessitates a huge number of cells in the primary samples, which confines its use in many genetic applications. ChIP-exo, which combines ChIP and lambda exonuclease digestion (exo), allows researchers to pinpoint a protein's genomic binding places with an almost single-base resolution by digesting contaminating DNA, which significantly minimizes background indication (Rhee and Pugh, 2012).

The Chromatin Conformation Capture (3C) method was established by Dekker et al. (2002), which involves giving cells with formaldehyde to cross-link protein-protein and protein–DNA interactions, formerly extracting and digesting the chromatin with a control enzyme. The appropriateness of the recombinant fragments was resolute by PCR, where the richness of the recombinant small fragments straightly correlates with the occurrence of interaction between the two interrelating regions (Sajan and Hawkins, 2012). This method allows for the detection of unidentified DNA region interactions with a specific section of interest, the Carbon Copy Chromosome Conformation Capture (5C) is a high-quantity adaptation of 3C that consequences in a library of all ligation goods that can be analyzed using microarray or DNA sequencing technologies, and the Combined Chromosome Conformation Capture C (Granada et al., 2017).

Epigenetics in Aquatic Animals

Many biological processes rely on epigenetic inheritance, including gene expression during early embryo growth, imprinting, and transposon muzzling (Triantaphyllopoulos et al., 2016). The epigenetic consequence happening the phenotype that can be transmitted down to successive cohorts and cannot be enlightened by variations in the main DNA sequence or Mendelian genetics is known as transgenerational epigenetic legacy (Daxinger and Whitelaw, 2010). Because of epigenetic reprogramming between cohorts in mammals, which expunges and rearranges the majority of epigenetic marks, it has only been recorded in mammals on a few occasions (Feng et al., 2010). In animals, still, the erasure of epigenetic markers is not a widespread norm. In zebrafish, *e.g.*, DNA methylation is not rearranged throughout initial development (Macleod et al., 1999), indicating that transgenerational epigenetic inheritance is possible. A recent zebrafish study corroborated this, revealing the inheritance of maternal DNA methylatome in progeny embryos (Jiang et al., 2013; Potok *et.*, 2013). Invertebrates and vertebrates have similar DNA methylation properties, but their methylation patterns differ, saying that DNA methylation in invertebrates acting a distinct part than in mammals (Zemach et al., 2010).

In eukaryotes, epigenetic variations to DNA and histones are required for connections between the surrounding and genetic information that govern gene expression and, in sequence, determine phenotypes. Describe the principles and procedures utilised in the analysis of high-throughput data, as well as the most often used apparatuses for epigenetic data analysis and epigenetic study resources. The advancement of next-group sequencing (NGS) has recently made it possible to distinguish genome-wide DNA methylation and histone modification patterns, which is what drives the epigenome notion (He and Song, 2017).

Though first proposed in 1998 (Handelsman et al., 1998), next-group sequencing (NGS) technologies, which have substantially increased the rapidity and price -efficiency of DNA sequencing since 2005, made the metagenomic technique easily applicable (Chen and pachter, 2005). Entire genome sequencing, genome variety analysis, metagenomics, epigenetics, discovery of non coding RNAs and protein-binding sites and gene-expression summarizing by RNA sequencing are all activities of NGS technologies (Metzker, 2010). The cohort Sequencing Simulator for Metagenomics (NeSSM) is a high-quantity metagenome sequencing simulation system (Jia et al., 2013).

Metagenomics and associated technology in the field of fisheries, as well as their use of modern genomics, approaches in aquatic animal health management, aquaculture dealing with microbial diversity, microcosms, antibiotic resistance genes, disease identification, and biofloc and probiotics (Kaviarasu and Sudhan, 2016). Gene expression designs can be permanently altered through epigenetic processes without disturbing the underlying DNA sequence (Feil and Fraga, 2012).

Epigenetics has been shown to have a crucial role in fish adaptation to environmental change (Aluru et al., 2011), germ and immature progressive biology (Moran and Perez Figueroa, 2011), and cancer (Moran and Perez Figueroa, 2011; Mudbhary and Sadler, 2011). Fish experimental models, as well as some invertebrate experimental models, are nowadays being used in universal biomedical investigations into epigenetics (Lu et al., 2011; Burga et al., 2011). During the process of embryonic development, epigenetic alterations of DNA happen over several processes and are expected to facilitate discrepancy into precise cell types. An epigenetic modification can be defined as a mitotically or meiotically heritable variation in the function of a gene lacking alterations in the gene sequence (Youngson and Whitelaw, 2008).Genes that are to be silenced from one of the parental alleles (i.e. expressed by only one allele) become methylated during embryonic development in a process called imprinting (Bird, 2002). There is only limited evidence of genomic imprinting in oviparous species, such as fish (Shimoda, 2007). However, DNA methylation reprogramming was observed during the early embryonic development of zebrafish (*Danio rerio*) in a recent study using an anti-5-methylcytosine antibody in immunohistochemistry and southwestern immunoblotting (Mackay, 2007). This reprogramming of DNA methylation found in zebrafish is similar to reprogramming during mammalian development (Santos and Dean, 2004). The role of DNA methylation in mediating gene expression is well documented. It has been proposed recently that the regulation of gene activity by DNA methylation can also be achieved by control of transcript splicing (Shukla et al., 2011).

The epigenetic regulation of gene expression during gonad differentiation could provide a basis for adaptive sex reversal in some genetic sex determination (GSD) organisms (Matsumoto and Crews, 2012), something that has been demonstrated to date in a few fish examples. For example, experiments in sea bass, *Dicentrarchus labrax*, suggested that the high temperature incubation during the early development stage increased the DNA methylation level of the promoter region of cyp19a1a (cytochrome P450, family 19, subfamily A, polypeptide 1a), leading to suppressed cyp19a1a expression and male development (Navarro-Martin et al., 2011).

Genetic and epigenetic interactions between redundant genes in polyploid fish have probably influenced their evolutionary fate, leading to their current impressive biological diversity (Le comber and Smith, 2004). Spontaneous polyploids have been observed in several phylogenetically distant orders, including both wild and farmed fish species (Thorgaard and Gall, 1979). In the vertebrates, polyploid species are not exclusive to fish, since they have been reported in different groups, from amphibians (Stock et al., 2002) to occasionally even in mammals (Gallardo et al., 1999). Polyploids can originate either from alterations of meiotic or mitotic processes in specimens within a species (autopolyploidy) or by reproductive contact among species (allopolyploidy) (Piferrer et al., 2009).

Manipulations during captive-rearing have been hypothesized to generate chromosomal abnormalities or heritable epigenetic changes, such as DNA methylation, that may affect individual fitness in salmonids (Araki et al., 2008). The genetic sources of variation can be associated with the presence of the extra maternal set of chromosomes and can involve simple gene dosage (additivity) between chromosome sets or positive or negative dosage compensation effects (heterosis), epigenetic mechanisms and transcriptional co-suppression (negative gene dosage compensation) (Piferrer et al., 2009).

Vitellogenesis is the production of yolky eggs in oviparous species, and involves transport of gene products from the liver to the ovary where proteins are deposited in the maturing oocytes (Stromqvist et al., 2010). They showed that the exposure to EE2 results in an alteration of DNA methylation levels in zebrafish which warrants studies of epigenetic changes in fish toxicological studies. They have for the first time shown sex and tissue differences in the level of DNA methylation in adult zebrafish in a franking region of the vitellogenin I gene. Further, following exposure to 100 ng EE2/L during 14 days the DNA methylation level they decreased in liver of both females and males. The results of them demonstrate the usefulness of the zebrafish as a model organism for studying epigenetic processes and possible epigenetic alterations following exposure to chemicals.

Pyrosequencing technology represents a tool to determine methylation levels of multiple CpG sites in specific genes of interest and this study shows its applicability to investigate pollutant-induced alterations of methylation levels in fish (Stromqvist et al., 2010).

The epigenetic mode, in which germ cell was conditionally specified and depended on inductive signals amongst embryonic cells at later embryonic stages, has been reported in the *M. musculus* (Saga, 2008) Strong *Locentrotus purpuratus* (Voronina et al., 2008) *Blatta germanica* (reviewed by Extavour and Akam) (Extavour and Akam, 2003) *Platynereis dumerilii* (Rebscher et al., 2007) and *Mnemiopsis leidyi* (reviewed by Extavour and Akam, 2003). The expression pattern of Fc-vasa-like, and the origin and migratory characteristics of germ cell suggest that the specification of germ cell in *F. chinensis* may begin at the limb bud stage and display an epigenetic mode (Feng et al., 2011). Characteristics of the origin and migration of the germ cell suggest that the specification of the germ cell of *F. chinensis* is epigenetically induced, which is different from those reported for other Crustaceans (Feng et al., 2011).

In zebrafish modulation of parental abiotic environment and nutrition confers increased resistance to the environmental stressor and alterations in cardiovascular parameters (stroke volume, heart rate, cardiac output and red blood cell concentration) to the subsequent generation (Ho, 2008). Research on steelhead trout (*Oncorhynchus mykiss*) found no evidence that the hatchery rearing environment generates global hyper- or hypo-methylation of the genome or generates differential methylated sites in comparison to the wild environment (Blouin et al., 2010).

Surprisingly, accelerated sequence evolution has been reported in African cichlid fish for most of the miRNA target sites (Loh et al., 2011). Due to their extremely rapid phenotypic diversification and speciation, cichlid fish are one of the most well-known model systems for the study of biological diversification (Kocher, 2004). With ~2,000 species, cichlids have formed spectacularly diverse and species-rich adaptive radiations showing impressive variation in body shape, coloration, and behavior (Verheyen et al., 2003).

Impact of the Environment on Fish Epigenetic Changes

The environmental changes causes impacts in both fish and other animal groups that when direct contact with temperature changes and water quality. This one is considered as heart of trans-generational inheritance approach exposed earlier. However, demonstrating the changes that occur in fish gametes and ovary epigenetic patterns by environmental changes is a real challenge.

Domestication

Domestication is considered as multi-step process because where a species collected from wild condition and admitted to a favorable environment for their growth, maturation, reproduction, and development. Domestication earnings the entire genetic background of a species and tries to adapt a specific population to different farming conditions (Teletchea and Fontaine, 2014). Fish culture is important process under epigenetic investigation. Hence, mechanisms of epigenetic are a response to these complete environmental changes deserve some specific attention. For example, during embryo incubation conditions can be influenced at early stages the ability of the fish to the culture at various environmental conditions and as demonstrated in mammals, modulation of gene expression as information is passed to the next generation through epigenetic memory in the gametes (C. Labbe et al, 2017).

Temperature

The effect of temperature on fish gametes and their early embryos is sex differentiation has been reported in several early studies. Epigenetics of sex differentiation has been reported in fish these mechanisms are different which is mainly based on ranging from genotypic to environmental sex determination models. In fish one of the most important organs is the gonads which are somatic tissues able to activated signaling pathways, activated or suppressed genes that can produce in completely various organs such as the ovary or testes. Therefore, it arises long after the establishment of the genotype at stage of the fertilization process while stressing the significance of epigenetic mechanisms (Munger and Capel, 2012). However, sex determination is mainly dependent on temperature which has been reported in many cultured fish species to induce masculinization via aromatase repression due to changes in the temperature range (Lance, 2009). The epigenetic process has been investigated in larval stages of European sea bass exposure to various temperatures. In this study promoter of gonadal aromatase (cyp19a) is double in juvenile males compared to females suppressing gene expression. In addition, the methylation status of the promoter occurs only in the gonads affected by the thermosensitive period of the experiment. However, methylation levels increased in females which induce gene silencing and masculinization due to exposure to high-temperature

(Navarro-Martin et al., 2011). This study provides the knowledge about the alter the epigenetic marks in a particular gene dependent on environmental changes and modifications of sex differentiation process promoted by temperature.

Nutrition

The progeny of mammals has a well-established effect on maternal nutrition. Moreover, both, starvation and overnutrition cause impacts that can be transmitted multi-generationally (Susiarjo and Bartolomei, 2014). The epigenome has components which are DNA methylation one of the most important components and proteins (histone modification) and RNA (non-coding RNAs) that other components act directly or indirectly (McGraw et al., 2013). Both nutritional conditions and dietary transgenerational effects are not only limited to the maternal factor on the sensitive period of pregnancy stage of mammals. This indicates that the programming of metabolic disease in the offspring is caused by the effect of paternal malnutrition (DelCurto et al., 2013). Fullston et al. (2013) reported that in mice paternal obesity induces epigenetic alterations on its spermatozoa and metabolic alterations in progeny and promoter of methylation status and gene expression levels can be changed by low protein diets, especially lipid and cholesterol biosynthesis happen in the progeny. However, much evidence has been reported in mammals specifically, rats and mice, that evidence are also the same reported in fish with few studies. Skaerven et al. (2014), investigated the Zebrafish experimentally feed with low levels B-vitamincytes which stimulated higher inclusion of lipids in the hepatocytes of F1 livers.

This study indicated the reversible condition of epigenetic changes, nutrients act as epigenetic regulatory factors which that convert or rescue the epigenetic status. Recently, many studies are focusing their attention on the potential correlation between nutrition, epigenome, and better larval performances for commercial interest in particular important species of *Senegalese sole* (Canada et al., 2014).

Xenobiotics

The gamete epigenetics is seriously affected by other environmental factors such as toxins. It is known that different abnormalities occur in the progeny due to parental exposure to an environment contaminated with pesticides or other toxic compounds. The leukemia conditions occur in the progeny during paternal exposure to polycyclic aromatic hydrocarbons has been reported by Castro-Jimenez and Orozco-Vargas., (2011) and paternal exposure to such compounds result in poor sperm quality (Jeng et al., 2013). Culture animals exposure to pollutants associated with heavy metals that disrupting spermatogenesis as well as other compounds which include the bisphenol A (BPA) and bis (2ethyl-hexyl) phthalate (DEHP), and dibutyl phthalate result in alteration of DNA methylation. The concentrations BPA and landfill leacheate has been estimated in river water less than 21 μg/L and less than 17.2 mg/L respectively (Crain et al., 2007). The experimental fish has been exposed to low concentrations of BPA that lead to testicular changes, decreasing mature spermatozoa count and increasing immature sperm stages. Carnevali et al. (2010) studied the fish exposed to DEHP which results in significant reduction of fecundity and impaired reproduction in zebrafish exposed to DEHP Corradetti et al. (2013). The decreased regeneration and downregulating genes downregulating genes that crucial for the regeneration process in zebrafish exposed to BPA demonstrated by Navarro et al. (2016). In a few studied genes downregulation is mainly associated with promoter demethylation. Lombo et al. (2015) investigated the increase in heart failures in the progeny up to the F2 from paternal zebrafish exposed to BPA while the decrease in remnant mRNA related to early development in spermatozoa from F0 and F1 individuals. The estrogenic effect of BPA has been reported in seabream (*Sparus aurata*) which results severely affect reproductive performance (Maradonna et al., 2014). The mature zebrafish follicles exposed to BPA lead to down-regulated oocyte maturation inducing signals and promoted apoptosis has been studied by Santangeli et al. (2016). They reported that these negative effects may be caused by epigenetic deregulation. Recently known that relationship between environmental contaminants and their effects cause the alterations in animal epigenome which is commonly accepted. In this method used for the focus on the epigenetic status of an organism as "foot-printing" to observe the contaminant exposure over the organism's lifetime. In this research area demonstrated in many fish species, but important model fish species is zebrafish (Kamstra et al., 2015).

Conclusion

In this chapter, we have considered that epigenetic activities play a major role in aquaculture for improving the production and diagnosis level, because it deals with the study of heritable changes in gene function (that cannot be explained by changes in the nucleotide sequence of DNA). We reviewed a number of the epigenetic modifications found in various aquatic organisms important for aquaculture development, with a focus on the potential use of this new understanding brought by epigenetics towards growing or improving aquatic animal production in this sector will reduce the cost. We can promote stable and effective breeding programs by understanding and controlling the important parameters such as sex differentiation, environmental adaptation and morphologic plasticity to expand the economically viable commercial aquaculture industries. Epigenetic mechanisms are very useful in identifying biomarkers of resistance to diseases or tolerance, stress and can potentially lead to improved diagnosis, etiology, and risk prediction in culture of economically important species. In future, it may help for innovation of new technologies and aquaculture production.

References

Aluru, N., Karchner, S.I. and Hahn, M.E. (2011). Role of DNA methylation of AHR1 and AHR2 promoters in differential sensitivity to PCBs in Atlantic Killfish, Fundulur heteroclitus. Aquatic Toxicology. 101, 288–294.

Antequera, F. and Adrian Bird. (1993). "Number of CpG islands and genes in human and mouse." Proceedings of the National Academy of Sciences. 90.24: 11995-11999.

Araki, H., Berejikian, B. A., Ford, M.J. and Blouin, M.S. (2008). Fitness of hatchery-reared salmonids in the wild. Evol. Appl. 1: 342–355.

Babu K, Zhang J, Moloney S, Pleasants T, McLean CA, Phua SH et al. (2012) Epigenetic regulation of ABCG2 gene is associated with susceptibility to xenobiotic exposure. Journal of Proteomics 75: 3410–3418.

Bird, A. (1986). CpG-rich islands and the function of DNA methylation. Nature. 321, 209-213.

Bird, A. (2002). DNA methylation patterns and epigenetic memory. Genes Dev.16:6–21.

Bishop KS, Ferguson LR (2015) The interaction between epigenetics, nutrition and the development of cancer. Nutrients 7: 922–947.

Blouin, M.S., Thuiller, V., Cooper, B., Amarasinghe, V., Cluzel, L., Araki, H. and Grunau, C. (2010). No evidence for large differences in genomic methylation between wild and hatchery steelhead. Can. J.Fish. Aquat. Sci. 67: 217-224.

Burga, A., Casanueva, M.O. and Lehner, B. (2011). Predicting mutation outcome from early stochastic variation in genetic interaction partners. Nature 480, 250–253.

Canada, P., Engrola, S., Conceicao, L.E.C., Teodosio, R., Mira, S., Sousa, V., Fernandes, J.M.O. (2014). A high inclusion of fish protein hydrolysate on Senegalse sole larval diet affects growth and is associated with altered expression of DNA methyltransferases. Proc EPICONCEPT Conference. Epigenetics and Periconception Environment. Ann Van Soom, Alirez Fazeli and Sofia Engrola Eds. (Vilamoura, Portugal, 1–3 October) 2014.

Carnevali, O., Tosti, L., Speciale, C., Peng, C., Zhu, Y., Maradonna, F., 2010. DEHP impairs zebrafish reproduction by affecting critical factors in oogenesis. PLoS One 5, e10201.

Castro-Jimenez, M.A., Orozco-Vargas, L.C., 2011. Parental exposure to carcinogens and risk for childhood acute lymphoblastic leukemia, Colombia, 2000–2005. Prev. Chronic Dis. 8, A106.

Catherine Labbe, Vanesa Robles, Maria Paz Herraez. (2017). Epigenetics in fish gametes and early embryo. Aquaculture. 472, 93–106.

Chen, K, and Pachter, L. (2005). Bioinformatics for whole-genome shotgun sequencing of microbial communities. PLoS Comput Biol. 1: 106-112.

Chen, P. and Li, G. (2010). Dynamics of the higher-order structure of chromatin. Protein Cell. 1 (11):967–971.

Corradetti, B., Stronati, A., Tosti, L., Manicardi, G., Carnevali, O., Bizzaro, D., 2013. Bis-(2-ethylexhyl) phthalate impairs spermatogenesis in zebrafish (Danio rerio). Reprod. Biol. 13, 195–202.

Crain, D.A., Eriksen, M., Iguchi, T., Jobling, S., Laufer, H., LeBlanc, G.A., Guillette Jr., L.J., 2007. An ecological assessment of bisphenol-A: evidence from comparative biology. Reprod. Toxicol. 24, 225–239.

Daxinger, L. and Whitelaw, E. (2012). Understanding transgenerational epigenetic inheritance via the gametes in mammals. Nat Rev Genet.13: 153–62.

Deaton, A.M. and Bird, A. (2011). CpG islands and the regulation of transcription. Genes Dev. 25, 1010–1022.

Dekker J, Rippe K, Dekker M, Kleckner N. (2002). Capturing chromosome conformation. Science. 295: 1306–1311.

DelCurto, H., Wu, G., Satterfield, M.C., 2013. Nutrition and reproduction: links to epige netics and metabolic syndrome in offspring. Current Opinion in Clinical Nutrition and Metabolic Care. 16, 385–391.

Ehrlich, M., Gama-Sosa, M.A. and Huang, LH. (1982). Amount and distribution of 5-methylcytosine in human DNA from different types of tissues of cells. Nucleic Acids Res. 10(8):2709–2721.

Extavour, C.G. and Akam, M. (2003). Mechanisms of germ cell specification across the metazoans: epigenesis and preformation. Development.130: 5869-5884

Feil, R. and Fraga, M.F. (2012). Epigenetics and the environment: emerging patterns and implications. Nature Reviews Genetics. 13:97-109.

Feng, S., Jacobsen, S.E. and Reik, W. (2010). Epigenetic reprogramming in plant and animal development. Science. 330(6004):622–627.

Feng, Z., Zhang, Z., Shao, M. and Zhu, W. (2011). Developmental expression pattern of the Fc-vasa-like gene, gonadogenesis and development of germ cell in Chinese shrimp, Fenneropenaeus chinensis. Aquaculture. 314: 202-209.

Fischle W, Tseng BS, Dormann HL, et al. Regulation of HP1-chromatin binding by histone H3 methylation and phosphorylation. Nature. 2005; 438(7071):1116–1122.

Friedman, R.C, Farh, K.K, Burge, C.B. and Bartel, D.P. (2009). Most mammalian mRNAs are conserved targets of microRNAs. Genome Res. 19(1):92–105.

Fullston, T., Teague, E.M.C.O., Palmer, N.O., DeBlasio, M.J., Mitchell, M., Corbett, M., Print, C.G., Owens, J.A., Lane, M., 2013. Paternal obesity initiates metabolic disturbances in two generations of mice with incomplete penetrance to the F-2 generation and alters the transcriptional profile of testis and sperm microRNA content. Faseb Journal 27, 4226–4243.

Fullwood MJ, Liu MH, Pan YF, Liu J, Han X, Mohamed YB et al. (2009). An oestrogen receptor a-bound human chromatin interactome. Nature. 462: 58–64.

Gallardo, M.H., Bickham, J.W., Honeycutt, R.L., Ojeda, R.A. and Kohler, N. (1999). Discovery of tetraploidy in a mammal. Nature. 401: 341.

Handelsman, J., Rondon, M.R., Brady, S.F., Clardy, J. and Goodman, R.M. (1998). Molecular biological access to the chemistry of unknown soil microbes: a new frontier for natural products. Chem Biol. 5: R245-R249.

He, Y. and Song, J. (2017). Bioinformatics Analysis of Epigenetics, in Bioinformatics in Aquaculture: Principles and Methods (ed Z. (. Liu), John Wiley & Sons, Ltd, Chichester, UK. doi: 10.1002/9781118782392.ch15.

Hewagama, A. and Richardson, B. (2009). The genetics and epigenetics of autoimmune diseases. J Autoimmun.33(1):3–11.

Ho, D.H. (2008). Morphological and physiological developmental consequences of parental effects in the chicken embryo (Gallus gallus domesticus) and the zebrafish larva (Danio rerio). Diss: University of North Texas.

Jablonka E, Raz G. (2009). Transgenerational epigenetic inheritance: prevalence, mechanisms, and implications for the study of heredity and evolution. The Quarterly Review of Biology. 84: 131–176.

Jeng, H.A., Pan, C.H., Lin, W.Y., Wu, M.T., Taylor, S., Chang-Chien, G.P., Zhou, G., Diawara, N., 2013. Biomonitoring of polycyclic aromatic hydrocarbons from coke oven emissions and reproductive toxicity in nonsmoking workers. J. Hazard. Mater. 244-245, 436–443.

Jia, B., Xuan, L., Cai, K., Hu, Z., Ma, L. and Wei, C. (2013). NeSSM: a nextgeneration sequencing simulator for metagenomics. PLoS ONE 8:e75448. doi: 10.1371/journal.pone.0075448.

Jiang, L., Zhang, J., Wang, J.J., Wang, L., Zhang, L., Li, G., Yang, X., Ma, X., Sun, X. and Cai, J. (2013). Sperm, but not oocyte, DNA methylome is inherited by zebrafish early embryos. Cell. 153: 773–784.

Jones, P.A. (2012). Functions of DNA methylation: islands, start sites, gene bodies and beyond. Nat. Rev.Genet.13, 484–492.

Kamstra, J.H., Loken, M., Alestrom, P., Legler, J. (2015). Dynamics of DNA Hydroxymethylation in Zebrafish. Zebrafish 12, 230–237.

Kaviarasu, D. and Sudhan, C. (2016). Review note on the application of metagenomics in emerging aquaculture systems and aquatic animal health management. International journal for innovative research in multidisciplinary field. vol-2, issue-8.

Kocher, T.D. (2004). Adaptive evolution and explosive speciation: the cichlid fish model. Nat Rev Genet. 5:288–298.

Kouzarides, T. (2007). Chromatin modifications and their function. Cell. 128(4):693–705.

Lance, V.A., 2009. Is regulation of aromatase expression in reptiles the key to understanding temperature-dependent sex determination? J Exp Zool A 311, 314–322.

Law, J.A. and Jacobsen, S.E. (2010). Establishing, maintaining and modifying DNA methylation patterns in plants and animals. Nat. Rev. Genet.11, 204–220.

Le Comber, S.C. and Smith, C. (2004). Polyploidy in fishes: patterns and processes. Biol J Linn Soc. 82: 431-442.

Li, B., Carey, M. and Workman, J.L. (2007). The role of chromatin during transcription. Cell. 128(4):707–719.

Li, G., Margueron, R., Hu, G., Stokes, D., Wang, Y.H. and Reinberg, D. (2010). Highly compacted chromatin formed in vitro reflects the dynamics of transcription activation in vivo. Mol Cell. 38(1):41–53.

Li ZH, Lu X, Gao Y, Liu SJ, Tao M, Xiao H et al. (2011). Polyploidization and epigenetics. Chinese Science Bulletin. 56: 245– 252.

Loh, Y.H., Yi, S.V. and Streelman, J.T. (2011). Evolution of microRNAs and the diversification of species. Genome Biol Evol. 3:55–65.

Lombo, M., Fernandez-Diez, C., Gonzalez-Rojo, S., Navarro, C., Robles, V., Herraez, M.P., 2015. Transgenerational inheritance of heart disorders caused by paternal bisphenol A exposure. Environ. Pollut. 206, 667–678 (Barking, Essex: 1987).

Lu, J.W., Hsai, Y. and Tu, H.C. (2011). Liver development and cancer formation in zebrafish. Birth Defects Research C: Embryo. 93, 157–172.

Luana Granada, Marco F.L. Lemos, Henrique N. Cabral, Peter Bossier and Sara C. Novais. (2017). Epigenetics in aquaculture – the last frontier. Reviews in Aquaculture. 0, 1–20.

MacKay, A.B., Mhanni, A.A., McGowan, R.A. and Krone, P.H. (2007). Immunological detection of changes in genomic DNAmethylation during early zebrafish development. Genome. 50: 778-785.

Macleod, D., Clark, V.H. and Bird, A. (1999). Absence of genome-wide changes in DNA methylation during development of the zebrafish. Nat Genet. 23:139–140.

Mahmoud SA, Poizat C (2013) Epigenetics and chromatin remodeling in adult cardiomyopathy. Journal of Pathology. 231: 147–157.

Maradonna, F., Nozzi, V., Dalla Valle, L., Traversi, I., Gioacchini, G., Benato, F., Colletti, E., Gallo, P., Di Marco Pisciottano, I., Mita, D.G., et al., 2014. A developmental hepatotoxicity study of dietary bisphenol A in Sparus aurata juveniles. Comp Biochem Physiol C Toxicol Pharmacol 166, 1–13.

Matsumoto, Y. and Crews, D. (2012). Molecular mechanisms of temperature dependent sex determination in the context of ecological developmental biology. Mol Cell Endocrinol. 354: 103–110.

McGraw, S., Shojaei Saadi, H.A., Robert, C., 2013. Meeting the methodological challenges in molecular mapping of the embryonic epigenome. Mol. Hum. Reprod. 19, 809–827.

Meda, F., Folci, M., Baccarelli, A. and Selmi, C. (2011). The epigenetics of autoimmunity. Cell Mol Immunol. 8(3):226–236.

Metzker, M.L. (2010). Sequencing technologies- the next generation. Nat. Rev. Genet.11, 31–46.

Mirbahai L, Yin G, Bignell JP, Li N, Williams TD, Chipman JK., (2011). DNA methylation in liver tumorigenesis in fish from the environment. Epigenetics. 6: 1319–1333.

Moore, L.D., Le, T. and Fan, G. (2013). DNA methylation and its basic function. Neuropsycho -pharmocology. 38, 23–38.

Moran, P. and Perez-Figueroa, A. (2011). Methylation changes associated with early maturation stages in Atlantic salmon. BMC Genetics. 12, 86, doi: 10.1186e1471-2156-12-86.

Mudbhary, R. and Sadler, K.C. (2011). Epigenetics, development, and cancer: Zebrafish make their mark. Birth Defects Research C Embryo Today. 93, 194–203.

Munger, S.C., Capel, B., 2012. Sex and the circuitry: progress toward a systems-level understanding of vertebrate sex determination. Wiley Interdiscip. Rev. Syst. Biol. Med. 4, 401–412.

Navarro, C., Valcarce, D.G., Riesco, M.F., Lombó, M., Cubría, C., Herráez, M.P., Robles, V., 2016. Effects of Bisphenol A on zebrafish caudal fin regeneration ability.

Navarro-Martin, L, Vinas, J., Ribas, L., Diaz, N., Gutierrez, A., Di, Croce, L. and Piferrer, F. (2011). DNA methylation of the gonadal aromatase (cyp19a) promoter is involved in temperature-dependent sex ratio shifts in the European seabass. PLoS Genet 7: e1002447.

Neumann, H., Hancock, S.M. and Buning, R. (2009). A method for genetically installing site-specific acetylation in recombinant histones defines the effects of H3 K56 acetylation. Mol Cell. 36(1):153–163.

Noordermeer, D., Duboule, D., 2013. Chapter Four - Chromatin Architectures and Hox Gene Collinearity. In: Edith, H. (Ed.), Current Topics in Developmental Biology. Academic Press, pp. 113–148.

Piferrer, F., Beaumont, A., Falguiere, J., Flajshans, M. and Haffray, P. (2009). Polyploid fish and shellfish: Production, biology and applications to aquaculture for performance improvement and genetic containment. Aquaculture. 293: 125156.

Portela, A. and Esteller, M. (2010). Epigenetic modifications and human disease.

Potok, M.E., Nix, D.A., Parnell, T.J. and Cairns, B.R. (2013). Reprogramming the maternal zebrafish genome after fertilization to match the paternal methylation pattern. Cell. 153: 759–772.

Rebscher, N., Zelada-Gonzalez, F., Banisch, T.U. and Raible, F. (2007). Vasa unveils a common origin of germ cell and of somatic stem cells from the posterior growth zone in the polychaete Platynereis dumerilii. Dev Biol. 306: 599-611.

Rhee HS, Pugh BF (2012) ChIP-exo method for identifying genomic location of DNA-binding proteins with near-single nucleotide accuracy. Current Protocols in Molecular Biology 100: 21.24.1–21.24.14.

Robertson, K.D. and Wolffe, A.P. (2000). DNA methylation in health and disease. Nat. Rev. Genet.1, 11–19.

Saga, Y. (2008). Mouse germ cell development during embryogenesis. Curr Opin Genet Dev. 18: 337-341.

Sajan SA, Hawkins RD. (2012). Methods for identifying higher-order chromatin structure. Annual Review of Genomics and Human Genetics. 13: 59–82.

Santangeli, S., Maradonna, F., Gioacchini, G., Cobellis, G., Piccinetti, C.C., Dalla Valle, L., Carnevali, O., 2016. BPA-induced deregulation of epigenetic patterns: effects on female zebrafish reproduction. Sci. Report. 6, 21982.

Santos, F. and Dean, W. (2004). Epigenetic reprogramming during early development in mammals. Reproduction. 127: 643-651.

Shao C, Li Q, Chen S, Zhang P, Lian J, Hu Q et al. (2014). Epigenetic modification and inheritance in sexual reversal of fish. Genome Research. 24: 604–615.

Sharma S, Kelly TK, Jones PA (2010) Epigenetics in cancer. Carcinogenesis 31: 27–36.

Shimoda, N. (2007). DNA methylation in zebrafish. In: Neumann HA (ed) Progress in DNA Methylation Research. Nova Science Publishers, Inc: 133-152.

Shukla, A., Chaurasia, P. and Bhaumik, S.R. (2009). Histone methylation and ubiquitination with their cross- talk and roles in gene expression and stability. Cellular and Molecular Life Sciences. 66, 1419–1433.

Skaerven, K.H., Aaners, H., Lie, K.K., Hamre, K., Dahl, J.A. (2014). Insufficient B-vitamin levels in the feed for zebrafish increases the lipid accumulation in the liver of the next generation. Proceedings of the EPICONCEPT Conference. Epigenetics and Periconception Environment. Ann Van Soom, Alirez Fazeli and Sofia Engrola Eds. (Vilamoura, Portugal, 1–3 October) 2014.

Stock, M., Lamatsch, D.K., Steinlein, C., Eppeln, J.T. and Grosse, W.R. (2002). A bisexually reproducing all-triploid vertebrate. Nat Genet. 30: 325-328.

Stromqvist, M., Tooke, N. and Brunstrom, B. (2010). DNA methylation levels in the 5'flanking region of the vitellogenin I gene in liver and brain of adult zebrafish (Danio rerio)—Sex and tissue differences and effects of 17a-ethinylestradiol exposure. Aquatic Toxicology. 98: 275-281.

Susiarjo, M., Bartolomei, M.S., 2014. Epigenetics. You are what you eat, but what about your DNA? Science 345, 733–734.

Teletchea, F., Fontaine, P., 2014. Levels of domestication in fish: implications for the sustainable future of aquaculture. Fish Fish. 15, 181–195.

Thorgaard, G.H. and Gall, G.A.E. (1979). Adult triploids in a rainbow trout family. Genetics 93: 961-973.

Triantaphyllopoulo, K.A., Ioannis, I. and Andrew J.B. (2016). Epigenetics and inheritance of phenotype variation in livestock. Epigenetics and Chromatin. 9:31.

Verheyen, E., Salzburger, W., Snoeks, J. and Meyer, A. (2003). Origin of the superflock of cichlid fishes from Lake Victoria, East Africa. Science. 300:325–329.

Voronina, E., Lopez, M., Juliano, C.E., Gustafson, E. and Song, J.L. (2008). Vasa protein expression is restricted to the small micromeres of the sea urchin, but is inducible in other lineages early in development. Dev Biol. 314: 276-286

Waddington, C.H. (1942). Canalization of development and the inheritance of acquired characters. Nature. 150, 3.

Wu, S.C. and Zhang, Y. (2010). Active DNA demethylation: many roads lead to Rome. Nat. Rev. Mol. Cell Biol.11, 607–620.

Yan, J., Zierath, J.R. and Barres, R. (2011). Evidence for non-CpG methylation in mammals. Exp. Cell Res.317, 2555–2561.

Youngson, N.A. and Whitelaw, E. (2008) Transgenerational epigenetic effects. Annu. Rev. Genom. Hum Genet. 9: 233-257.

Zemach, A., McDaniel, I.E., Silva, P., and Zilberman, D. (2010). Genome-wide evolutionary analysis of eukaryotic DNA methylation. Science, 328(5980), 916–919.

Zentner, G.E. and Henikoff, S. (2014). High-resolution digital profiling of the epigenome. Nat. Rev. Genet.15, 814–827.

Zhang J, Poh HM, Peh SQ, Sia YY, Li G, Mulawadi FH et al. (2012). ChIA-PET analysis of transcriptional chromatin interactions. Methods. 58: 289–299.

2

Integrated Multi Trophic Aquaculture A Balanced System for Sustainable Aquaculture

[1]Kalidoss Radhakrishnan, [2]P Dheeran, Rajendiran [3]Rajeshkannan, [1]P Seenivasan, [4]N Raghuvaran, [5]Radhakrishnan Naveenkumar, [1]Abhilash Thapa and [6]Kanagaraja Anantharaja

[1]Fisheries Economics Extension and Statistics Division, ICAR-Central Institute of Fisheries Education, Mumbai, India

[2]Division of Aquaculture, ICAR-Central Institute of Fisheries Education, Mumbai India

[3]Fish Genetics and Biotechnology Division, ICAR-Central Institute of Fisheries Education Mumbai, India

[4]Fish Nutrition, Biochemistry and Physiology Division, ICAR-Central Institute of Fisheries Education, Mumbai, India

[5]Aquatic Environment and Health Management Division, ICAR-Central Institute of Fisheries Education, Mumbai, India

[6]Regional Research Centre of ICAR-Central Institute of Freshwater Aquaculture Bengaluru India

---

Abstract

Intensive aquaculture productions in open water release particulates and dissolved waste into the environment. There is also a growing concern about nutrient uptake from wastes of aquaculture, as well as promoting more sustainable aquaculture development. In order to meet the demands of the world, aquaculture is striving to establish production strategies that are sustainable and minimize the impact on the environment. The integrated multi-trophic aquaculture (IMTA) is one of the production strategies that promotes synergistic interactions between species by combining trophic levels and utilizing complementary ecological services. Additionally, the IMTA based product's nutritional profile makes it very useful as a natural ingredient in aquafeed compositions and as a potential food supplement for human nutrition. In spite of numerous studies on IMTA, we have little information about the collective comprehensive development responses of fed, extractive species and the spatial scale at which they can be cultivated. The purpose of this review is to provide comprehensive knowledge and overview of types of species that are cultivated, their combinations, metagenomics, economics, and how microbes play a significant role in elements cycling, nutrients retention, energy flow, reducing environmental impacts, and improving farmed species health in IMTA system. Of the various species reared, Penaeus vannamei showed a better growth in fed species and U. rigida, Perna viridis and Holothuria tubulosa are in extractive species. IMTA system reared species has the capability to mitigate the negative impacts of aquaculture pollutants and the environmental surrounding. Metagenomics approach in various study revealed IMTA has ability to maintain antibiotic resistance and bacterial community structure than conventional culture system. More willing to pay for IMTA produced product will encourage new venture to enter in the farming practice. Development of BMP guidelines and regulatory governance further strengthen existing IMTA farming practices to achieve environmentally sustainable aquaculture. As part of this review, we aimed to consolidate the peer-reviewed data on IMTA to fill any research knowledge gaps by analyzing data at various trophic levels in a sustainable ecosystem.

Keywords: Biomitigation; Economic stability; Nutrient retention; Metagenomics; Social acceptability

Introduction

According to the United Nations Department of Economic and Social Affairs, the world population increase from 7.7 to 9.7 billion by 2050; therefore, a growing demand for seafood and the production gap is anticipated due to stagnating fish supply from capture fisheries. The production gap may be curtailed by intensification of aquaculture practices rather than horizontal expansion. Intensive farming practices such as cage and pen culture in open water have gained significant attention and are related to the efficient utilization of untapped potential water bodies and of existing resources. However, adoption of such intensive farming may lead to creating several environmental issues. For instance, the uneaten excess feed and fish faeces in intensive farming increase the nutrient concentration in water (eutrophication) and further, it deteriorates the water quality and may cause oxygen depletion that affects the health condition of the fish. Increased nutrient loading in fish farming is substantial (Wang et al., 2012). Nitrogen (N_2) that can negatively impact the benthic environment by altering the sediment chemistry with knock-on effects on benthic biodiversity (Giles, 2008; Hargrave, 2010) the selection of monitoring parameters is surrounded by uncertainty, which impedes the development of cost-effective monitoring procedures. Studies examining benthic impacts of fish farms have come to different conclusions concerning the severity of impacts. Previous attempts to quantitatively review these studies have had only limited success due to the uncertainty in the data originating from discrepancies in sampling and analytical protocols as well as different temporal and spatial scales covered in the studies. The main objective of this study was to review publications on fish farm benthic impacts and to develop a Bayesian network (BN). Studies aimed to address the issue of nutrient loading around the farming area by improving the digestibility of fish feed and computerized feed management system, but it failed to eliminate the nutrient pollution associated with fish farming.

In this context, there is a necessity for development of a sustainable new aquaculture system that ensures environmental sustainability, economic stability, societal acceptability, low production cost with low risk, and multiple species culture by improving total production. The Ecosystem Approach for Aquaculture (EAA) promotes the efficient use of nutrient resources, and opportunities for diverse products, hereby, integrated aquaculture is an imperative practical way to implement the EAA (Soto & Food and Agriculture Organization of the United Nations, 2009)in the marine environment, it has been much less reported. However, in recent years the idea of integrated aquaculture has been often considered a mitigation approach against the excess nutrients/ organic matter generated by intensive aquaculture activities particularly in marine waters. In this context, integrated multitrophic aquaculture (IMTA. Integrated aquaculture and more specifically, Integrated Multi-trophic Aquaculture (IMTA) is an emerging concept having beneficial environmental effects to bio mitigate waste production and utilize resources effectively by farming multiple species that will increase the resilience of the operation with reduced risks. The idea behind integrated multitrophic aquaculture is to farm species from different trophic levels and with harmonious ecosystem functions that can be farmed in proximity so that uneaten food and wastes, nutrients and by-products from one species can be recycled into fertilizer, feed, and energy for the other species, and that can be used to exploit the synergistic interactions between species (Chopin et al., 2001). The IMTA approach is regarded as a feasible method to reduce the negative impacts of aquaculture waste. The IMTA concept was developed to increase the sustainability of intensive aquaculture systems. Several research works were conducted globally about various aspects of IMTA system and we commence with a review of the existing knowledge species selection criteria, different species combination and their growth performance (fed species, extractive species), nutritional retention efficiencies, nutritional profile, sustainability, economics and marketing challenges of IMTA system.

IMTA System: Definition and Its Components

IMTA is a practice in which byproducts from one species are recycled to become inputs for another. IMTA is the culture of fed aquaculture species (fish/shrimp) along with inorganic extractive species (seaweed) and organic extractive species (oyster/mussel), bottom feeder (sea cucumber/polychaete) to create a balanced ecosystem for environmental remediation (bio-mitigation), economic stability (diversified species culture at low cost with low risk) and social acceptability (better management practices, improved regulatory governance, and appreciation of differentiated and safe products) (Chopin et al., 2001).

IMTA systems consist of fed species, and extractive (organic, inorganic and deposit feeder) species. Feed is the only source of nutritional energy supplied to fed species. The species, size, and feed composition are the few of variables that affect the amount and form of the nutrients (dissolved and particulate nutrients) discharged by fed species to extractive species. The organic (shellfish/bivalves or herbivorous fish) and inorganic (seaweeds or aquatic weeds) extractive species collectively maintain the ecological balance. The organic extractive species such as bivalves which is a filter feeder and generally grown next to mesh fish cages. This reduces nutrient loadings by filtering and incorporate particulate wastes (fish feed and faeces). The inorganic extractive species such as seaweeds and aquatic plants (act as bio-filtration) are reduced the environmental impact by absorbing inorganic nutrients waste. The sea cucumber, sea urchin, lobster and polychaetes which are the deposit feeders, it absorbs the organic and inorganic nutrients that generated by the organic extractive species (Granada et al., 2016).

Economic Stability

Output

Economic stability of IMTA is accomplished through augmenting production which relies heavily on the species selected. Growth of the species varies; Specific Growth Rate (SGR) is commonly used to compare with other species that helps to identify the suitable species – fed species and extractive species – based on growth because it is directly proportional to the output of that species. However, we have summarized (see supplementary file) the specific growth rate of fed species and extractive species reared in IMTA system for better understanding which may be useful in species selection. The SGR lies between 1.43 (*Pseudoplatystoma punctifer*) to 6.48 (*Penaeus vannamei*) under fed species list. The seaweeds, *Ulva rigida*, *Gracilaria conferta* and *Hypnea musciformis* are grown faster in IMTA system compared to algae in the seawater (Ashkenazi, Israel, & Abelson, 2019; da Silva, Castilho-Barros, & Henriques, 2022; Neori et al., 2003) their performance poses a dilemma: TAN (Total Ammonia Nitrogen). For instance, among the inorganic extractive species *U. rigida* had the highest SGR of 18.27% day^{-1}. A recent study detailly discussed about the importance of sea weed and extractive species in world aquaculture production (Chopin & Tacon, 2021). It is anticipated that fed-species aquaculture production is expected to grow 58.96 million tonnes by 2025 in order to achieve this growth average rate of 7.7% year^{1} including the supply of feed ingredient inputs (Tacon et al., 2022). Mussels and sea cucumber are commonly nurtured as an organic extractive species (Rejeki et al., 2016; Shpigel et al., 2017). (Yokoyama, 2013; Tolon et al., 2017) study found that it grows faster in IMTA system than the normally cultured system. Growth rate of fish directly influenced by the quality of the diet (Israel et al., 2019). examined digestibility of seabream faecal waste in reared in different trophic species such as grey mullet (*Mugil cephalus*), sea urchin (*Paracentrotus lividus*), and sea cucumber (*Actinopyga bannwarthi*) in IMTA system. The sea cucumber and sea urchins had better digestibility of seabream waste but not in mullets (Israel et al., 2019; Lupatsch et al., 2003). *Sparus aurata*, farm located in the Eastern Mediterranean Sea. To achieve this, digestibility trials were conducted with three species representing different trophic levels, the grey mullet, *Mugil cephalus*, the sea urchin, *Paracentrotus lividus,* and the sea cucumber, Actinopyga bannwarthi using the particulate farm effluents as feed. To prepare the feeds, seabream faecal waste was collected by means of a large conical bag attached below a fish cage. The farm wastes were dried and pelleted after adding chromic oxide indigestible marker.

In addition to this, sea weeds produced in IMTA system has been utilized to replace fish meal due to amino acid profile and protein quality of seaweeds – *Porphyra dioica*, *Porphyra umbilicalis*, *Gracilaria vermiculophylla*, and *Ulva rigida* (Machado et al., 2020) – are imperative (Table 1). The protein quality relies on the amino acid composition, and chemical parameters including the amino acid score and essential amino acid index. (Shpigel et al., 2017a) study reported that amino acids, leucine, threonine, and valine concentration was higher in *Porphyra* species followed by *Gracilaria vermiculophylla.* The essential amino acids more than 90% is indicator of high-quality protein that observed in *Saccharina latissima* than food crops such as rice, wheat, oats and soyabeans. The aspartic acid and glutamic acid are, non-essential amino acids, reported higher in *Porphyra dioica* conchocelis. (Shpigel et al., 2017a) noticed that *Ulva lactuca* produced in IMTA had high concentration of protein (26-37%) and lipid (2-2.75%) than that of it wild produced. Another observed that the combination of *Ulva lactuca* and poultry meal has the capability to replace the fishmeal and it reduced the nitrogen input in aquatic habitat and feed cost ($0.25 per kg). Further, (Laramore, Baptiste, Wills & Hanisak, 2018) study found that *Ulva lactuca* replaced commercial pellet feed up to 25% for the juvenile production of *P. vannamei* without compromising the growth (Table 1). We could be able to infer from the reviewed literature is that *Penaeus vannamei, Ulva lactuca* and *Actinopyga bannwarthi* can rear in the IMTA system which produces better output. The culture of molluscs and seaweed are increasingly recognized for its ecosystem services (Naylor et al., 2021).

Table 1: Amino acid and nutrient profile of different species cultured in the IMTA system

Author name	Species	CRUDE PROTEIN (% ds)	EAA% (mg/g ds)	NEAA% mg/g ds)	∑TAA mg/g ds)	AAS (%)	EAAI (%)	LAA	EAA/ non-EAA	TN	NPN
(Machado et al., 2020)	*Porphyra dioica (blades)*	23.7	36.84	93.28	13.57	48.7	90.8	Trp		4.74	1.12
	P. dioica (Conchocelis)	26.73	39.78	85.49	16.17	87.1	114.2	Met		5.35	0.88
	P. umbilicalis (blades)	23.11	38.45	93.38	13.88	56.5	96.5	Met		4.62	0.64
	P. umbilicalis (Conchocelis)	23.97	38.65	89.94	13.2	93.4	115.7	Met		4.79	0.46
	Gracilaria vermiculophylla	13.38	40.14	80.09	3.36	58.4	101.9	Met		2.68	0.44
	Ulva rigida	10.19	40.79	55.03	4.95	90.8	123.4	Met		2.04	0.22
		PROTEIN (g/100 g dw)			∑AA						
(Vieira et al., 2018)	*Porphyra sp (Apr-Jul)*	27.4	44.3	33.7	38.8				1.32		
	Porphyra sp (Oct-Nov)	28.2	39.9	39.2	39.9				1.04		
	Gracilaria (Apr-Jul)	24.7	51.2	28.8	35.5				1.74		

	Gracilaria (Oct-Nov)	24.4	49.7	33.8	39.6				1.47		
	C. crispus (Apr-Jul)	19.5	46.7	29.5	31.2				1.58		
	C. crispus (Oct-Nov)	19.1	42	38.2	38.9				1.1		
	O. pinnatifida (Apr-Jul)	24.3	47.9	29.9	39.4				1.6		
	O. pinnatifida (Oct-Nov)	22.8	41	35.6	39.6				1.15		
	A. nodosum (Apr-Jul)	6.9	39.2	36.9	36.6				1.06		
	A. nodosum (Oct-Nov)	9.4	37.7	37.9	36.1				1		
	F. spiralis (Apr-Jul)	11.8	38.7	36	40.2				1.08		
	F. spiralis (Oct-Nov)	11.7	37.7	37.1	39.6				1.02		
	S. polyschides (Apr-Jul)	12.4	49	29.4	37.7				1.67		
	S. polyschides (Oct-Nov)	11.8	47.8	28.8	38.2				1.66		
	U. pinnatifida (Apr-Jul)	16.5	37.2	42.6	44.2				0.87		
	U. pinnatifida (Oct-Nov)	19.5	40.2	40	42.3				1.01		

	Ulva spp. (Apr-Jul)	20.5	44.5	33.6	41.3				1.32		
	Ulva spp. (Oct-Nov)	23.3	43.6	33.4	49.9				1.3		
(Bancarosa et al., 2017)	*C. crispus*	19.3			12.8					3.1	41.1
	F. lumbricalis	13.1			8.7					2.1	37.6
	M. stellatus	15.2			11.2					2.4	34.3
	P. palmate	16.2			12.4					2.6	33.2
	P. dioica	31			24.2					5	31.5
	P. purpurea	18			15.9					2.9	23.4
	P. umbilicalis	24			17.7					3.8	35.5
	C. rupestris	19.5			13.9					3.1	39.4
	U. intestinalis	14.8			13.1					2.4	24.7
	U. lactuca	22.6			17.5					3.6	33.5
	A. esculenta	14.2			11.8					2.3	32.8
	A. nodosum	4.5			3.5					0.7	37.3
	C. flagelliformis	10.2			7.4					1.6	39.4
	F. serratus	4.8			4.6					0.8	22
	F. spiralis	5.8			4.6					0.9	33.8
	F. vesiculosus	5.4			4.3					0.9	33.3
	H. siliquosa	7.3			5.4					1.2	39.9
	H. elongate	8.7			5.9					1.4	43.5
	L. digitata	9.6			7.7					1.5	31.8

	P. canaliculata	7.3			4.8					1.2	45.8
	S. latissima	12			9.8					1.9	31.5
			EAA (% dry weight)	NEAA (% dry weight)							
(Gressler et al., 2010)	*C. crispus*		3.5	4.1	7.6				0.9		
	F. lumbricalis		4.2	4.9	9.1				0.9		
	M. stellatus		5.5	5.8	11.3				0.9		
	P. palmate		2.9	3.8	6.7				0.8		

Lowering Cost by Enhancing Production

By increasing production output, minimizing waste and efficient utilization of resources in the IMTA system would help to reduce the production cost and accomplish better economic profitability. It generally measures through profitability indicators such as Net Present Value (NPV), Internal Rate of Return (IRR), Pay Back Period (PBP), and Benefit-Cost Ratio (BCR). Profitability is commonly determined by type of species combination nurtured in IMTA system, number of species culture in IMTA system and geographical location (temperate or tropical). Most of the studies reported that integrating shellfish and/or seaweeds with existing salmon/shrimp monoculture operations increase economic profits of the farm (Bergamo et al., 2021). The salmon produced with seaweeds generated the revenue of USD 34000 year $^{-1}$. Production of salmon with mussel had generated a better NPV (USD 2.63 million) followed by salmon monoculture (USD1.7 million) and mussel monoculture (USD 0.650 million) (Whitmarsh, Cook, & Black, 2006) (Table 2). IMTA in ponds showed a better net income and benefit-cost ratio than the shrimp monoculture ponds (Balasubramanian et al., 2018)farming of species from different trophic levels and with complimentary ecosystem function, is regarded as a suitable approach to develop a sustainable aquaculture system. In order to establish an IMTA system, a study was carried out in Sindhudurg District, Maharashtra, India for selected tropical brackish-water species. Two equal sized pens (250 m2. Integrated production of *P. Perna*, *N. nodosus*, and *K. alvarezii* showed a higher IRR, and NPV and lesser payback period (da Silva et al., 2022)Nodipecten nodosus scallop, and *Kappaphycus alvarezii* algae. IMTA with mullets and tiger shrimp as fed species and oyster and water spinach as extractive species resulted in 3.35, 3.48, and 1.6-fold increases in total production, net income, and BCR, respectively (Biswas et al., 2020).

We could observe from other studies that number of species reared in the IMTA system also remarkably affect the economics returns. For instance, when three species (salmon–mussel–kelp farm) reared in IMTA also showed 24% higher NPV (USD 3296037) than salmon monoculture (USD 2664112) at 5% discount rate for a period of 10-years (Ridler et al., 2007). It was also reported that IMTA as three species (*Salmo salar*, *Mytilus edulis*, and *Saccharina latissima*) had better NPV and four species (S*almo salar*, *Mytilus edulis*, and *Saccharina latissimi* and *Strongylocentrotus droebachiensis*) compared to salmon monoculture (*Salmo salar*) (Carras et al., 2020). These findings are consistent with the another study (Bunting & Shpigel, 2009) competition for resources and conflict. Consequently, a new paradigm of ecologically-sound, socially responsible and economically viable aquaculture development based on systems-thinking, resource use efficiency and joint analysis with stakeholders is needed. Horizontal integration, combining aquaculture production systems to optimise resource use efficiency constitutes a promising approach in this regard. Research shows that an array of horizontally integrated systems are technically viable, however, few studies have combined this with consideration of the managerial, financial and economic demands. A bioeconomic modelling approach was employed here to assess the broader implications of adopting horizontally integrated land-based marine aquaculture in temperate and warm water settings. The temperate system, integrating fish, microalgae and shellfish culture and a polishing lagoon was developed on the Atlantic coast of France. Modelling outcomes predicted that when all costs were considered this approach failed to generate a positive Internal Rate of Return (IRR that IMTA three species (shrimp, oyster and seahorse) also found to be economically profitable, yielding an IRR of 131.1%, benefit-cost US$ 20.72, payback periods of 1.1 years and a breakeven point of 18% (Fonseca et al., 2017). It is also reported that warm water systems (sea urchins, shrimp, and halophyte) had better economic profitability from the IMTA system compared to temperate water systems (fish, microalgae, shellfish) (Bunting & Shpigel, 2009) competition for resources and conflict. Consequently, a new paradigm of ecologically-sound, socially responsible and economically viable aquaculture development based on systems-thinking, resource use efficiency and joint analysis with stakeholders is needed. Horizontal integration, combining aquaculture production systems to optimise resource use efficiency constitutes a promising approach in this regard. Research shows that an array of horizontally integrated systems are technically viable, however, few studies have combined this with consideration of the managerial, financial and economic demands. A bioeconomic modelling approach was employed here to assess the broader implications of adopting horizontally integrated land-based marine aquaculture in temperate and warm water settings. The temperate system, integrating fish, microalgae and shellfish culture and a polishing lagoon was developed on the Atlantic coast of France. Modelling outcomes predicted that when all costs were considered this approach failed to generate a positive Internal Rate of Return (IRR. It reduces the environmental costs by creating environmental and social benefits (Bergamo et al., 2021; Carras et al., 2020) through ecosystem goods and services (Yu et al., 2017). Diversification of product and changing market conditions are low-cost option that can boost economic returns while mitigating the risks associated with small-scale monoculture systems (Bergamo et al., 2021)

Table 2: Discount and non-discount measure of the IMTA system

S. No	Authors	Species	BCR	NPV	IRR %	BEP %	PP (Years)
1.	(Whitmarsh et al., 2006)	Integrated salmon mussel culture		$2.63 million			
2.	(Ridler et al., 2007)	Salmon–mussel–kelp		$32,96,037			
3.	(Carras et al., 2020)	Salmon, blue mussel, kelp		$33,974,817			
		If 10 % premium		$40,538,598			
4.	(Fonseca et al., 2017)	Shrimp–oyster IMTA with seahorse		$4,74,530.02	131.10	18.7	1.1
5.	(Bunting and Shpigel 2009)	Fish, oysters and clams		€41,685 (5%) & €40,546 (20%)	Negative (10 years)		34.7
		If 20% premium		€24,523 (5%) & €601 (20%)	19.40		4.1
6.	(Balasubr ama nian et al., 2018)	*Penaeus indicus,* Chanoschanos, Etroplussuratensis, *Mugil cephalus, Crassostrea madrasensis*					
7.	(da Silva et al., 2022)	P. Perna, N. nodosusand K. alvarezii		$97,240.88 (NPV) & $13,211.92 (ANPV) (6%)	41.13 - 50.71		3.35 to 2.94
8.	(Biswas et al., 2020)	*Ipomoea aquatica, Crassostrea cuttackensis*					
9.	(Bergamo et al., 2021)	*Pernaperna,* Rachycentron canadum		1,23,594 (%)	24		1.04

Product Diversification

Species from different trophic levels co-exist in the IMTA system (fed species and extractive species – organic, inorganic and deposit feeders). IMTA systems may be open-water, land-based either from the marine water, brackish water, and freshwater systems that includes variety of species (Neori et al., 2004). Combinations including fish + seaweed + shellfish, shellfish + shrimp, seaweed + shrimp, and fish + shrimp have been used in several IMTA systems (Troell et al., 2009) industry and policy makers as a promising opportunity for large-scale expansion of the aquaculture industry. Simultaneously, there has also been increased interest in both landbased and nearshore aquaculture systems which combine fed aquaculture species (e.g. finfish. The fed species cultured in IMTA system from marine water species (*Sparus auratus, Dicentrarchus labrax, Colossoma macropomum, Gadus morhua, Trachinotus blochii, Pseudoplatystoma punctifer, Epinephelus lanceolatus, Pagellus bogaraveo, Penaeus vannamei*, and *Penaeus monodon*) freshwater species (*Clarias batrachus, Oreochromis niloticus, Cyprinus carpio, Oncorhynchus mykiss, Salmo salar, Macrobrachium borellii, Macrobrachium amazonicum, Oncorhynchus tshawytscha, Tinca tinca,* and *Rutilus rutilus*) and brackishwater species (*Mugil cephalus, Chanos chanos, Liza parsia* and *Exopalaemon carinicauda*).

Extractive organic marine water species (*Mytilus edulis, Mytilus galloprovincialis, Perna viridis Mimachlamys nobilis, Meretrix lusoria* and *Jassa spp*), brackish water species (*Sinonovacula constricata, Tegillarca granosa, Crassostrea gasar, Crassostrea cuttackensis*), extractive inorganic seaweed species (*Osmundea pinnatifida, Saccharina latissima, Halimione portulacoides, Gracilaria bursa-pastoris, Saccharina japonica, Sargassum hemiphyllum, Ulva rigida, Halopithys incurva, Fucus spiralis, Codium intertextum, Dermocorynus dichotomus, Hypnea spinella, Laurencia dendroidea, Caulerpa racemose, Treptacantha abies-marina, Macrocystis pyrifera, Ulva lactuca, Porphyra, Pyropia yezoensis, Gracilaria tikvahiae, Gracilaria lemaneiformis, Gracilaria heteroclada, Caulerpa lentillifera, Eucheuma denticulatum, Sarcocornia neri, Palmaria palmata* and *Enteromorpha spp*), microalgae (*Chlorella vulgaris, Platymonas helgolandica*), diatoms (*Chaetoceros debilis, Navicula parva*), deposite feeders includes seacucumber (*Apostichopus japonicu, Holothuria tubulosa and Cucumaria frondose*), sabellid polychaetes, sponges, sea urchin (*Paracentrotus lividus*), tunicates (*Styela clava*). Most cultured brackish water fed species in IMTA includes *Sparus auratus* and *Dicentrarchus labrax* and whereas in case of cold-water fisheries *Salmo salar* is most commonly cultured. According to the researchers the most commonly cultured extractive organic species includes (*Mytilus edulis*), extractive inorganic species (*Saccharina latissima*), deposit feeder (*Holothuria tubulosa*), because of their growth performances and market demand.

Risk Reduction

Risk reduction in IMTA system refers to an entrepreneur/producer can reduce the financial losses by implementing risk management techniques.

The production risk, marketing risk and managing risk are common risks in aquaculture and is also affect the IMTA production. Some unexcepted risk such as COVID-19 also affect IMTA production these kinds of risks very rare and some papers specially focused on this aspect (Sarà, Zenone, & Tomasello, 2009), therefore we have not included those aspect here. However, apart from these we focused the risk which is pertaining to IMTA system such as animal health and fish escape and environmental impacts that are the key challenges that may weaken the economic profitability and production.

Animal Health

It is significant to monitoring the possibilities of disease transmission within the IMTA reared species by a harmful biological agent including parasite, bacteria, virus, and fungal infection. Few studies suggested that shellfish are key species for spreading the diseases. In IMTA system, organic extractive species are susceptible bioaccumulate organism in particular bivalves (mussel, clam, and oyster) may increase the possibilities of disease risk by filtering the pathogens from fed species faecal matters and environment (Molloy, Pietrak, Bricknell, & Bouchard, 2013) shellfish may also increase disease risk on farms by serving as reservoirs for important finfish pathogens such as infectious pancreatic necrosis virus (IPNV. It is observed that mussels are releasing viable *Vibrio anguillarum* in faecal matter at high concentration (Pietrak, Molloy, Bouchard, Singer, & Bricknell, 2012) and it may also have high risk to transmit the Infectious Pancreatic Necrosis Virus (IPNV) to cocultured species. All these studies suggested that the sea cucumber also may increase the pathogen risk level to IMTA species by serving as disease vectors (Carrier). Invertebrates generally act as an intermediate host in the parasite life cycle, it may be responsible for parasitic infections in cultured species in IMTA. However, the precautionary measures (quarantine practices and selection of specific pathogen-free or resistant species) are highly advisable actions to check the cultivable species health prior to release into IMTA system.

Environmental Impacts – Nutrient Retention

In any farming practices, the nutrient from waste (uneaten feed and fish excreta) are crucial factors that create negative impact such as eutrophication, deterioration of water quality, altering aquatic habitat (plankton and microbial community), etc. Unlike other aquaculture activities, the nutrients are efficiently utilized in IMTA system by nurturing extractive species at different tropic level. For instance, uneaten fish waste and fish excreta are taken as an input by the extractive species in IMTA system, which is technically called nutrient retention. It is the measure of proportion of nutrient remaining for extractive species that obtained from the waste of fed species or high to low trophic level species in IMTA system. The nutrient retention efficiency is quantified by summarizing the eco-physiological response of the fed species and extractive species. The extractive species are categorized into three, first autotrophic species, which consumes inorganic nutrients, second filter feeder which consumes particulate organic matter (POM) and third deposit feeder, which scavenging on POM that settles on the bottom (Soto & Food and Agriculture Organization of the United Nations, 2009) in the marine environment, it has been much less reported. However, in recent years the idea of integrated aquaculture has been often considered a mitigation approach against the excess nutrients/organic matter generated by intensive aquaculture activities particularly in marine waters. In this context, integrated multitrophic aquaculture (IMTA).

Fed Species

Nutrient retention efficiency of the fed species are influenced by the type and size of species reared, feeding level, composition of diets, and water temperature (Islam, 2005; Schneider, Sereti, Eding, & Verreth, 2005). The nitrogen, phosphorus and carbon are the imperative nutrients and 13-43%, 18-36% and 14-38% was observed in marine species respectively (Nederlof et al., 2022). It is related to fecal and feed stability that is determined by feed composition and environmental factors.

Extractive Species

The nutrients, which are not utilized by fed species become input for extractive species. The waste nutrients from fed species are categorized into two fractions, namely organic and inorganic. The fish excretes inorganic N (NH_3/NH_4^+), P (PO_4^{3-}) and C (CO_2). The ammonia (NH_4^+) is transformed into nitrate (NO_3^-) by nitrifying bacteria in aerobic condition with intermediate product of NO_2^-. These three forms of nitrogen referred as dissolved inorganic nitrogen. The POM was produced by aggregating uneaten feed and faeces which converted into dissolved organic matter (5–45% N, 42–54% P and 5–44% C) by extractive species (Nederlof et al., 2022) (Figure 2).

The dissolved inorganic nitrogen is predominantly intake by the seaweeds and microalgae. Its efficiency various with species (A.B. Jones, Preston, & Dennison, 2002). The inorganic P and C efficiencies ranges 3-18% (Hernández et al., 2006; A. B. Jones et al., 2002) and 4–8% respectively (Table 3). The bivalves and oysters – filter feeders – are directly (feed) and indirectly (plankton) removed the particulate organic matter (POM) from the water column. Balance method is generally applied in closed system to estimate the bivalve waste extraction efficiencies. The *Crassostrea gigas* has the assimilation efficiency of 56% when it cultured along with fed species (Lefebvre et al., 2000). In three species IMTA system (fish-microalgae-bivalve system), the bivalves are consumed the microalgae up to 100%, and the microalgae assimilated 67% of TAN-N and 47% of PO_4-P from the fish (Jones & Iwama, 1991). Deposit feeders – sea cucumbers and polychaetes – which are eliminating the POM in IMTA system. The sea cucumbers mitigate aquaculture waste that can extract 0.1–20% OM, 3–10% organic C, 7–16% organic N and 21–25% organic P from sediments abundant with aquaculture waste. The polychaetes has higher waste extraction efficiencies compared to sea cucumbers, which shows 20–85% OM, 40–91% organic C and 30–91% organic N of the aquaculture waste fed.

Table 3: Scaling of a conceptual four-species IMTA system; biomass (tonnes wet weight) and area (m^2) required per extractive species for maximum retention of waste (salmon farm fed 1 tonne of commercial feed) (Nederlof et al., 2022).

	Biomass (tonnes wet weight)			**Area (m^2)**		
	Nitrogen	**Phosphorus**	**Carbon**	**Nitrogen**	**Phosphorus**	**Carbon**
Seaweed	4–23	3–18	4–8	389–23091	356–15680	465–7571
Mussels	0.08–0.11	0.3–0.6	0.3	11–41	42–207	98
Oyster	0.13–0.15	0.6–0.7	0.3	26–30	123–140	54
Sea cucumber	0.7–0.9	ND	2–3	664–930	ND	2366–2821
Polychaete	0.6–1	ND	1–3	1842–3542	ND	4225–9526

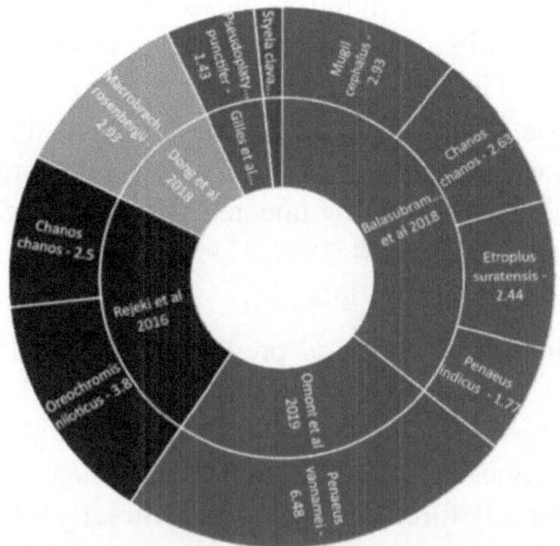

Figure 1: Specific growth rate reported for different fed species in the IMTA system

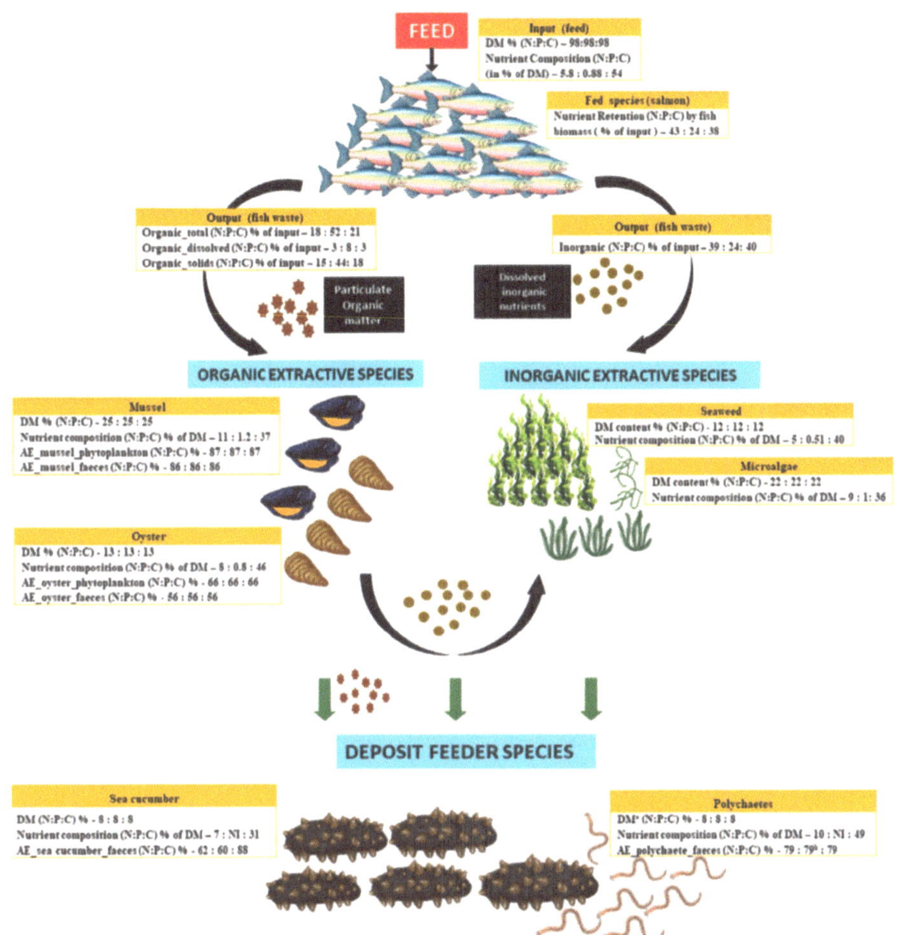

Figure 2: Non-retained nutrients (waste) from salmon culture utilized by the different extractive (data source: (Nederlof et al., 2022). DM-dry matter; AE-assimilation efficiency; NI- No information. The superscript “a” indicates that no information is available for polychaetes; therefore, the same data were used as for sea cucumbers. The superscript “b” indicates that no data available for phosphorus assimilation efficiency; therefore, the same data were used as for nitrogen and carbon efficiency.

Retention Potential of IMTA Systems

In four species IMTA system (salmon–kelp–mussel–polychaete), the nutrient retention potential has greater in theoretical maximum (94% N, 79% P and 94% C) than the closed (65–75% N, 65% P and 45–75% C) and open IMTA system 50% N, 40% P and 40-50% C (Figure 3). Moderate nutrient retention efficiency has observed in closed IMTA system which is related to fish feed is the only nutrient input. The lowest nutrient retention efficiency has noticed in open water IMTA system and is mainly related exposure time to waste nutrients capture efficiency, ambient nutrient, temporal issues and spatial design. For instance, salmon reared in closed IMTA system has better nutrient retention efficiency than the IMTA open system. Similar to this nutrient recycled by the extractive species are greater in closed IMTA system (22%–32% N, 41% P and 7%–37% C) than that of the open IMTA system (7% N, 16% P and 2%–12% C). This is suggesting that closed-loop systems are allowing control of nutrient-rich waste (Chopin et al., 2001) but open-water IMTA lacks this fine control with the dilution of waste occurring by natural seawater movement (currents).

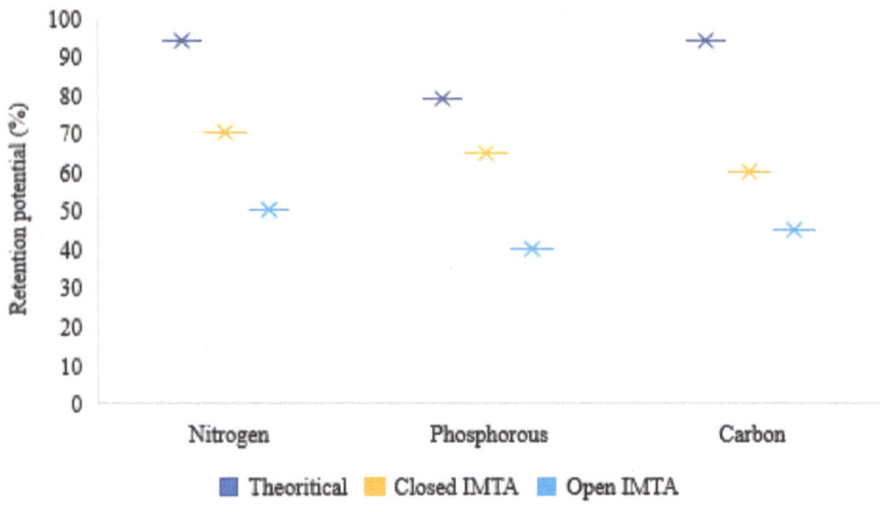

Figure 3: Nutrient retention potentials of different types of IMTA system (fish–seaweed–bivalve–deposit feeder)

Environmental Remediation

Biomitigation

IMTA is a biomitigation strategy to mitigate the negative impacts of aquaculture pollutants on the aquatic ecosystem as well as the environmental surrounding. The biomitigation services depend on the selection of extractive organism species, selected trophic niche, feeding behaviour (deposit feeding, filter feeding, autotrophic) and their ecological functions. The most commonly cultured extractive organisms in the IMTA system are seaweed, bivalve and sea cucumber.

Although the majority of studies employed seaweed such as *Gracilaria lichenoides*, *G. lemaneiformis*, and *G. chilensis*, there were only few studies related to efficacy of *Gracilaria verrucosa* bioremediation in the field (Buschmann et al., 2008; Huo et al., 2012; Mao et al., 2009; Troell et al., 1997; Xu et al., 2008; Yang et al., 2006; Zhou et al., 2006) fertilizers, plant growth control products, human food or animal fodder and feed additives. These multiple uses of algae offer a number of possibilities for coupling this activity to salmon, abalone and filter-feeder farming. In this context, different experiments carried out in Chile have demonstrated that Gracilaria chilensis and Macrocystis pyrifera have great potential in the development of an integrated aquaculture strategy. The present Integrated Multi-Trophic Aquaculture (IMTA. With fish, scallops, or shrimp species culturing with microalgae in the IMTA system, *Gracilaria* sp. can effectively remove excess amount of nutrients (such as N and P) than the other species (Buschmann et al., 2008; Hernåndez et al., 2006; Huo et al., 2011; Jones et al., 2002; Mao et al., 2009; Neori et al., 2004; Troell et al., 1997; Yang et al., 2006; Zhou et al., 2006) fertilizers, plant growth control products, human food or animal fodder and feed additives. These multiple uses of algae offer a number of possibilities for coupling this activity to salmon, abalone and filter-feeder farming. In this context, different experiments carried out in Chile have demonstrated that *Gracilaria chilensis* and *Macrocystis pyrifera* have great potential in the development of an integrated aquaculture strategy. The present Integrated Multi-Trophic Aquaculture (IMTA. A field cultivation experiment indicated that *G. lemaneiformis* removed most of the nutrients in the co-culture system, and its mean absorption rates of N and P were 10.64 and 0.38umol g^{-1} dry weight h^{-1}, respectively (Zhou et al., 2006) including dissolved inorganic nitrogen and phosphorus. In China, fish mariculture in coastal waters has been increasing since the last decade. However, there is no macroalgae commercially cultivated in north China in warm seasons. To exploit fish-farm nutrients as a resource input, and at the same time to reduce the risk of eutrophication, the high-temperature adapted red alga Gracilaria lemaneiformis (Bory et al., 2011) reported that the kelps (*Alaria esculenta* and *Saccharina latissima*) co-cultured in the vicinity of salmon cages could remove 2.3–4.4 kg of dissolved inorganic nitrogen (DIN). In addition to higher bioremediation efficiency, *Gracilaria* (Rhodophyta) are having high commercial value by manufacturing agar-agar, used as feed ingredient and human consumption. Bivalves are filter feeder which consumes particulate organic matter. Mussels can capture ambient seston and transform it into pseudofeces (settling particles), which can have no effect on the quantity of organic matter in the environment (Filgueira et al., 2017) the evaluation of shellfish−finfish synergy requires a combined study of biological and physical processes, which can be achieved by the implementation and coupling of mathematical models. A highly configurable mathematical model was developed that can be applied at the apparent spatial scale of IMTA sites. The model tracks different components of the seston, including feed wastes, fish faeces, shellfish faeces, natural detritus and phytoplankton. Based on the characterization of these fluxes, a hypothetical IMTA site was used to explore different spatial arrangements for evaluating finfish–shellfish farm mitigation efficiency. The site was modelled following a factorial design, which tested 2 levels of background seston concentrations, 3 farm designs, 2 hydrodynamic conditions and 2 levels of aquaculture intensity. The model predicts that mitigation efficiency is highly dependent on the background environmental conditions, obtaining maximum mitigation under oligotrophic conditions that stimulate shellfish filtration activity. The dominance of vertical fluxes of particulate matter triggered by the high settling velocity of finfish aquaculture wastes suggests that suspended shellfish aquaculture cannot significantly reduce organic loading of the seabed. Consequently, this suggests that waste mitigation at IMTA sites should be best achieved by placing organic extractive species (e.g. deposit feeders).

The ability of sea cucumbers to bioremediate IMTA system is confirmed by their consumption and assimilation of aquaculture waste as well as their ability to lower its organic and nutritional content (MacDonald et al., 2011; Nelson et al., 2012; Robinson, & Barrington, 2011; Nelson et al., 2012). Sea cucumbers reportedly eliminated around 70% of the particulate organic carbon (POC) from Atlantic salmon farms (Cubillo et al., 2016; Ren et al., 2012). It needs less space than seaweed to recover nutrients, although as they occupy distinct trophic niches. The average deposition rate in finfish farms, a sea cucumber's net solid uptake (0.645 kg m^{-2} yr^{1}) is insufficient to alter the biological condition of the sediment. When *H. scabra* is grown at high densities and at its maximal ingestion rate, it offers better bioremediation capability for aquaculture effluent. Because sea cucumber waste extraction efficiency was low, it might not be feasible to remove fish waste in large numbers using a commercial scale. When beginning with a more evenly distributed production of sea cucumbers and fish (1.3:1), sea cucumbers can remove 0.73 percent of the annual faeces load from finfish under simulated conditions. Additional research is required in order to thoroughly assess the ecological consequences of IMTA systems on the local ecological level because sea cucumbers might not be capable of handling waste mitigation on their own. Regardless of the niche taken into account, the objective of 100 percent bioremediation appears unrealistic for open water IMTA systems; size of culture area and biomass of extractive organisms based IMTA system needed to be designed in future.

Metagenomics Application in Biomitigation

Metagenomics is defined as the application of massive sequencing technologies to genetic material regarded to microbes and planktons that are obtained directly from the environmental samples (water or soil) from a particular ecosystem (Chithira et al., 2021). Bacteroidetes, Planctomycetes, Firmicutes, Cyanobacteria, and Actinobacteria in the metagenome retrieved from the sediment sample. Unclassified bacteria also contributed a significant portion of the metagenome. Two potential shrimp pathogens viz Vibrio harveyi and Acinetobacter lwoffii detected in the sediment sample show the risk associated with the pond. Microbes that play essential roles in nutrient cycling and mineralization of organic compounds such as Bacteroidetes, Planctomycetes, Gammaproteobacteria, Firmicutes, Cyanobacteria, and Actinobacteria could also be identified. The present study provides preliminary data with respect to the microbial community present in the sediments of a shrimp culture system and emphasizes the application of metagenomics in exploring the microbial diversity of aquaculture systems, which might help in the early detection of pathogens within the system and helps to develop pathogen control strategies in semi-intensive aquaculture systems.The metagenomics approach is applied to explore microbial diversity to understand their effects on fish production in the IMTA system (Figure 4). Studies have been reported that the phylum of Proteobacteria, Bacteroidetes, Chloroflexi, Cyanobacteria, Actinobacteria, Firmicutes, and Planctomycetes are the dominant group in sediment and the Proteobacteria, Bacteroidetes, Cyanobacteria, Actinobacteria, Verrucomicrobia, Firmicutes, Tenericutes, Cloacimonetes and Lentisphaerae are dominant group in water (Deng et al., 2019; Liu et al., 2022; Qiao et al., 2020; Ying et al., 2018) energy flow and farmed-species health. The aim of this study was to evaluate how feed types, fresh frozen fish diet (FFD (Table 4). However, Proteobacteria is the most dominating phylum in both water (average 40.1%) and sediment (61.9%) followed by Cyanobacteria (21.9%) in water and Bacteroidetes (13.3%) in sediment (Deng et al., 2019). These microbial communities are mainly involved in different activities such as N-cycle, fermenting carbohydrates, photosynthesis, and improving the growth of aquatic animals. Different genes are expressed in microbes that are involved in nitrogen and sulfur fixation (Deng et al., 2019). As compared to water, sediments are more abundant in N acquisition genes responsible for denitrification, nitrification, and Dissimilatory Nitrate Reduction to Ammonium (DNRA) that include nirK, nirS, norB, nosZ, nirB, and nrfA genes and genes specific to S-dependent ANAMMOX (Anaerobic Ammonium Oxidation) sulfate reductase (dsrA). A relatively lower level of expression of genes specific to assimilatory nitrate reduction (narB and nirA) is observed in sediment than in water. The correlation of functional gene relationship with environmental properties shows that the pH and salinity of the water shows a positive associationship with relative abundance of N-cycling gene but P and NH_4^+ of the sediment shows a negative relationship with a relative abundance of N-cycling gene (Liu et al., 2022). Another study concludes that IMTA has the capacity to maintain antibiotic resistance and bacterial community structure than the conventional farming system which indicates that it is a sustainable farming practice (Ying et al., 2018)

Phytoplankton has a significant role in aquatic habitat to maintain the water quality by uptake of nutrients in photosynthesis and serve as a food source for cultured aquatic animals. In addition to these, it also hinder the growth of pathogenic bacteria(Martins, Odebrecht, Jensen, D'Oca, & Wasielesky, 2016). In order to establish a balanced and healthy environmental condition, we need to know the phytoplankton community structure of aquatic ecosystems. The phytoplankton community structure is identified in IMTA using 23S rDNA gene primers by using a high-throughput sequencing method. Here, studies identify that 63.98% of Eukaryota reported when sequences were annotated in water samples. The Cyanophyta is the highest (30.88%) followed by Chlorophyta (29.41%), Bacillariophyta (16.18%) and Dinophyta (5.88%). Cryptophyta, Ochrophyta and Haptophyta. In general, Bacillariophyta and Cyanophyta are the major dominant phylum of phytoplankton in the IMTA system (Qiao et al., 2020) integrated multi-trophic aquaculture (IMTA (Table 5). To the best of our knowledge, we have noticed very lesser studies related to other omic technologies which were not covered in this study.

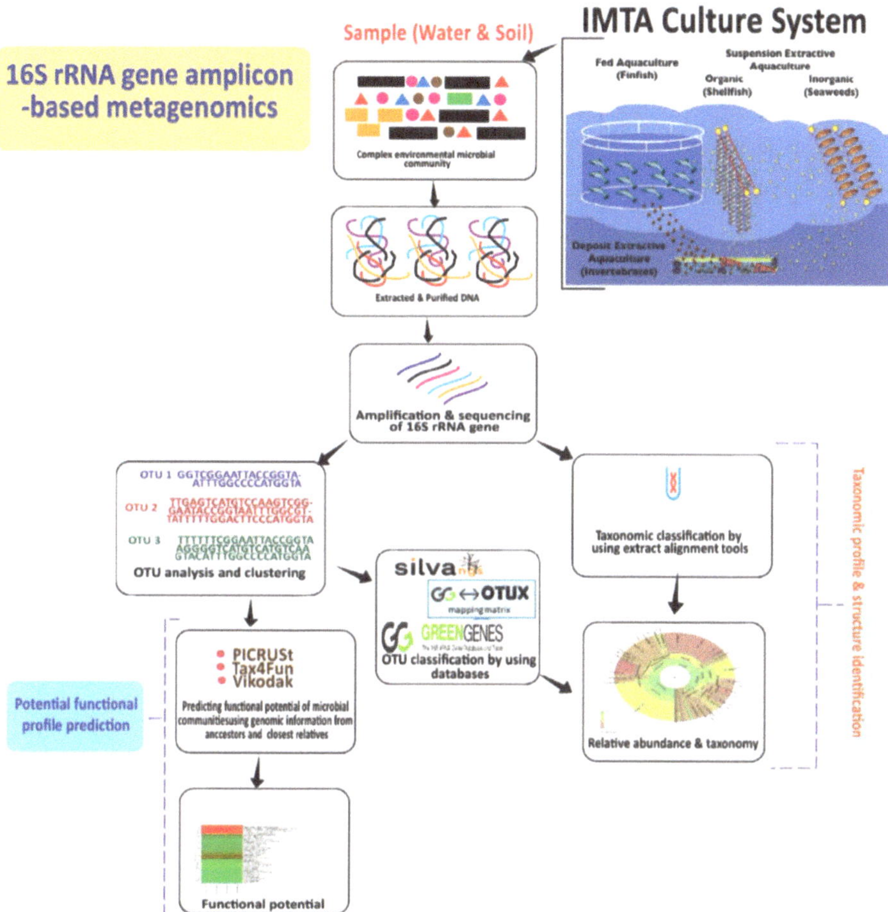

Figure 4: Identification of microbial community and phytoplankton diversity by using metagenomic approach in the IMTA system.

Table 4: Microbial community (Phylum) reported in the IMTA system

Name of the author	Culture systems	Sources	Cyano bacteria	Actino-bacteria	Proteo bacteria	Bactero idetes	Verruco microbia	Chloro flexi	Firmi cutes	Teneri cutes	Cloaci monetes	Lentis phaerae	Plancto mycetes
(Liu et al., 2022)	shrimp-crab pond (SC)	water	√	√	√	√	√						
		sediment	√	√	√	√		√		√			
	shrimp-crab-clam pond (SCC)	water	√	√	√	√	√						
		sediment	√	√	√	√			√	√			
(Deng et al. 2019)	Fresh Frozen Fish Diet (FFD) IMTA pond	water	√		√	√				√	√	√	
		sediment			√	√							
	Formulated Diet (FD) IMTA pond	water	√		√						√	√	
(Ying et al., 2018)		sediment			√	√							
	IMTA	water		√	√	√			√				
	traditional aquaculture	sediment			√			√					√
		water		√		√							√
(Qiao et al., 2020)	IMTA pond (Model I)	water	√		√	√	√		√				
	IMTA pond (Model II)	water	√		√	√	√		√				

Table 5: Phytoplankton community (Phylum) in water reported in the IMTA system

Name of the author	Culture systems	Sampled Months	Cyano phyta	Chloro phyta	Bacillario	Dino phyta	Crypto phyta	Ochro phyta	Hapto phyta	Eugleno phyta	Raphidophyta phyta
(Qiao et al. 2020)	IMTA pond (Model I)	September		√							√
		October			√					√	
	IMTA pond (Model II)	September			√						
		October		√						√	

Social Acceptability

Better Management Practices

It is an effective way of preventing source of pollution to accomplishing better growth without altering the aquatic habitat in open water system. The product such as fish/shrimp/sea cucumber/seaweed produced from the IMTA system may need awareness for better social acceptability which is produced in the eco-friendly manner. Since IMTA in the premature stage, the guidelines yet to be developed to achieve sustainable aquaculture production in IMTA system for that we have reviewed several better management practices guideline of other culture system. In this context, we have suggested the following aspects that may be considered to develop the guidelines of BMP for IMTA system.

- Site selection: selected site should be free from pollutants, cyclone prone area, nursery/fish breeding ground, marine protected area. Easy to transportation.
- Construction: Use adequate cage construction to avoid animal entanglement in water and more threatened species.
- Water Quality: Regular cage cleaning is essential to maintain better water quality
- Feed management: Provision of optimum feed to improve the feed utilization that also may consider the quantum of extractive species leads to reduce nutrient accumulation at the IMTA system site.
- Environmental management: Protect natural biodiversity, especially endangered species, by minimizing ecological impacts. If invasive species are reared, proper care should be taken which reduce disruption to aquatic ecosystems.
- Fish Health: Develop and employ biosecurity practices and quarantine protocols
- Escape: Develop and regularly update an escapes reduction and mitigation plan. And proper cage design can minimize the possibility of escape,
- Permitting: Permission needed to be obtained from the regulatory agencies prior to start the IMTA system in the open water. Comply with all local, territorial and federal regulations,
- Recordkeeping and Reporting: Proper record keeping is prerequisite for better decision making in future. Provide information on a timely basis to regulatory agencies,

Regulatory Governance

Aquaculture has a negative social impact due to a lack of appropriate planning and a failure to fully understand how aquaculture is interconnected with and affects natural systems. Aquaculture regulations are generally designed for a single species/group of species and can stymie a holistic approach by failing to consider species interactions. In order for IMTA to be expanded and implemented, while some serious impeding regulations must be communicated and changed into enabling and flexible regulations, so that it won't be continue to be nonessential and regulatory hurdles (Chopin et al., 2013).

The changing patterns of marine utilization pose significant challenges to governance and management in terms of acceptance within society and at various political levels. IMTA are practiced in coastal areas. The majority of IMTA practiced in coastal areas, which are subject to intense competition, inevitably leading to disputes, potential market chains, and revenue flow considerations needed to be addressed. From a socioeconomic standpoint, a broader participation strategy for its implementation would be preferable, because several different types of seafood products are produced in the same area, each with its own set of issues in terms of technology, biology, markets, and socioeconomic benefits to the participating actors. IMTA's experience, thus, far has been limited, particularly in terms of scope and set-up to become commercially robust and viable. Regulatory impediments to the enlargement and implementation of IMTA should be removed while more attention to be paid in relation to eutrophication and stocking density. Collaborative efforts may provide novel avenues for further investigating IMTA's potential for aquaculture developments by allowing local ownership in decision-making improve the social acceptability (Buck et al., 2018) offshore aquaculture can be seen as a possible step towards the large-scale expansion of marine food production. Integrated multi-trophic aquaculture (IMTA).

Appreciation of Differentiated and Safe Products

The IMTA is developed to create a balanced system in terms of environmental sustainability, economic viability, and social acceptability. It is well documented that IMTA reduces the negative environmental effect but how do people react toward the purchase of IMTA-produced salmon and how much they would be willing to pay for it. In this context, market analysis studies were conducted in New York (Shuve, 2009), Eastern Canada (Barrington et al., 2010), Canada (Martínez-Espiñeira et al., 2015), and U.S. West Coast (Yip et al., 2017). We could observe from all these studies that more than 50% of respondents was willing to pay extra amount 10% for IMTA produced salmon and 36% for IMTA produced oysters than conventional farming practices (Table 6) and is related to the less environmental impact from the aquaculture industry (Knowler et al., 2020).

Table 6: Willing to pay higher for IMTA-labeled products with different combinations of fish reared in the IMTA system

S.No	Authors	Species	WTP	
			Percentage of respondent agreed to pay of the total respondent	Either Percent or amount willing to pay higher for IMTA-labeled products
1.	(Barrington et al., 2010)	kelps, mussels, and salmon	50	10%
2.	(Shuve, 2009)	IMTA model	88	
		IMTA - FW mussel - 10% premium	38	
		IMTA - FW mussel - 20% premium	18	
3.	(Knowler et al., 2020)	IMTA –oysters		24 to 36 %
4.	(Martínez-Espiñeira et al., 2015)	IMTA salmon		CAD 280 million per year
5.	(Yip et al., 2017)	IMTA salmon	63.5	

Challenges

IMTA system is in the infancy stage in the aquaculture industry; therefore, it faces social, economic, and environmental challenges (Barrington et al., 2010). The social conflicts may create a negative perception about the development of IMTA system in open water bodies (Ridler et al., 2007). Further, uncertainties about the economic viability may hinder the development of IMTA technology (Klinger & Naylor, 2012). The market adaptability and consumer acceptance are considered major challenges faced by the development of the IMTA system (Klinger and Naylor, 2012)

Further, IMTA also has various environmental challenges that are influenced by the water current, type of species cultured, etc. The coastal current depends on the seasons and location, which have an impact on tidal waters (Soto & Food and Agriculture Organization of the United Nations, 2009) in the marine environment, it has been much less reported. However, in recent years the idea of integrated aquaculture has been often considered a mitigation approach against the excess nutrients/organic matter generated by intensive aquaculture activities particularly in marine waters. In this context, integrated multitrophic aquaculture (IMTA. When there is close proximity of different species in IMTA, there can be possible to increase pathogen exposure which may transmit the disease and parasites to the wild from diseased IMTA species (Krkošek et al., 2007; Toranzo et al., 2005), posing a health risk to consumers (Klinger and Naylor, 2012).

Summary

Studies have conducted to evaluate the production performance of different – fed and extractive – species combination (product diversification), nutrient retention, biomitigation (environmental remediation), and regulatory measures. The published literature shown that *Sparus auratus* and *Dicentrarchus labrax* were commercially nurtured fed species; similarly, *Mytilus edulis*, *Saccharina latissimi* and *Holothuria tubulosa* under extractive species category. The SGR of the fed species reared in the IMTA system showed a better growth performance in *Penaeus vannamei* and similarly in case of extractive species *U. rigida*, *Perna viridis* and *Holothuria tubulosa*. Despite, IMTA-produced *U. lactuca* has the ability to replace the around 25% fishmeal without altering the growth and survival of the fed species with better FCR. It is also noticed that three species IMTA system showed a remarkably significant output than other species combination and monoculture. The product diversification is not alone boosting the economic benefits but also mitigate the risks associated with culture system. The *Ulva* sp. and *Gracilaria* sp. has an ability to intake up to 100% of nutrient generated by fed species. The species reared in the IMTA system has an ability to mitigate the negative impacts of aquaculture pollutants on the aquatic ecosystem and the environmental surrounding. The metagenomics study of the IMTA system suggests it has potential to maintain antibiotic resistance and bacterial community structure than conventional system which pave the way for sustainable farming. Consumers are willing to pay more for IMTA produced product which may encourage new venture to get into this farming system however, development of guidelines for BMP and regulatory governance for IMTA system will enhance the aquaculture production.

Future Challenges and the Way Forward

The IMTA multi-crop multifariousness strategy (fish, invertebrates, seaweeds and microorganisms) is the way to reduce economic risk and prepare for future climate change consequences. IMTA system economic viability can be increased by stocking larger size fish, cultivating edible seaweeds with commercial value, and building marketing channels for mussels and oysters. The presence of extractive species in spatially widespread locations has been found to significantly reduce the organic loading in fish cage. High throughput sequencing approach will help to understanding the interaction between microbial community and IMTA reared species. There is a possibility of disease transmission from deposit feeder/organic extractive species to co-culture species needed to be addressed. There is a growing interest in open-water IMTA, and more studies are anticipated to improve IMTA processes, and which will result in more sustainable IMTA systems in the future. As a bio-mitigation option, IMTA should be considered as a part of future long-term sustainable development programmes.

Conflict of Interest

The authors declare no competing interests.

Acknowledgement

We are grateful to the Director, ICAR-Central Institute of Fisheries Education, Mumbai, India for providing the logistical facilities. Authors would like to acknowledge Mr. Himansu Shankar Nage, Fish Genetics and Biotechnology Division, ICAR-Central Institute of Fisheries Education, Mumbai, India for assistance in creating the graphical illustration in the present study. The authors did not receive support from any organization for the submitted work. The authors written this paper during their leisure time and institute has no influence on the content presented in the manuscript.

References

Ashkenazi, D. Y., Israel, A., & Abelson, A. (2019). A novel two-stage seaweed integrated multi-trophic aquaculture. Reviews in Aquaculture, 11(1), 246–262. https://doi.org/10.1111/raq.12238

Balasubramanian, C. P., Mhaskar, S. S., Sukumaran, K., Panigrahi, A., Vasagam, K., Kumararaja, P., ... Vasusdevan, N. (2018). Development of integrated multi-trophic aquaculture (IMTA) for tropical brackishwater species in Sindhudurg District, Maharashtra, west coast of India. Indian Journal of Fisheries, 65(1). https://doi.org/10.21077/ijf.2018.65.1.70128-10

Barrington, K., Ridler, N., Chopin, T., Robinson, S., & Robinson, B. (2010). Social aspects of the sustainability of integrated multi-trophic aquaculture. Aquaculture International, 18(2), 201–211. https://doi.org/10.1007/s10499-008-9236-0

Bergamo, G. C. A., Olier, B. S., de Sousa, O. M., Kuhnen, V. V., Pessoa, M. F. G., & Sanches, E. G. (2021). Economic feasibility of mussel (Perna perna) and cobia (Rachycentron canadum) produced in a multi-trophic system. Aquaculture International, 29(5), 1909–1924. https://doi.org/10.1007/s10499-021-00762-x

Biswas, G., Kumar, P., Ghoshal, T. K., Kailasam, M., De, D., Bera, A., Vijayan, K. K. (2020). Integrated multi-trophic aquaculture (IMTA) outperforms conventional polyculture with respect to environmental remediation, productivity and economic return in brackishwater ponds. Aquaculture, 516, 734626. https://doi.org/10.1016/j.aquaculture.2019.734626

Buck, B. H., Troell, M. F., Krause, G., Angel, D. L., Grote, B., & Chopin, T. (2018). State of the Art and Challenges for Offshore Integrated Multi-Trophic Aquaculture (IMTA). Frontiers in Marine Science, 5, 165. https://doi.org/10.3389/fmars.2018.00165

Bunting, S. W., & Shpigel, M. (2009). Evaluating the economic potential of horizontally integrated land-based marine aquaculture. Aquaculture, 294(1–2), 43–51. https://doi.org/10.1016/j.aquaculture.2009.04.017

Buschmann, A. H., Varela, D. A., Hernández-González, M. C., & Huovinen, P. (2008). Opportunities and challenges for the development of an integrated seaweed-based aquaculture activity in Chile: Determining the physiological capabilities of Macrocystis and Gracilaria as biofilters. Journal of Applied Phycology, 20(5), 571–577. https://doi.org/10.1007/s10811-007-9297-x

Carras, M. A., Knowler, D., Pearce, C. M., Hamer, A., Chopin, T., & Weaire, T. (2020). A discounted cash-flow analysis of salmon monoculture and Integrated Multi-Trophic Aquaculture in eastern Canada. Aquaculture Economics & Management, 24(1), 43–63. https://doi.org/10.1080/13657305.2019.1641572

Chithira, M. S., Aishwarya, P. V., Mohan, A. S., & Antony, S. P. (2021). Metagenomic analysis of microbial communities in the sediments of a semi-intensive penaeid shrimp culture system. Journal of Genetic Engineering and Biotechnology, 19(1), 136. https://doi.org/10.1186/s43141-021-00237-9

Chopin, T., & Tacon, A. G. J. (2021). Importance of Seaweeds and Extractive Species in Global Aquaculture Production. Reviews in Fisheries Science & Aquaculture, 29(2), 139–148. https://doi.org/10.1080/23308249.2020.1810626

Chopin, T., Buschmann, A. H., Halling, C., Troell, M., Kautsky, N., Neori, A., … Neefus, C. (2001). Integrating seaweeds into marine aquaculture systems: a key toward sustainability. Journal of Phycology, 37(6), 975–986. https://doi.org/10.1046/j.1529-8817.2001.01137.x

Chopin, T., MacDonald, B., Robinson, S., Cross, S., Pearce, C., Knowler, D., … Hutchinson, M. (2013). The Canadian Integrated Multi-Trophic Aquaculture Network (CIMTAN)—A Network for a New Era of Ecosystem Responsible Aquaculture. Fisheries, 38(7), 297–308. https://doi.org/10.1080/03632415.2013.791285

Cubillo, A. M., Ferreira, J. G., Robinson, S. M. C., Pearce, C. M., Corner, R. A., & Johansen, J. (2016). Role of deposit feeders in integrated multi-trophic aquaculture—A model analysis. Aquaculture, 453, 54–66. https://doi.org/10.1016/j.aquaculture.2015.11.031

da Silva, E. G., Castilho-Barros, L., & Henriques, M. B. (2022). Economic feasibility of integrated multi-trophic aquaculture (mussel Perna perna, scallop Nodipecten nodosus and seaweed Kappaphycus alvarezii) in Southeast Brazil: A small-scale aquaculture farm model. Aquaculture, 552, 738031. https://doi.org/10.1016/j.aquaculture.2022.738031

Deng, Y., Zhou, F., Ruan, Y., Ma, B., Ding, X., Yue, X., Yin, X. (2019). Feed Types Driven Differentiation of Microbial Community and Functionality in Marine Integrated Multitrophic Aquaculture System. Water, 12(1), 95. https://doi.org/10.3390/w12010095

Fei, X. Solving the coastal eutrophication problem by large scale seaweed cultivation. 8.

Filgueira, R., Guyondet, T., Reid, G., Grant, J., & Cranford, P. (2017). Vertical particle fluxes dominate integrated multi-trophic aquaculture (IMTA) sites: Implications for shellfish-finfish synergy. Aquaculture Environment Interactions, 9, 127–143. https://doi.org/10.3354/aei00218

Fonseca, T., David, F. S., Ribeiro, F. A. S., Wainberg, A. A., & Valenti, W. C. (2017). Technical and economic feasibility of integrating seahorse culture in shrimp/oyster farms. Aquaculture Research, 48(2), 655–664. https://doi.org/10.1111/are.12912

Giles, H. (2008). Using Bayesian networks to examine consistent trends in fish farm benthic impact studies. Aquaculture, 274(2–4), 181–195. https://doi.org/10.1016/j.aquaculture.2007.11.020

Granada, L., Sousa, N., Lopes, S., & Lemos, M. F. L. (2016). Is integrated multitrophic aquaculture the solution to the sectors' major challenges? - A review. Reviews in Aquaculture, 8(3), 283–300. https://doi.org/10.1111/raq.12093

Hargrave, B. (2010). Empirical relationships describing benthic impacts of salmon aquaculture. Aquaculture Environment Interactions, 1(1), 33–46. https://doi.org/10.3354/aei00005

Hernández, I., Pérez-Pastor, A., Vergara, J. J., Martínez-Aragón, J. F., Fernández-Engo, M. Á., & Pérez-Lloréns, J. L. (2006). Studies on the biofiltration capacity of Gracilariopsis longissima: From microscale to macroscale. Aquaculture, 252(1), 43–53. https://doi.org/10.1016/j.aquaculture.2005.11.048

Huo, Y., Wu, H., Chai, Z., Xu, S., Han, F., Dong, L., & He, P. (2012). Bioremediation efficiency of Gracilaria verrucosa for an integrated multi-trophic aquaculture system with Pseudosciaena crocea in Xiangshan harbor, China. Aquaculture, 326–329, 99–105. https://doi.org/10.1016/j.aquaculture.2011.11.002

Huo, Y., Zhang, J., Xu, S., Tian, Q., Zhang, Y., & He, P. (2011). RETRACTED: Effects of seaweed Gracilaria verrucosa on the growth of microalgae: A case study in the laboratory and in an enclosed sea of Hangzhou Bay, China. Harmful Algae, 10(4), 411–418. https://doi.org/10.1016/j.hal.2011.02.003

Huo, Y. Z., Xu, S. N., Wang, Y. Y., Zhang, J. H., Zhang, Y. J., Wu, W. N., … He, P. M. (2011). Bioremediation efficiencies of Gracilaria verrucosa cultivated in an enclosed sea area of Hangzhou Bay, China. Journal of Applied Phycology, 23(2), 173–182. https://doi.org/10.1007/s10811-010-9584-9

Islam, Md. S. (2005). Nitrogen and phosphorus budget in coastal and marine cage aquaculture and impacts of effluent loading on ecosystem: Review and analysis towards model development. Marine Pollution Bulletin, 50(1),48–61. https://doi.org/10.1016/j.marpolbul.2004.08.008

Israel, D., Lupatsch, I., & Angel, D. L. (2019). Testing the digestibility of seabream wastes in three candidates for integrated multi-trophic aquaculture: Grey mullet, sea urchin and sea cucumber. Aquaculture, 510, 364–370. https://doi.org/10.1016/j.aquaculture.2019.06.003

Jones, A. B., Preston, N. P., & Dennison, W. C. (2002). The efficiency and condition of oysters and macroalgae used as biological filters of shrimp pond effluent: Efficiency and condition of biofilters of shrimp pond effluent A B Jones et al. Aquaculture Research, 33(1), 1–19. https://doi.org/10.1046/j.1355-557X.2001.00637.x

Jones, T. O., & Iwama, G. K. (1991). Polyculture of the Pacific oyster, Crassostrea gigas (Thunberg), with chinook salmon, Oncorhynchus tshawytscha. Aquaculture, 92, 313–322. https://doi.org/10.1016/0044-8486(91)90037-8

Klinger, D., & Naylor, R. (2012). Searching for Solutions in Aquaculture: Charting a Sustainable Course. Annual Review of Environment and Resources, 37(1), 247–276. https://doi.org/10.1146/annurev-environ-021111-161531

Knowler, D., Chopin, T., Martínez-Espiñeira, R., Neori, A., Nobre, A., Noce, A., & Reid, G. (2020). The economics of Integrated Multi-Trophic Aquaculture: Where are we now and where do we need to go? Reviews in Aquaculture, raq.12399. https://doi.org/10.1111/raq.12399

Krkošek, M., Ford, J. S., Morton, A., Lele, S., Myers, R. A., & Lewis, M. A. (2007). Declining Wild Salmon Populations in Relation to Parasites from Farm Salmon. Science, 318(5857), 1772–1775. https://doi.org/10.1126/science.1148744

Laramore, S., Baptiste, R., Wills, P. S., & Hanisak, M. D. (2018). Utilization of IMTA-produced Ulva lactuca to supplement or partially replace pelleted diets in shrimp (Litopenaeus vannamei) reared in a clear water production system. Journal of Applied Phycology, 30(6), 3603–3610. https://doi.org/10.1007/s10811-018-1485-3

Lefebvre, S., Barille, L., & Clerc, M. (2000). Pacific oyster žCrassostrea gigas/ feeding responses to a fish-farm effluent. 14.

Liu, Q., Li, J., Shan, H., & Xie, Y. (2022). Metagenomic Insights into the Structure of Microbial Communities Involved in Nitrogen Cycling in Two Integrated Multitrophic Aquaculture (IMTA) Ponds. Journal of Marine Science and Engineering, 10(2), 171. https://doi.org/10.3390/jmse10020171

Lupatsch, I., Katz, T., & Angel, D. L. (2003). Assessment of the removal efficiency of fish farm effluents by grey mullets: A nutritional approach: Bioremediation by grey mullets. Aquaculture Research, 34(15), 1367–1377. https://doi.org/10.1111/j.1365-2109. 2003. 00954.x

MacDonald, B. A., Robinson, S. M. C., & Barrington, K. A. (2011). Feeding activity of mussels (Mytilus edulis) held in the field at an integrated multi-trophic aquaculture (IMTA) site (Salmo salar) and exposed to fish food in the laboratory. Aquaculture, 314(1–4), 244–251. https://doi.org/10.1016/j.aquaculture.2011.01.045

Machado, M., Machado, S., Pimentel, F. B., Freitas, V., Alves, R. C., & Oliveira, M. B. P. P. (2020). Amino Acid Profile and Protein Quality Assessment of Macroalgae Produced in an Integrated Multi-Trophic Aquaculture System. Foods, 9(10), 1382. https://doi.org/10.3390/foods9101382

Mao, Y., Yang, H., Zhou, Y., Ye, N., & Fang, J. (2009). Potential of the seaweed Gracilaria lemaneiformis for integrated multi-trophic aquaculture with scallop Chlamys farreri in North China. Journal of Applied Phycology, 21(6), 649–656. https://doi.org/10.1007/s10811-008-9398-1

Martins, T. G., Odebrecht, C., Jensen, L. V., D'Oca, M. G., & Wasielesky, W. (2016). The contribution of diatoms to bioflocs lipid content and the performance of juvenile Litopenaeus vannamei (Boone, 1931) in a BFT culture system. Aquaculture Research, 47(4), 1315–1326. https://doi.org/10.1111/are.12592

Martínez-Espiñeira, R., Chopin, T., Robinson, S., Noce, A., Knowler, D., & Yip, W. (2015). Estimating the biomitigation benefits of Integrated Multi-Trophic Aquaculture: A contingent behavior analysis. Aquaculture, 437, 182–194. https://doi.org/10.1016/j.aquaculture. 2014.11.034

Molloy, S. D., Pietrak, M. R., Bricknell, I., & Bouchard, D. A. (2013). Experimental Transmission of Infectious Pancreatic Necrosis Virus from the Blue Mussel, Mytilus edulis, to Cohabitating Atlantic Salmon (Salmo salar) Smolts. Applied and Environmental Microbiology, 79(19), 5882–5890. https://doi.org/10.1128/AEM.01142-13

Nations, U. Population. Retrieved July 15, 2022, from United Nations website: https://www.un.org/en/global-issues/population

Nederlof, M. A. J., Verdegem, M. C. J., Smaal, A. C., & Jansen, H. M. (2022). Nutrient retention efficiencies in integrated multi-trophic aquaculture. Reviews in Aquaculture, 14(3), 1194–1212. https://doi.org/10.1111/raq.12645

Nelson, E. J., MacDonald, B. A., & Robinson, S. M. C. (2012). The absorption efficiency of the suspension-feeding sea cucumber, Cucumaria frondosa, and its potential as an extractive integrated multi-trophic aquaculture (IMTA) species. Aquaculture, 370–371, 19–25. https://doi.org/10.1016/j.aquaculture.2012.09.029

Neori, A., Chopin, T., Troell, M., Buschmann, A. H., Kraemer, G. P., Halling, C., Yarish, C. (2004). Integrated aquaculture: Rationale, evolution and state of the art emphasizing seaweed biofiltration in modern mariculture. Aquaculture, 231(1–4), 361–391. https://doi.org/10.1016/j.aquaculture.2003.11.015

Neori, A., Msuya, F. E., Shauli, L., Schuenhoff, A., Kopel, F., & Shpigel, M. (2003). A novel three-stage seaweed (Ulva lactuca) biofilter design for integrated mariculture. Journal of Applied Phycology, 15(6), 543–553. https://doi.org/10.1023/B:JAPH.0000004382.89142.2d

Pietrak, M. R., Molloy, S. D., Bouchard, D. A., Singer, J. T., & Bricknell, I. (2012). Potential role of Mytilus edulis in modulating the infectious pressure of Vibrio anguillarum 02β on an integrated multi-trophic aquaculture farm. Aquaculture, 326–329, 36–39. https://doi.org/10.1016/j.aquaculture.2011.11.024

Qiao, L., Chang, Z., Li, J., & Chen, Z. (2020). Phytoplankton community structure in an integrated multi-trophic aquaculture system revealed by morphological analysis and high-throughput sequencing. Applied Ecology and Environmental Research, 18(3), 3907–3933. https://doi.org/10.15666/aeer/1803_39073933

Reid, G. K., & Chopin, T. (2011). Spatial Modelling of Integrated Multi-Trophic Aquaculture (IMTA) Shellfish: Workshop Discussions and Developments. 9.

Rejeki, S., Wisnu Ariyati, R., & Lakhsmi Widowati, L. (2016). Application of integrated multi tropic aquaculture concept in an abraded brackish water pond. Jurnal Teknologi, 78(4–2). https://doi.org/10.11113/jt.v78.8213

Ren, Y., Dong, S., Qin, C., Wang, F., Tian, X., & Gao, Q. (2012). Ecological effects of co-culturing sea cucumber Apostichopus japonicus (Selenka) with scallop Chlamys farreri in earthen ponds. Chinese Journal of Oceanology and Limnology, 30(1), 71–79. https://doi.org/10.1007/s00343-012-1038-6

Ridler, N., Wowchuk, M., Robinson, B., Barrington, K., Chopin, T., Robinson, S., … Boyne-Travis, S. (2007). Integrated multi − trophic aquaculture (imta): A potential strategic choice for farmers. Aquaculture Economics & Management, 11(1), 99–110. https://doi.org/10.1080/13657300701202767

Sarà, G., Zenone, A., & Tomasello, A. (2009). Growth of Mytilus galloprovincialis (mollusca, bivalvia) close to fish farms: A case of integrated multi-trophic aquaculture within the Tyrrhenian Sea. Hydrobiologia, 636(1), 129–136. https://doi.org/10.1007/s10750-009-9942-2

Schneider, O., Sereti, V., Eding, E. H., & Verreth, J. A. J. (2005). Analysis of nutrient flows in integrated intensive aquaculture systems. Aquacultural Engineering, 32(3–4), 379–401. https://doi.org/10.1016/j.aquaeng.2004.09.001

Shpigel, M., Guttman, L., Shauli, L., Odintsov, V., Ben-Ezra, D., & Harpaz, S. (2017). Ulva lactuca from an Integrated Multi-Trophic Aquaculture (IMTA) biofilter system as a protein supplement in gilthead seabream (Sparus aurata) diet. Aquaculture, 481, 112–118. https://doi.org/10.1016/j.aquaculture.2017.08.006

Shuve, H. (2009). Survey Finds Consumers Support Integrated Multitrophic Aquaculture—Effective Marketing Concept Key. 3.

Soto, D., & Food and Agriculture Organization of the United Nations (Eds.). (2009). Integrated mariculture: A global review. Rome: Food and Agriculture Organization of the United Nations.

Tacon, A. G. J., Metian, M., & McNevin, A. A. (2022). Future Feeds: Suggested Guidelines for Sustainable Development. Reviews in Fisheries Science & Aquaculture, 30(2), 135–142. https://doi.org/10.1080/23308249.2020.1860474

Tolon, M. T., Emiroglu, D., Gunay, D., & Ozgul, A. (2017). Sea cucumber (Holothuria tubulosa Gmelin, 1790) culture under marine fish net cages for potential use in integrated multi-trophic aquaculture (IMTA). INDIAN J. MAR. SCI., 46(04), 9.

Toranzo, A. E., Magariños, B., & Romalde, J. L. (2005). A review of the main bacterial fish diseases in mariculture systems. Aquaculture, 246(1–4), 37–61. https://doi.org/10.1016/j.aquaculture.2005.01.002

Troell, M., Joyce, A., Chopin, T., Neori, A., Buschmann, A. H., & Fang, J.-G. (2009). Ecological engineering in aquaculture—Potential for integrated multi-trophic aquaculture (IMTA) in marine offshore systems. Aquaculture, 297(1–4), 1–9. https://doi.org/10.1016/j.aquaculture.2009.09.010

Troell, Max., Halling, C., Nilsson, A., Buschmann, A. H., Kautsky, N., & Kautsky, L. (1997). Integrated marine cultivation of Gracilaria chilensis (Gracilariales, Rhodophyta) and salmon cages for reduced environmental impact and increased economic output. Aquaculture, 156(1–2), 45–61. https://doi.org/10.1016/S0044-8486(97)00080-X

Wang, X., Olsen, L., Reitan, K., & Olsen, Y. (2012). Discharge of nutrient wastes from salmon farms: Environmental effects, and potential for integrated multi-trophic aquaculture. Aquaculture Environment Interactions, 2(3), 267–283. https://doi.org/10.3354/aei00044

Whitmarsh, D. J., Cook, E. J., & Black, K. D. (2006). Searching for sustainability in aquaculture: An investigation into the economic prospects for an integrated salmon–mussel production system. Marine Policy, 30(3), 293–298. https://doi.org/10.1016/j.marpol.2005.01.004

Xu, Y., Fang, J., & Wei, W. (2008). Application of Gracilaria lichenoides (Rhodophyta) for alleviating excess nutrients in aquaculture. Journal of Applied Phycology, 20(2), 199–203. https://doi.org/10.1007/s10811-007-9219-y

Yang, Y.-F., Fei, X.-G., Song, J.-M., Hu, H.-Y., Wang, G.-C., & Chung, I. K. (2006). Growth of Gracilaria lemaneiformis under different cultivation conditions and its effects on nutrient removal in Chinese coastal waters. Aquaculture, 254(1–4), 248–255. https://doi.org/10.1016/j.aquaculture.2005.08.029

Ying, C., Chang, M.-J., Hu, C.-H., Chang, Y.-T., Chao, W.-L., Yeh, S.-L., … Hsu, J.-T. (2018). The effects of marine farm-scale sequentially integrated multi-trophic aquaculture systems on microbial community composition, prevalence of sulfonamide-resistant bacteria and sulfonamide resistance gene sul1. Science of The Total Environment, 643, 681–691. https://doi.org/10.1016/j.scitotenv.2018.06.204

Yip, W., Knowler, D., Haider, W., & Trenholm, R. (2017). Valuing the Willingness-to-Pay for Sustainable Seafood: Integrated Multitrophic versus Closed Containment Aquaculture: Valuing the Willingness-to-Pay for Sustainable Seafood. Canadian Journal of Agricultural Economics/Revue Canadienne d'agroeconomie, 65(1), 93–117. https://doi.org/10.1111/cjag.12102

Yokoyama, H. (2013). Growth and food source of the sea cucumber Apostichopus japonicus cultured below fish cages - Potential for integrated multi-trophic aquaculture. Aquaculture, 372–375, 28–38. https://doi.org/10.1016/j.aquaculture.2012.10.022

Yu, L. Q. J., Mu, Y., Zhao, Z., Lam, V. W. Y., & Sumaila, U. R. (2017). Economic challenges to the generalization of integrated multi-trophic aquaculture: An empirical comparative study on kelp monoculture and kelp-mollusk polyculture in Weihai, China. Aquaculture, 471, 130–139. https://doi.org/10.1016/j.aquaculture.2017.01.015

Zhou, Y., Yang, H., Hu, H., Liu, Y., Mao, Y., Zhou, H., Zhang, F. (2006). Bioremediation potential of the macroalga Gracilaria lemaneiformis (Rhodophyta) integrated into fed fish culture in coastal waters of north China. Aquaculture, 252 (2–4), 264–276. https://doi.org/10.1016/j.aquaculture. 2005.06.046

3

COVID-19 Pandemic Impact on Fisheries & Aquaculture

[1]Hari Prasad Mohale and [1]Sanjay Chandravanshi

[1]*Department of Fisheries Biology and Resource Management Fisheries College and Research Institute (TNJFU), Thoothukudi, Tamil Nadu*

Introduction

The impact of COVID-19 on the fisheries and food industries of vary, and the situation is rapidly evolving. Fisheries products that are highly dependent on international trade suffered quite early in the development of the pandemic from the restrictions and closures of global markets, whereas fresh fish and shellfish supply chains were severely impacted by the closure of the food service sectors (e.g. hotels, restaurants and catering facilities, including school and work canteens). Although COVID-19 does not fish, the fisheries sector is still have indirect impact in the pandemic caused changing consumer demands, market access or logistical problems related to transportation and border restrictions (Purkait et al., 2020). This will in turn have a damaging the effect on fisheries and aquaculture farmers in livelihoods, as well as on their food security of the populations that rely heavily of the fish for protein and essential micronutrients. The processing sector also faced closures due to reduced/lost consumer demand. This has had a significant impact, especially on women, who form the majority of the workforce in the post-harvest sector (Avtar et al., 2021).

The lockdowns implemented by some countries have resulted in logistical difficulties in seafood trade, particularly in relation to transportation and border restrictions. The salmon fish industries in particular have to suffer from the increased in air freight costs and they have to cancellation of the flights. The tuna industries have to also reported that the movement of restrictions for professional seafarers in which that including at-sea fishery observers as well as the marine personnel in the ports thereby the preventing crew changes and repatriation of seafarers "Most of the seafood-consuming countries in the Mediterranean belt including Italy, Spain, Greece, Portugal and France are suffering. Europe is a large consumer of squid, cuttle fish and octopus besides shrimp, the mainstay of Indian seafood export (Kumaran et al., 2021).

Some shortages of seeds, feeds and related aquaculture items (e.g. vaccines) have also been reported, due to restrictions on transportation and travel of personnel, with particular impacts on the aquaculture industry. As a result of the drop in demand, and resulting price drops, the capture fisheries production on the some countries has been brought to a halt or significantly reduced, which may positively influence wild fish stocks in the short term. In the aquaculture they are growing evidence that unsold produced that will result in the increase of the live fish stocks and therefore have higher costs for feeding and also greater risk for the fish mortality (Workie et al., 2020).

Safe to eat Fishery and Aquaculture Products

Fish and fisheries product are a key component to a healthy diet and are safe to eat. Misleading perceptions in some countries have led to decreased consumption of these products. Yet, coronavirus cannot infect aquatic animals (finfish, reptiles, amphibians and invertebrates such as crustaceans and molluscs), therefore these animals do not play an epidemiological role in spreading COVID-19 to humans. While they are no evidence of the viruses that caused respiratory and illnesses are transmitting to food and food packaging industries fishery and aquaculture products can become contaminated if handled by people who are infected with COVID-19 and are who are not following good hygiene practices. For this reason, as before COVID-19 is the important to emphasize the need to implement robust hygiene practices to protect fishery and aquaculture products from contamination (Love et al., 2021).

COVID-19 pandemic affect local / global fish food chain

The fisheries products are among the most traded food products in the world, with 38 percent of fish/seafood entering international trade. At the same time, fishing and fish farming are important at local level for the livelihoods of many fish-dependent communities, as well as for low-income countries and small island developing states. Measures to contain the spread of COVID-19 (e.g. closure of food services, cessation of tourism, reduction of transport services, trade restrictions, etc.) have caused disruption in both domestic and international supply chains (Ruiz-Salmón et al., 2021).

If one of these producers – buyer- seller links is broken by the disease or containment measures, the outcome will be a cascading chain of disruption that will affect the sector's economy. The fact that live, fresh or chilled fish, which represent 45 percent of fish consumed, are highly perishable products presents additional logistical challenges. Furthermore, widespread containment measures can have a notable impact on nations that trade significant amounts of seafood, reducing foreign incomes or threatening food security. Keeping the supply chain open is fundamental to avoid a global food crisis (Kharbikar et al., 2020).

Key Impact on Global and Local Seafood-Dependent Economies and Livelihoods

At a local level, fishers and fish workers are adapting by changing fishing gears, targeting different species or selling their products to the domestic market. Some fishers, fish farmers and fish workers are selling directly to the consumer. While these innovations will support communities, especially women operating in the post-harvest sector, domestic markets have limits both in terms of demand and price (Agrawal et al., 2020). In short terms possible disruptions to economies and livelihoods could come from labour shortages (travel barriers, labour lay-offs, etc.); direct boat-to-consumer sales; aquaculture input shortages (feed, seed, vaccines); as well as fishing (e.g. bait, ice, gear, etc.); competition for sourcing and transport services (something which is already happening in the agricultural sector); and a lack of finance and cash flow (delayed payment of past orders) (Minahal et al., 2020).

Varied impacts in Aquaculture production with uncertainties for the future

Effects on aquaculture production will vary. Due to the market disruptions aquaculture farmer cannot to be sell and their harvest, they must keep large quantities of live fish that need to be fed for an indeterminate period. This increases costs, expenditures and risks. Some farmed species for export (e.g. pangasius) have been reportedly affected by the closure of international markets (China, European Union). Shellfish aquaculture (e. g. oysters) is affected mainly because of the closure of foodservices (e.g. tourism, hotels and restaurants) and retailers (e.g. European Union) In addition, due to a wide range of restrictions by different countries on cargo movements and airport clearing etc, hatchery operators and brood stock traders may find it difficult to trade brood stock for seed production, which could cause a sharp decline in production. Small scale aquaculture on the other hand may benefit from reduced competition with fish imports. The aquaculture production capacity may be also affecting by the difficulty in sourcing inputs (seed and feed) and also finding the labour due to lockdown (Ragumaran et al., 2021).

Implication for the most vulnerable

The pandemic have to create in the unprecedented social, economic and health crisis with the impacts on the most of vulnerable group including also women (like harvesters, processors and vendors), migrant fishers, fish workers, ethnic minorities and crew members. Many individuals are not registered; operate in the informal labour market with no labour market policies, including no social protection and no access to relief package/aid. These conditions might exacerbate the secondary effects of COVID-19, including poverty and hunger (Meharoof et al., 2020). The small-scale fisheries sector is trying to make ends meet, to continue fishing and provide locally-caught fresh fish, but it is experiencing great difficulties due to the closure of markets, limited storage facilities, falling wholesale fish prices and new sanitary requirements and physical distancing measures. Because of these difficulties, many activities have been reduced. The reduction of fishing and fish farming activities will reduce the amount of fish available for processing and trade. Furthermore, mobility restrictions will adversely affect the transfer of fish to markets (Nguyen et al., 2021). In the current situation, migrant fishers and fish workers, including ethnic minorities, are unable to return to their native villages due to lockdowns. They require immediate assistance including food and transportation (where movement restrictions permit) to reach their villages (Das et al., 2021).

Consequence for management

While the closing of fishing operations will offer respite for the some overexploit fish populations similar constraints will also to be apply in science and management support of the operations. For example, fish assessment surveys may be reduced or postponed, obligatory fisheries observer programmes may be temporarily suspended, and the postponement of science and management meetings will delay implementation of some necessary measures and the monitoring of management measures (Aura et al., 2020).

Support the fisheries and aquaculture

On 2 April 2020 , the European Commission adopted a set of ambitious proposals to mitigate the socio-economic impact of the coronavirus in the fishery and aquaculture sectors. This initiative introduces additional measures and provides flexibility to the rules governing expenditure under the European Maritime & Fisheries Fund (EMFF). The package of the specific or temporary measures including: Support of the fisheries temporary cessation of fishing activities with due to coronavirus; Support to aquaculture farmers for the suspension or reduction of production due to coronavirus; Support to producer organizations for the temporary storage of fishery and aquaculture products;

A more flexible reallocation of financial resources within the operational programme of each Member State and a simplified procedure for amending operational programmes with respect to the introduction of the new measures (Loh et al., 2021). The proposed measures, once approved by the European Parliament and the Council, will be eligible retroactively as of 1 February 2020 and will be available until 31 December 2020. This proposal of strengthens and the Coronavirus. The response investment initiative proposed by Commission on 13 March 2020, the revise state aid rules under the new Temporary Framework, adopted on 19 March 2020, which aimed to bring immediate relief to the seafood sector (Bhowmik et al., 2021).

Support for fishers

To mitigate the significant socio-economic consequences of the coronavirus outbreak and the need for liquidity in the economy, the EMFF could grant a financial compensation to fisheries for the temporary on cessation of their fishing activities. The EU will be pay up to 75% for this compensation and the rest to be borne for member States. The support temporary for cessation of the fishing activities and caused by coronavirus outbreak will not be subject to the financial capping applicable to the other cases of temporary cessation, thus allowing Member States to grant support on the basis of need. Vessels that have already reached the maximum six month duration of EMFF support for temporary cessation under Article 33 of the EMFF Regulation will nevertheless be eligible for support under the Coronavirus measures until the end of 2020 (Cooke et al., 2021).

Support for aquaculture farmers

The proposal gives the possibility to grant financial compensation to aquaculture farmers for the temporary suspension or reduction of production, where it is the consequence of the coronavirus outbreak. This compensation will be calculated on the basis of income foregone. The EU have to pay up to 75% for this compensation and the rest to be borne for member States (Kiruba-Sankar et al., 2021).

Major impacts of Covid-19 on the fisheries harvest and post-harvest in India'

The ongoing Covid-19 lockdown has dealt a massive blow to the fisheries sector, especially the harvest and post-harvest sector. ICAR- Central Institute of Fisheries Technology (CIFT), Kochi said, "Fisheries sector provides livelihood to around 16 million people in the country apart from its contribution to food and nutritional security and as a major forex earner. On 2018-19 the seafood exports brought in the $6.7 billion to exchequer (Asante et al., 2021). According to the director of ICAR-CIFT, Kochi will be the supporting in government through the Covid-19 impact assessment study conducted by its scientists in Kerala, Andhra Pradesh, Gujarat and Maharashtra.

The comprehensive reported on the Study of the major impacts of Covid-19 in the fisheries harvest and post-harvest in India (Bhendarkar et al., 2021). Along with the suitable recommendations has been to submit the ICAR of New Delhi, Ministry of Agriculture and Farmers' Welfare for effective policy implications to keep national food supply chains alive in terms of food and nutritional security. The findings will help mitigate the fallout of this pandemic on the food system and protect vulnerable communities in the fisheries sector during the looming food crisis (Gopal et al., 2020).

COVID- 19 impact on the livelihood of marine fishing communities in India

The outbreak of COVID-19, the resultant total lockdown in India has greatly affected to the livelihoods of all fishing communities of India. Total lockdown may help to be arrest and the spread of corona virus; however, quick and effective intervention is required for fishers to minimize the disruptive effect on the livelihoods of vulnerable population particularly on food systems, storage and market chains, both locally and regionally (Yusoff et al., 2021). Fisheries in India are an important sector of food and nutritional security.

More than nine million active fishers directly and indirectly depend on fisheries and aquaculture for their livelihood of which 80% are small scale fishers. It employs over 14 million people and contributes to 1.1 per cent of the Indian GDP. The east coast of India has covers four maritime states as Tamil Nadu, Andhra Pradesh, Odisha and West Bengal and also the Union Territories of Puducherry and Andaman and Nicobar Islands. Fishing is mainly carried out with traditional fishing crafts, motorized boats and small mechanized crafts. The overall east coast regions produce 25 per cent of total Indian marine fish landing.

Small scale fishers in India have issues in three areas: pricing, marketing and organization. Many of these are long term needs but there are a few that are immediate and related to the coronavirus (Selvam et al., 2020). Complete lockdown in the harbours and the landing centres has mostly affected to the fisher-folks' day-to-day earnings in all coastal districts. Small scale fisheries especially are responsible for providing fish as a significant source of protein at low cost for consumers.

This is particularly important for marginalized communities and lack of fish in the diet will have considerable impact on nutrition security of these people (Khan, et al., 2020). In some villages near Chennai, small scale fishers fishing near shore areas are struggling to market their catch. Due to physical distancing norms, only few fisherwomen are able to buy fish from the fishermen in the landing centres. Since the time allotted to sell the fish is very short, they are forced to sell their catch at a low price. For example, if the fish rate was INR 500 per kilogram before COVID-19 lock down, the rate now is just INR 300 to 350 (Sunny et al., 2021).

References

Agrawal, S., Jamwal, A. and Gupta, S., 2020. Effect of COVID-19 on the Indian economy and supply chain.

Asante, E.O., Blankson, G.K. and Sabau, G., 2021. Building Back Sustainably: COVID-19 Impact and Adaptation in Newfoundland and Labrador Fisheries. Sustainability, 13(4), p.2219.

Aura, C.M., Nyamweya, C.S., Odoli, C.O., Owiti, H., Njiru, J.M., Otuo, P.W., Waithaka, E. and Malala, J., 2020. Consequences of calamities and their management: The case of COVID-19 pandemic and flooding on inland capture fisheries in Kenya. Journal of Great Lakes Research, 46(6), pp.1767-1775.

Avtar, R., Singh, D., Umarhadi, D.A., Yunus, A.P., Misra, P., Desai, P.N., Kouser, A., Kurniawan, T.A. and Phanindra, K.B.V.N., 2021. Impact of COVID-19 lockdown on the fisheries sector: A case study from three harbors in western India. Remote Sensing, 13(2), p.183.

Bhendarkar, M.P., Gaikwad, B.B., Ramteke, K.K., Joshi, H.D., Ingole, N.A., Brahmane, M.P. and Gupta, N., 2021. Anticipating the impact of the COVID-19 lockdowns on the Indian fisheries sector for technological and policy reforms.

Bhowmik, J., Selim, S.A., Irfanullah, H.M., Shuchi, J.S., Sultana, R. and Ahmed, S.G., 2021. Resilience of small-scale marine fishers of Bangladesh against the COVID-19 pandemic and the 65-day fishing ban. Marine Policy, 134, p.104794.

Cooke, S.J., Twardek, W.M., Lynch, A.J., Cowx, I.G., Olden, J.D., Funge-Smith, S., Lorenzen, K., Arlinghaus, R., Chen, Y., Weyl, O.L. and Nyboer, E.A., 2021. A global perspective on the influence of the COVID-19 pandemic on freshwater fish biodiversity. Biological Conservation, 253, p.108932.

Das, B.K., Roy, A., Som, S., Chandra, G., Kumari, S., Sarkar, U.K., Bhattacharjya, B.K., Das, A.K. and Pandit, A., 2021. Impact of COVID-19 lockdown on small-scale fishers (SSF) engaged in floodplain wetland fisheries: evidences from three states in India. Environmental Science and Pollution Research, pp.1-12.

Das, B.K., Roy, A., Som, S., Chandra, G., Kumari, S., Sarkar, U.K., Bhattacharjya, B.K., Das, A.K. and Pandit, A., 2021. Impact of COVID-19 Pandemic Lockdown on Small Scale Fishers (SSF) Engaged in Floodplain Wetland Fisheries: Evidences from Three States in India.

Gopal, N., Edwin, L. and Ravishankar, C.N., 2020. COVID-19 Throws The Indian Fisheries Sector Out Of Gear.

Khan, I., Shah, D. and Shah, S.S., 2021. COVID-19 pandemic and its positive impacts on environment: an updated review. International Journal of Environmental Science and Technology, 18(2), pp.521-530.

Kharbikar, H.L., Radhika, C., Naitam, R.K., Daripa, A., Malav, L. and Raghuvanshi, M.S., 2020. Consequences of COVID-19 pandemic and lockdown on food and agribusiness sector in india. Food and scientific reports, 1(6), pp.13-18.

Kiruba-Sankar, R., Saravanan, K., Haridas, H., Praveenraj, J., Biswas, U. and Sarkar, R., 2022. Policy framework and development strategy for freshwater aquaculture sector in the light of COVID-19 impact in Andaman and Nicobar archipelago, India. Aquaculture, 548, p.737596.

Kumaran, M., Geetha, R., Antony, J., Vasagam, K.K., Anand, P.R., Ravisankar, T., Angel, J.R.J., De, D., Muralidhar, M., Patil, P.K. and Vijayan, K.K., 2021. Prospective impact of Corona virus disease (COVID-19) related lockdown on shrimp aquaculture sector in India–a sectoral assessment. Aquaculture, 531, p.735922.

Loh, H.C., Looi, I., Ch'ng, A.S.H., Goh, K.W., Ming, L.C. and Ang, K.H., 2021. Positive global environmental impacts of the COVID-19 pandemic lockdown: a review. GeoJournal, pp.1-13.

Love, D.C., Allison, E.H., Asche, F., Belton, B., Cottrell, R.S., Froehlich, H.E., Gephart, J.A., Hicks, C.C., Little, D.C., Nussbaumer, E.M. and da Silva, P.P., 2021. Emerging COVID-19 impacts, responses, and lessons for building resilience in the seafood system. Global Food Security, p.100494.

Meharoof, M., Gul, S. and Qureshi, N.W., 2020. Indian seafood trade and COVID-19: Anticipated impacts and economics. Food Sci. Rep.

Minahal, Q., Munir, S., Komal, W., Fatima, S., Liaqat, R. and Shehzadi, I., 2020. Global impact of COVID-19 on aquaculture and fisheries: A review. Int. J. Fish. Aquat. Stud, 8, pp.42-48.

Nguyen, X.P., Hoang, A.T., Ölçer, A.I. and Huynh, T.T., 2021. Record decline in global CO2 emissions prompted by COVID-19 pandemic and its implications on future climate change policies. Energy Sources, Part A: Recovery, Utilization, and Environmental Effects, pp.1-4.

Purkait, S., Karmakar, S., Chowdhury, S., Mali, P. and Sau, S.K., 2020. Impacts of novel coronavirus (COVID-19) pandemic on fisheries sector in India: A Minireview. Ind. J. Pure App. Biosci, 8(3), pp.487-492.

Ragumaran, M., Mohan Raj, V., Susan George, S.R. and Mathu Mitha, C., 2021. Impact of Covid 19 pandemic on seafood exports from India. Intern. J. Zool. Invest, 7(2), pp.808-813.

Ruiz-Salmón, I., Fernández-Ríos, A., Campos, C., Laso, J., Margallo, M. and Aldaco, R., 2021. The fishing and seafood sector in the time of COVID-19: Considerations for local and global opportunities and responses. Current Opinion in Environmental Science & Health, 23, p.100286.

Selvam, S., Jesuraja, K., Venkatramanan, S., Chung, S.Y., Roy, P.D., Muthukumar, P. and Kumar, M., 2020. Imprints of pandemic lockdown on subsurface water quality in the coastal industrial city of Tuticorin, South India: a revival perspective. Science of the Total Environment, 738, p.139848.

Sunny, A.R., Sazzad, S.A., Prodhan, S.H., Ashrafuzzaman, M., Datta, G.C., Sarker, A.K., Rahman, M. and Mithun, M.H., 2021. Assessing impacts of COVID-19 on aquatic food system and small-scale fisheries in Bangladesh. Marine policy, 126, p.104422.

Workie, E., Mackolil, J., Nyika, J. and Ramadas, S., 2020. Deciphering the impact of COVID-19 pandemic on food security, agriculture, and livelihoods: A review of the evidence from developing countries. Current Research in Environmental Sustainability, p.100014.

Yusoff, F.M., Abdullah, A.F., Aris, A.Z. and Umi, W.A.D., 2021. Impacts of COVID-19 on the Aquatic Environment and Implications on Aquatic Food Production. Sustainability, 13(20), p.11281.

4

Nutraceuticals as Functional Feed Additives: Roles and Scope in Aquaculture

Raghuvaran N

Fish Nutrition Biochemistry and Physiology Division, ICAR-CIFE Mumbai Maharashtra

Abstract

Aquaculture being one of the chief contributors to animal protein production has grown meteorically in recent years. The higher demand from the sector has resulted in the rapid intensification of aquaculture practices. The culture practices with higher stocking densities and unregulated use of antibiotics in the last decades have not only increased the chance of disease outbreak but also the risk of biomagnification to higher trophic levels. Thus, to ensure nutritional security alternatives are being looked out to ensure higher production without any potential health risks. Thus, dietary intervention with health-benefiting products such as nutraceuticals has gained much interest in food production sectors. Nutraceuticals are health-benefiting products that augment the higher growth rate and health of cultured organisms. Nutraceuticals are of many types and are derived from many different sources. The scope of nutraceuticals as functional feed additives are immense in aquaculture. This review briefly discusses the present status and different roles of these substances as functional additives in aquaculture.

Keywords: Nutraceuticals, Functional feeds, Stress mitigation, Growth promoters, Immuno stimulants

Introduction

Rapid increase in world population in last few decades has challenged the food producing sector (Agriculture, poultry, dairying and aquaculture) massively to meet the global food demand. The population is expected to be 9.9 billion by 2050 which means the food producing sector have to increase production by 60% further more than the current production (FAO, 2020). According to FAO (2016) approximately 17% animal protein and 7% total protein consumed by humans were provided by fish. Accordingly, per capita fish consumption was gradually increasing worldwide at 1.5% rate annually from 9.0 kg in 1961 to 20.5 kg in 2017 (FAO, 2018; Fajardo et al., 2022). Aquaculture which accounts for 47% of total fish production is expected to overtake capture fisheries by 2024 and reach 52% by 2029 (FAO, 2020). Further vertical expansion of aquaculture is not possible due to land and water availability. Hence horizontal expansion of available resources with more intensification in aquaculture practices is the way forward. But further intensification in aquaculture attracts more problems in the form of new diseases, metabolic stress to the cultured species due to higher stocking densities. The usage of antibiotics and drugs is not a wise option due to bioaccumulation in the cultured organisms and water. Hence, the aquaculture sector needs more technological innovation for disease control, efficient drug use, water quality improvement and production of desired tailored fish, (Aklakur et al., 2015; Fajardo et al., 2022; Nasr-Eldahan et al., 2021). Nutraceuticals are products derived from food sources that are purported to provide extra health benefits, in addition to the basic nutritional value found in foods. Dietary intervention with these growth promoting substances have broad and unlimited scope. In addition to their growth and immunity enhancing capability, they are also used for improving the culture water quality. A wide variety of substances such as probiotics, prebiotics, amino acids, fatty acids, vitamins, minerals, nucleotides, yeast products, fortified enzymes, carotenoids and herbal extracts are being used as nutraceuticals. Thus, the use of nutraceuticals in aquaculture sector has gained much interest in recent years specially for improving growth, immunity and for mitigating stress in the intensified culture medium with high stocking densities. In this review, we will see a detailed review of the usage and roles of nutraceuticals used in aquaculture.

Evolution of use of nutraceuticals

The word nutraceutical is a combination of the words nutrition and pharmaceutical. Originally, this concept was recognized for human health improvement but nowadays it gets an importance in animal and fish nutrition. The term "nutraceuticals" is a food or food product that produces health and medicinal benefits, including prevention and treatment of disease defined by Dr. Stephen D. Felice in 1989. Day by day the use of nutraceuticals becomes more popular than drugs and dietary supplements due to their potential nutritional and therapeutic benefits. The nutraceuticals have been reported to have immune-boosting and growth enhancing ability in fish. In aquaculture, some of the feed additives are also included in the list of nutraceuticals like amino acids, vitamins, minerals etc (Brower, 1998).

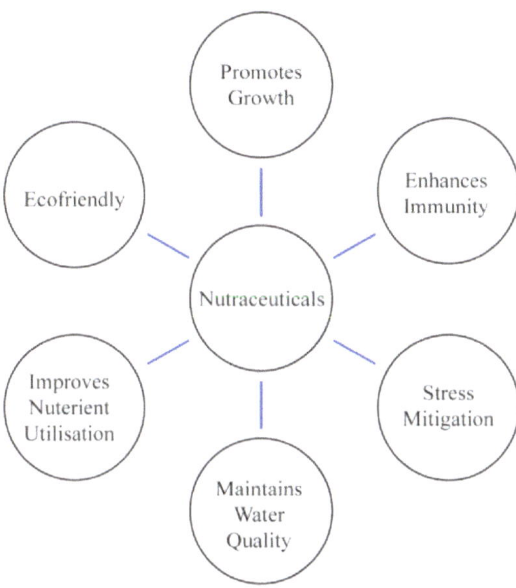

Figure 1: Different roles of nutraceuticals

Nutraceuticals in Aquaculture

Exogenous Enzymes

Exogenous enzymes used as fish feed additives have become an effective way to improve animal performance, and attracted extensive attention from feed producers and aquaculture researchers (Deguara, 1998; Ghomi et al., 2012; Hlophe-Ginindza et al., 2016; Hung et al., 2015; Kolkovski et al., 1993; KuzMina & Golovanova, 2004; Li et al., 2009; Liu et al., 2018; Xavier et al., 2012).. Exogenous enzymes are widely used as additives in fish feed all over the world. Following enzymes are commonly used:

Phytases

To increase feed utilisation and efficiency enzymes such as phytases are commonly used. The major problem in incorporating plant based ingredients in aquafeed is because of the form of phosphorous (phytate) present. The major plant ingredients used in feed such as soyabean meal, rice bran, wheat, maize, cotton seed meal and other ingredients all contains phosphorous in the form of phytate. This can be overcome by supplementing exogenous phytases in fish feed which will convert the phytate phosphorous into digestible form. Apart from adding into complete feeds, these enzymes can also be used individually to treat feed ingredients prior to mixing. This will be particularly beneficial in feeds which are extruded and subjected to higher processing temperatures. The application of phytase enzyme has been studied in several species such as *Oreochromis niloticus* (Portz and Liebert, 2004; Goncalves et al., 2005; Liebert and Portz, 2005; Silva et al., 2005), *Labeo rohita* (Verlhac-Trichet et al., 2014; Baruah et al., 2007a; Baruah et al., 2007b), *Ictalurus punctatus* (Eya and Lovell, 1997; Li and Robinson, 1997), *Ctenophayngodon idella* (Liu et al., 2013b; Liu et al., 2014), *Pagrus major* (Biswas et al., 2007a; Laining et al., 2012), *Clarias gariepinus* (Van Weerd et al., 1999), *Pangasius pangasius* (Debnath et al., 2005), *Pangasianodon hypophthalmus* (Hung et al., 2014), *Cyprinus carpio* (Nwanna and Schwarz, 2007), *Carassius auratus* (Liu et al., 2012), *Tandanus tandanus* (Huynh & Nugegoda, 2011), *Piaractus mesopotamicus* (Furuya et al., 2008), *Litopenaeus vannamei* (Suprayudi et al., 2012) and it is found that phytase supplementation have a positive effect on phosphorus as well as overall nutrient utilisation.

Proteases

Proteases are enzymes that are responsible for hydrolysis of peptide bonds located in the middle (endopeptidase) or at the C or N terminus (exopeptidase). This enzyme hydrolyze the complex proteins into smaller peptides and amino acids enhancing their digestibility. Researchers have reported that supplementing exogenous protease have a significant effect on the feed conversion ratio (FCR), weight gain and ANPU (Apparent net protein utilisation). This combination of exogenous protease and endogenous digestive enzymes allows easy digestion of complex protein molecules present in plant feedstuffs. Studies have been conducted on several fish species such as *Litopenaeus vannamei* (Li et al., 2016), *Carassius auratus gibelio* (Shi et al., 2016), *Oreochromis niloticus* (Hassaan et al., 2019) where supplementation of exogenous proteases improved growth and nutrient utilisation apart from improving metabolic activity. Zheng et al. (2020) have also reported that acid proteases facilitate fish growth. Recently, Sharma et al. (2021) found that addition of bromelain in spirulina based diets improved growth and served as an economical alternative to the highly priced fish meal in fish feed.

Lipases

Lipases are a family of carboxylic ester hydrolytic enzymes, which hydrolyze the ester bonds present in triglycerides sequentially to form glycerol and fatty acids. Lipases and phospholipase A2 are the two types of lipolytic enzymes that have been mostly studied in fish metabolism (Iijima et al., 1998; Zambonino and Cahu, 2007). They play an important role in modulating fish adipose tissues, ultimately affecting carcass yield and flesh quality of farmed fish species (Weil et al., 2013). The oral cavity of fish larvae often contains lipases (Murray et al., 2003; Srivastava et al., 2002), but bile salt-activation is usually required to activate their function (Iijima et al., 1998; Murray et al., 2003). Because juvenile fish have a better ability to digest phospholipids than triglycerides, and the pancreatic lipase of juveniles is non-linear in the digestive level of dietary triglycerides (Cahu et al., 2003), lipases have been added as a feed component to increase the level of lipid digestion. Studies on lipases as an exogenous enzyme added to fish feed and their effects on fish performance are still relatively few and primarily conducted using mixtures with other enzymes. Ghomi et al. (2012) found that adding lipase to the feed improved SGR, final body weight and quality of the meat in addition to the n-3 essential fatty acids content in *Huso huso* fingerlings. Similar findings were reported by Zamini et al. (2014) in Caspian salmon (*Salmo trutta caspius*). Hence, the use of exogenous lipases can have a series of beneficial effects in fish fed with low protein high lipid diet. More studies on the use of these enzymes, in combination with microbial products are needed.

Cellulases

Cellulases are enzymes that hydrolyze β-1,4 glycosidic bonds in the polymer to release glucose units allowing the use of cellulase as a source of carbohydrates to provide energy to the body (Barr et al. 1996). Cellulase activity has also been detected in the GI tract of some fish (Stickney and Shumway 1974), and it has been demonstrated that the cellulase activity is mainly contributed by the gastrointestinal microbial community of the fish rather than by the fish itself (Lindsay and Harris 1980; Saha and Ray 1998). Studies in fish species such as *Dicentrarchus labrax* (Dias et al., 1998), *Orchorhyrchus mykiss* (Bromley and Adkins 1984; Hansen and Storebakken, 2007), *Oreochrornis niloticus* (Amirkolaie et al., 2005) fed with high cellulase have resulted in low intestinal absorption and decreased growth. Whereas Zhou et al. (2013) reported that addition cellulose resulted in improved growth performance in grass carp (*Ctenophayngodon idella*).

Hemicellulases

Hemicellulases are a group of enzymes involved in the breakdown and hydrolysis of galactans, xylans, mannans, and arabans (Chadha et al., 2019). To reduce the feed cost non starch polysaccharides such as wheat, bran and granins are widely used in feed production. As fishes have limited capacity to digest such NSP's these enzymes are gaining attention in aquafeed preparation. Currently, xylanases and glucanases are widely used as feed additives in aquaculture industry. Studies on fish species such as *Salmo salar* (Jacobsen et al., 2018), *Bidyanus bidyanus* (Stone 2003), *O. niloticus* (Maas et al., 2018, 2020) and *Litopenaeus vannamei* (Qiu and Davis, 2017) with addition of xylanases and glucanases have shown positive results in terms of improved growth rate and nutrient utilisation.

Probiotics

Probiotics are live microbes introduced into the GI tract of animals via feed or culture water. It promotes good health by enhancing intestinal microbial balance. They generally act by competing with the harmful bacteria for adhesion sites and nutrients and produces inhibitory compounds leading to enhanced digestion, absorption and gut health (Balcazar et al., 2006). Two of the most important probiotic candidate species Lactic acid bacteria (LAB) and *Bacillus* spp both belong to the phylum *Firmicutes*. These probiotics can tolerate acidic pH and colonize the gut resulting in positive impact on digestion and feed intake. Many species of LAB such as *C. maltaromaticum*, *L. curvatus*, *L. sakei*, and *L. plantarum* have been isolated from salmonid intestine. Several species of LAB have been demonstrated to reduce the adhesion of pathogens such as *V. anguillarum, Y. ruckeri*, *A. hydrophila* and *A. salmonicida.* Even *Vibrio alginolyticus* despite their pathogenicity in some fishes have been used as probiotics to inhibit *V. parahaemolyticus* in *Litopenaeus vannamei.* Several probiotics species such as *Micrococcus luteus (*Abd El-Rhman et al., 2009), *S. cerevisiae* (Abdel-Tawwab et al., 2008; Boonanuntanasarn et al., 2019), *Pediococcus acidilactici* (Ferguson et al., 2010; Castex et al., 2010), *B. thuringiensis* (Ghanei-Motlagh et al., 2021), *L. plantarum* AH 78 (Hamdan et al., 2016), *B. subtilis* C-3102 (He et al., 2013), *L. rhamnosus* ATCC 7469 (Hooshyar et al., 2020), *Lactococcus lactis* (Kim et al., 2013), *L. acidophilus* (Venkat et al., 2004) have been reported to enhance growth performance and immunity in a wide variety of fish species.

Prebiotics

The concept of prebiotics was introduced by Glenn Gibson and Marcel Roberfroid in 1955 and described prebiotic as a non-digestible food ingredient that beneficially affects the host by serving as a food to the beneficial microorganisms in the host digestive system. In 6th ISAPP meeting 2008, prebiotics was defined as selectively fermented ingredient that results in specific changes activity and composition of gut microbiota conferring benefits upon host health. It can be of any compound, substrate, long chain sugar, nutrient, or fibre. They are resistant to the acidic pH in stomach and can be fermented by gut microbiota. The commonly used prebiotics in aquaculture include β-glucan, inulin, mannan oligosaccharide (MOS), galactooligosaccharide (GOS), fructooligosaccharides (FOS), arabinoxylan oligosaccharide (AXOS) and oligosaccharides. Generally, these prebiotics are applied in mixture to yield better results. In many fish species mixture of prebiotics had been applied namely *Sciaenops ocellatus (*FOS + GOS + MOS + GGM), *Litopenaeus* *vannamei (*inulin + FOS + GOS), *Eriocheir sinensis (*β-glucan + inulin + MOS) and have yielded positive results. Among all the combination of β-glucan and MOS have been quite successful and resulted in better growth and immunity in many species such as *Oreochromis niloticus* (Ismail et al., 2019), *Channa striata* (Munir et al., 2016), *Eriocheir sinensis* (Lu et al., 2019), *Salmo trutta caspius* (Jami et al., 2019).

Nucleotides

Nucleotides are intracellular compounds that plays key roles in nearly all biochemical processes in the body and are building blocks of nucleic acids. It consists of a nitrogenous base, a sugar (either ribose or deoxyribose) and one to three phosphate groups. Normally the nucleotides are produced through the salvage pathway or direct de novo synthesis from amino acids in the body. But both these processes are complex and under stressful conditions such as high stocking densities they are deemed as conditionally essential. The exogenous application of nucleotides in aquaculture as dietary supplements have shown promise in enhancing feeding and disease resistance in the intensively cultured animals. Supplementation of nucleotides have been found to promote hyperplasia, hypertrophy and upregulation of growth related genes such as GH, IGFs, MRFs and myostatin (Asaduzzaman et al., 2019). They are also mainly used as functional additives in low fish meal based alternative protein source diets where nucleotide supplementation has been found to increase feeding rate and utilisation of plant based sources. Nucleotides such as IMP, AMP and GMP acts as flavour enhancers and are commonly used as chemoattractants in aquafeed (Lin et al., 2007; Lin et al., 2009). The nucleotides have been found to enhance growth and feed utilization in many different species such as Tilapia, *O. niloticus* (Ramadan & Atef, 1991, Asaduzzaman et al., 2017; Reda et al., 2018), Atlantic salmon, *salmo salar* (Burrells et al., 2001), grouper (Lin et al., 2009), rainbow trout, *O. mykiss* (Tahmasebi-Kohyani et al., 2011), Sea cucumber, *Apostichopus japonicas* (Wei et al., 2015), cray fish (Safari et al., 2015). Exogenous application of nucleotides have also been reported to improve non-specific immune components such as lysosome activity, complement activity, total serum protein and bactericidal activity (Hossain et al., 2017, Kader et al., 2018). But the exact mechanism by which nucleotide supplementatio n influences growth and immunity is not completely understood. Thus, more researches are needed in this regard in future to know about its mechanism of action.

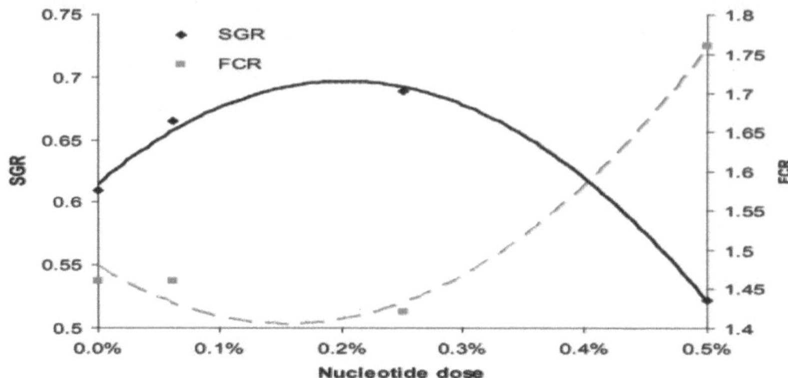

Figure. 2: Growth and nutrient utilization of rainbow trout fed diets with increasing doses of nucleotides (adopted from Ringo et al. 2012)

Acidifiers

Free organic acids or their salts and a few inorganic acids such as phosphoric acids are considered as acidifiers. They are generally short chain fatty acids or short chain organic acids which are volatile and weak acids with one or more carboxyl groups in their structure. The most commonly used organic acids and their salts include acetic acid (sodium acetate), formic acid (their salts potassium formate, calcium formate), lactic acid (in form of calcium lactate), citric acid, butyric acid (in their salt form as sodium butyrate).

Mechanism of action

Inclusion of acidifiers in feed improves gut morphology by lowering the pH of stomach and also act as antimicrobial agents resulting in extended shelf life of the feeds. Lowering the pH in stomach helps in stimulation, secretion and activation of inactive digestive enzymes such as pepsinogen to active pepsin in stomach (Yúfera et al., 2012). In this way, the acidifiers reduce the effect of pathogens especially the gram negative bacteria like *E. coli* and *Salmonella* by pH reduction and improves the immunity of fish. Mroz et al. (2000) have reported that inclusion of organic acids in the diet have positive effect on protein hydrolysis. Acidifiers in feed can counteract the antinutritional factors present in plant protein sourced feeds such as phytic acid and improves the bioavailability of minerals such as calcium, magnesium, phosphorus, copper and zinc (Malicki et al., 2004; Baruah et al., 2005). Overall, this results in better nutrient digestibility especially the protein and enhanced absorption, utilisation and retention of amino acids in the body (Sardar et al., 2020). Organic acids also have good amount of energy stored in the form of chemical bonds. Hence, they can act as a good source of energy. The short chain fatty acids are absorbed via intestinal epithelia by passive diffusion and enters the TCA cycle directly for ATP production.

Table 1: Effect of acidifiers on growth performance and immunity in different fish species

Acidifiers form	Species	Effects	References
Formic acid and calcium propionate	Mrigal (*Cirrhinus mrigala)*	Improved growth performance and immunity	Kumar et al. (2017)
Sodium propionate	European seabass (*Dicentrarchus labrax)*	Enhanced growth, immunity and gut health	Wassef et al. (2020)
Sodium butyrate	European seabass (*Dicentrarchus labrax)*	Improved growth, immunity and intestinal microbiota, villi area and goblet cells	Abdel-Mohsen et al. (2018)
Potassium diformate	Turbot (*Scophthalmus maximus*)	Enhanced growth performance	Fuchs et al. (2015)
Formic acid, propionic acid and calcium propionate	Nile tilapia (*O. niloticus*)	Increase in final BW, weight gain, specific growth rate and feed conversion ratio and protection against *Aeromonas* infection	Reda et al. (2016)
Sodium alginate	Common carp (*Cyprinus carpio)*	Enhanced resistance against *Edwardsiella tarda*	Fujiki et al. (1994)
Poly-hydroxy-butyrate (PHB)	European sea bass (*Dicentrarchus labrax*)	Improved growth performance, intestinal bacterial community	De Schryver et al. (2010)
Citric acid	Rohu *(Labeo rohita*)	Improves the action of phytase, nutrient digestibility and growth performance	Baruah et al. (2007)
Citric acid	Beluga *(Huso huso*)	Enhanced growth performance and phosphorus digestibility	Khajepour and Hosseini (2012)

Carotenoids

Carotenoids are a group of 800 fat soluble pigments found mainly in plants, algae, animals, fungi and photosynthetic bacteria. Out of this, only plants, bacteria, algae and fungi can synthesize carotenoids. Animals obtain their source from the diet. They are hydrophobic, lipophilic and insoluble in water. In high value species such as salmon, trout, shrimp and lobsters a bright and appropriate colour is considered an essential quality. Hence, in this species many natural and synthetic colourants are added in their diet to enhance the colouration and freshness. In animals, carotenoids are the second most widely occurring natural pigments after melanin.

Types and Sources of Carotenoids

Carotenoids are present in many different forms such as α-Carotene, β-carotene, β-cryptoxanthin, lutein, zeaxanthin, and lycopene. Among these, *β*-carotene is the most abundant pigment found in wide variety of fruits, vegetables and has the highest provitamin A activity. Lycopene is a red pigment found only in vegetables in which tomato is considered the most important source. Both lutein and zeaxanthin are present in green and dark green leafy vegetables like spinach, sprouts and broccoli. Astaxanthin is a reddish-pink pigment found in aquatic animals such as shrimp, crab and lobster including their processing wastes. They are also synthesized in large quantities by micro algae such as *Chlorella vulgaris*, *Haematococcus pluvialis* and *Phaffia rhodozyma*.

Functions of Carotenoids

Carotenoids are mainly added in fish diets to enhance the flesh colour and quality. But apart from this, they are also found to have several biochemical functions in the body. Carotenoids play an important role in the photosynthetic process and prevents damage against light and oxygen. They are strong antioxidants which scavenge ROS products such as singlet molecular oxygen and peroxyl radicals generated during the process of lipid peroxidation (Stahl and sies, 2003). They also play an important role in protection of cellular membranes and lipo proteins against oxidative damage. Carotenoids also act as precursor of vitamin A in the fish body. Pigments such as astaxanthin, canthaxanthin and zeaxanthin were found to be precursors of vitamin A_1 and A_2 in rainbow trout (*O. mykiss*). They are mobilized from muscle to gonads during the reproductive phase which confirms their role in reproduction (Chavarria and Flores., 2013). Supplementation of astaxanthin has resulted in better fecundity, egg quality and larval survival rate under captive conditions (Sawanboonchun et al., 2008). Astaxanthin directly affects the liver glycogen and increases its concentration. Thus, there is great potential and scope for carotenoids in aquaculture. It will help the industry to shift away from synthetic ingredients and colourants.

Micro Algae

Micro algae are diverse group of autotrophic organisms that grows rapidly utilizing the light energy and produce more biomass per hectare than vascular plants. Globally, algae have been used as a food source and for treatment of various diseases for hundreds of years. Algae can form numerous compounds that are being used as nutraceuticals presently and have graet potential to be intensely exploited. Different microalgae species such as *Anabaena, Chlorella, Nostoc, Spirulina, Chlamydomonas, Scenedesmus, Synechococcus, Parietochloris,* and *Porphyridium* are potential sources of different vitamins namely Vit. A (Retinol), Vit. B1 (Thiamine), Vit. B2 (Riboflavin), Vit. B3 (Niacin), Vit. B6 (Pyridoxine), B9 (Folic acid), B12 (Cobalamin), Vit. C (L-Ascorbic acid), D, E (Tocopherol), and B7 (Biotin) and minerals such as Potassium, zinc, copper, selenium, sodium, calcium, magnesium and nitrogen. Some of the important micro algae which are explored are:

Chlorella

Chlorella is a green unicellular algae composed of 55-70% protein, 1-4% chlorophyll, 9-18% dietary fiber, numerous vitamins and minerals. It contains all the essential amino acids needed by the fishes. Extracts of chlorella have been found to have antiinflammatory, antitumour, antioxidant, antimicrobial and anticataract properties. It is also reported to reduce blood pressure, lower cholesterol, enhances immune system and fastens wound healing (Guzman et al., 2003). However, the cost of production is quite high and challenges its potential uses in aquaculture.

Spirulina

Spirulina is a prokaryotic cyano bacteria rich in nutrients such as B vitamins, chlorophyll, phycocyanin, vitamin E, omega 6 fatty acids and numerous minerals (Gershwin and Belay., 2008). It consists of 60-70% protein and 10 times more beta carotene than carrots per unit mass. It is widely used in treating various diseases such as diabetes, hyper tension, high blood pressure and have also been reported to have anticancer properties (Tang and suter, 2011). It positively affects the cholesterol metabolism by increasing HDL levels and leads to healthy cardio vascular functions.

Haematococcus

Haematococcus is a unicellular, green alga that is widely used as a nutraceutical component in aquaculture and pharmaceuticals production. Among all, *Haematococcus pluvialis* is widely cultivated and used for its astaxanthin producing capacity. It is largest natural source of astaxanthin (1.5-3% of dry weight) which have antioxidant capacity 10 times more than beta carotene and 1000 times more than vitamin E (Lorenz et al., 2000; Guerin et al., 2003). This astaxanthin possesses anti oxidative, anti-cancer and anti-microbial activities.

Dunaliella

Dunaliella (*D. salina* widely used) a unicellular, green alga that is being used in nutraceutical production as it contains high amounts of beta carotene, glycerol and protein. It produces numerous carotenoid pigments with beta carotene being the dominant of all and small amounts of alpha carotene, lycopene and lutein (Murthy et al., 2005). Beta carotene from *Dunaliella* positively influences intra cellular communication and enhance immune response against many diseases.

Amino Acids

The protein content (i.e the ideal amino acids in the fish diet has shown promise to mitigate stress caused by temperature (Kumar et al., 2011), salinity (Li et al., 2011), pH (Qi et al., 2020) and crowding (Abdel-Tawwab, 2012). This can be due to increased need of essential amino acids under stressful conditions for synthesis of immunoproteins, antibodies and energy. The roles of different amino acids as nutraceuticals is discussed below.

Aromatic Amino Acids

Tryptophan is the major aromatic amino acid that has shown promise as stress mitigator. Being the precursor of serotonin and melatonin it positively affects the stress mitigating capacity by lowering the production of cortisol and increases antioxidant capacity (Hoseini et al., 2012; Ciji et al., 2015). Serotonin also stimulates the lymphocyte proliferation and regulates immunity (Hoseini et al., 2016). Researchers have reported that tryptophan decreases cortisol and glucose level in many species such as Common carp, *Cyprinus carpio* (Hoseini et al., 2012), Mrigal, *Cirrhinus mrigala* (Tejpal et al., 2009), Meagre, *Argyrosomus regius* (Gonzalez-Silvera et al., 2018; Herrera et al., 2020), Atlantic cod, *Gadus morhua* (Basic et al., 2013; Herrera et al., 2017), Rohu, *Labeo rohita* (Akhtar et al., 2013; Kumar et al., 2014) and Rainbow trout, *Oncorhynchus mykiss* (Hoseini et al., 2020). Other aromatic amino acids such as phenylalanine and tyrosine are also involved in stress alleviation via synthesis of thyroid hormones (T3 & T4) and catecholamines such as epinephrine, norepinephrine and dopamine (Herrera et al., 2019).

Sulphur Amino Acids

The Sulphur containing amino acids include methionine, cysteine and homocysteine. Of this, the methionine acts as the precursor for synthesis of antioxidant glutathione. Also being a methyl donor it is involved in the synthesis of polyamines (Machado et al., 2015). Taurine a derivative of methionine is also involved in osmoregulation, antioxidation and feeding stimulation (Li et al., 2016). Studies on cysteine have also shown that its supplementation reduces oxidative stress (Xie et al., 2016).

Basic Amino Acids

The Major basic amino acids include arginine and histidine. Arginine is the precursor of nitric oxide (NO) and polyamines (Hoseini et al., 2020). It modulates the antioxidant defense and other immune reactions. Likewise, histidine being rich in haemoglobin plays a critical role in haematopoiesis and stress tolerance (Bahabadi et al., 2018).

Essential Fatty Acids and Phospholipids

Normally, fishes require high amount of essential fatty acids as energy source under stressful conditions. PUFA's such EPA and arachidonic acid serves as the precursor for synthesis of eicosanoids such as prostaglandins, thromboxanes and leukotrienes. This prostaglandin produced from arachidonic acid modulates he HPI (Hypothalamic-pituitary-interrenal) axis and regulates cortisol level in fish under stress (Bell et al., 2003; Anholt et al., 2004). Phospholipids enhance the membrane fluidity (Zhao et al. 2013), antioxidant capacity (Gao et al., 2014; Cai et al., 2016) and facilitates digestion of lipids. Soy lecithin a source of phospholipid has shown promise in enhancing stress tolerance in fishes.

Vitamins

Among the vitamins, vitamin C (L-ascorbic acid) and vitamin E (α-tocopherol) are major stress mitigating vitamins widely studied in fishes. Vitamin C modulates lysozyme activity, complement system, antioxidant capacity and proliferation of natural killer cells (Gao et al., 2013; Chen et al., 2015). It also regulates RBC production and haemoglobin synthesis by facilitating iron absorption by intestine and scavenges free radicals generated during cellular activity. Likewise, vitamin E modulates heat shock proteins (HSP) expression (Cheng et al., 2018), cortisol level (Montero et al., 2001) and increases total serum protein, lysozyme, complement and respiratory burst activity (Liu et al., 2014). The roles of other vitamins such as vitamin A and D have not been much explored in fish. Among the B complex vitamins Vitamin B1 (Thiamine), Vitamin B2 (Riboflavin), Vitamin B9 (Folic acid), Vitamin B12 (Cyanocobalamin) are essential for production of erythrocytes (Barros et al., 2009). Choline a water soluble vitamin regulates HPI axis against environmental stress whereas myoinositol is a major intracellular osmolyte and a constituent of cell membrane. It prevents free oxygen radicals and prevents oxidative damage (Jiang et al., 2015).

Minerals

Most of the studies on minerals in fish are to determine their basic requirements. Studies on their roles in stress mitigation is very limited except for a few trace minerals such as selenium, copper, zinc and manganese. Among these, selenium is the major stress mitigating mineral studied as it plays major role in glutathione peroxidase system. It is also involved in secretion of growth hormone indirectly from the pituitary gland (Ibrahim et al., 2021). In addition, selenium also acts a cofactor for synthesis of digestive enzymes and thyroid hormones aiding in better growth and nutrient utilization (Shenkin, 2006). The selenium requirement increases under stress resulting rapid utilization of body selenium. Thus, it becomes essential to supplement them in diet of fishes cultured intensively. In selenium, organic form is better compared to the inorganic form in terms of utilization and digestibility. Selenium in form of nanoparticles is widely studied in different species. Selenium nanoparticles have a particle size of 1–100 nm by reduction of selenite or selenite. These nanoparticles have higher bioavailability, surface activity, catalytic efficiency, specific surface area, and particle dispersion compared to other forms. The effects of dietary selenium nanoparticles studied in different fishes is given in the Table 2. Zinc, copper and manganese are the other potential trace minerals involved in stress mitigation as they function as cofactors for several enzymes such as superoxide dismutase (SOD) and antioxidant defense system. Zinc also acts as neuro modulators and plays a key role in Na^+/K^+ ATPase activity. Kumar et al. (2016) have reported that dietary zinc improves tolerance to high temperature stress and lead toxicity. Studies on nutraceutical potential of copper and manganese are very scarce.

Table. 2: The effects of orally taken SeNPs diets on growth performance and immunity in fish.

Fish	**Senps (Size)**	**Senps (Dose)**	**Growth Indices**	**Effect**	**References**
Crucian Carp *(Carassius auratus gibelio)*	30–45 nm	0.5, 5, 10 µg/g	IFN, WG, PER	+	Tian et al. (2020)
African Sharptooth *Catfish (Clarias gariepinus)*	90 nm	2 mg/kg, 4 mg/kg	WG, PER, RBC, HCT	+	Owalabi et al. (2020)
Common Carp *(Cyprinus carpio)*	30–45 nm	0.5, 1, mg/kg	FI, WG, Lysozyme activity	+	Ashouri et al. (2015)
European Seabass *(Dicentrarchus labrax)*	60 ± 20 nm	0.25, 0.5, 1 mg/kg	WG, FCR, SGR, RBC, WBC, PCV	+	Abd El Kader et al. (2020)
Grass Carp *(Ctenophary ngodon idella)*	-	0.3, 0.6 mg/kg 0.9, 1.2 mg/kg	SGR, FBW, IGF-1, A/G, RBC	+	Yu et al. (2020)

Nile Tilapia *(Oreochromis niloticus)*	30–45 nm	0.7 mg/kg	PER, SGR, WG, GLO, LYM	+	Neamat-Allah et al. (2019)
Striped Catfish *(Pangasianodon hypophthalmus)*	205 nm	1 mg/kg	WG, PER, SGR	+	Kumar et al. (2018a)
Yellowfin Seabream *(Acanthopagrus latus)*	-	1, 1.5 mg/ kg	WG	+	Kienarsi et al. (2020)
Red Seabream *(Pagrus major)*	38.7 nm	0.5, 1, 2 mg/ kg	FW, WG, SGR	+	Dawood et al. (2019a)

Herbal and Animal Extracts

Many studies have been done on use of herbal and animal extracts as nutraceuticals in aquaculture. These extracts have a wide variety of bioactive compounds such as polyphenols, terpenoids, essential oils, glycosides, alkaloids, organic acids and polysaccharides. Many herbal extracts of plants including moringa, *Moringa oleifera* (Gbadamosi et al., 2016; Khalil & Kormi, 2017); bamboo, *Melocanna baccifera* (Khan et al., 2018); oak, *Quercus castaneifoli* (Paray et al., 2020); rosemary, *Rosmarinus officinalis* (Yousefi et al., 2019); green tea, *Camellia sinensis* (Hwang et al., 2013) have been successfully used to combat various infectious agents and environmental stressors. Invertebrate extracts such as Ete from marine tunicate (*Ecteinascidia turbinate)*, Hde from abalone (*Haliotis discus hannai)* and firefly squid (*Watasenia scintillans)* extract are also reported to have immunostimulatory effects against pathogens.

Conclusion

Aquaculture being the future of fish protein source is increasingly becoming intensive and occurrence of diseases is inevitable. Many different strategies have been tried and dietary interventions to augment growth appears more promising and feasible. Thus, the use of nutraceuticals can be a better strategy to combat stress, improve growth and dietary nutrient utilization. But, most of the studies pertaining to the potential of these nutraceuticals have been done on laboratory scale. Hence, more research should focus on testing these nutraceuticals on field level to develop novel feed additives and ingredients to boost aquaculture production without compromising the growth and health of the cultured animals.

References

Abd El-Rhman, A.M., Khattab, Y.A. and Shalaby, A.M., 2009. Micrococcus luteus and Pseudomonas species as probiotics for promoting the growth performance and health of Nile tilapia, Oreochromis niloticus. Fish & Shellfish Immunology, 27(2), pp.175-180.

Abdel-Mohsen, H.H., Wassef, E.A., El-Bermawy, N.M., Abdel-Meguid, N.E., Saleh, N.E., Barakat, K.M. and Shaltout, O.E., 2018. Advantageous effects of dietary butyrate on growth, immunity response, intestinal microbiota and histomorphology of European Seabass (Dicentrarchus labrax) fry. Egyptian Journal of Aquatic Biology and Fisheries, 22(4), pp.93-110.

Abdel-Tawwab, M., 2012. Effects of dietary protein levels and rearing density on growth performance and stress response of Nile tilapia, Oreochromis niloticus (L.). International Aquatic Research, 4(1), pp.1-13.

Abdel-Tawwab, M., Abdel-Rahman, A.M. and Ismael, N.E., 2008. Evaluation of commercial live bakers' yeast, Saccharomyces cerevisiae as a growth and immunity promoter for Fry Nile tilapia, Oreochromis niloticus (L.) challenged in situ with Aeromonas hydrophila. Aquaculture, 280(1-4), pp.185-189.

Akhtar, M.S., Pal, A.K., Sahu, N.P., Ciji, A., Meena, D.K. and Das, P., 2013. Physiological responses of dietary tryptophan fed Labeo rohita to temperature and salinity stress. Journal of Animal Physiology and Animal Nutrition, 97(6), pp.1075-1083.

Aklakur, M., Asharf Rather, M. and Kumar, N., 2016. Nanodelivery: an emerging avenue for nutraceuticals and drug delivery. Critical reviews in food science and nutrition, 56(14), pp.2352-2361.

Amirkolaie, A.K., Leenhouwers, J.I., Verreth, J.A. and Schrama, J.W., 2005. Type of dietary fibre (soluble versus insoluble) influences digestion, faeces characteristics and faecal waste production in Nile tilapia (Oreochromis niloticus L.). Aquaculture research, 36(12), pp.1157-1166.

Asaduzzaman, M., Ikeda, D., Abol-Munafi, A.B., Bulbul, M., Ali, M.E., Kinoshita, S., Watabe, S. and Kader, M.A., 2017. Dietary supplementation of inosine monophosphate promotes cellular growth of muscle and upregulates growth-related gene expression in Nile tilapia Oreochromis niloticus. Aquaculture, 468, pp.297-306.

Ashouri, S., Keyvanshokooh, S., Salati, A.P., Johari, S.A. and Pasha-Zanoosi, H., 2015. Effects of different levels of dietary selenium nanoparticles on growth performance, muscle composition, blood biochemical profiles and antioxidant status of common carp (Cyprinus carpio). Aquaculture, 446, pp.25-29.

Balcázar, J.L., De Blas, I., Ruiz-Zarzuela, I., Vendrell, D., Calvo, A.C., Márquez, I., Gironés, O. and Muzquiz, J.L., 2007. Changes in intestinal microbiota and humoral immune response following probiotic administration in brown trout (Salmo trutta). British journal of nutrition, 97(3), pp.522-527.

Barr, B.K., Hsieh, Y.L., Ganem, B. and Wilson, D.B., 1996. Identification of two functionally different classes of exocellulases. Biochemistry, 35(2), pp.586-592.

Barros, M.M., Ranzani-Paiva, M.J.T., Pezzato, L.E., Falcon, D.R. and Guimarães, I.G., 2009. Haematological response and growth performance of Nile tilapia (Oreochromis niloticus L.) fed diets containing folic acid. Aquaculture Research, 40(8), pp.895-903.

Baruah, K., Pal, A.K., Sahu, N.P., Debnath, D., Nourozitallab, P. and Sorgeloos, P., 2007. Microbial phytase supplementation in rohu, Labeo rohita, diets enhances growth performance and nutrient digestibility. Journal of the world Aquaculture Society, 38(1), pp.129-137.

Baruah, K., Pal, A.K., Sahu, N.P., Jain, K.K., Mukherjee, S.C. and Debnath, D., 2005. Dietary protein level, microbial phytase, citric acid and their interactions on bone mineralization of Labeo rohita (Hamilton) juveniles. Aquaculture Research, 36(8), pp.803-812.

Baruah, K., Sahu, N.P., Pal, A.K., Debnath, D., Yengkokpam, S. and Mukherjee, S.C., 2007. Interactions of dietary microbial phytase, citric acid and crude protein level on mineral utilization by rohu, Labeo rohita (Hamilton), juveniles. Journal of the World Aquaculture Society, 38(2), pp.238-249.

Baruah, K., Sahu, N.P., Pal, A.K., Jain, K.K., Debnath, D. and Mukherjee, S.C., 2007. Dietary microbial phytase and citric acid synergistically enhances nutrient digestibility and growth performance of Labeo rohita (Hamilton) juveniles at sub-optimal protein level. Aquaculture Research, 38(2), pp.109-120.

Basic, D., Schjolden, J., Krogdahl, Å., von Krogh, K., Hillestad, M., Winberg, S., Mayer, I., Skjerve, E. and Höglund, E., 2013. Changes in regional brain monoaminergic activity and temporary down-regulation in stress response from dietary supplementation with L-tryptophan in Atlantic cod (Gadus morhua). British journal of nutrition, 109(12), pp.2166-2174.

Bell, J.G., McEvoy, L.A., Estevez, A., Shields, R.J. and Sargent, J.R., 2003. Optimising lipid nutrition in first-feeding flatfish larvae. Aquaculture, 227 (1-4), pp.211-220.

Biswas, A.K., Kaku, H., Ji, S.C., Seoka, M. and Takii, K., 2007. Use of soybean meal and phytase for partial replacement of fish meal in the diet of red sea bream, Pagrus major. Aquaculture, 267(1-4), pp.284-291.

Boonanuntanasarn, S., Ditthab, K., Jangprai, A. and Nakharuthai, C., 2019. Effects of microencapsulated Saccharomyces cerevisiae on growth, hematological indices, blood chemical, and immune parameters and intestinal morphology in striped catfish, Pangasianodon hypophthalmus. Probiotics and antimicrobial proteins, 11(2), pp.427-437.

Bromley, P.J. and Adkins, T.C., 1984. The influence of cellulose filler on feeding, growth and utilization of protein and energy in rainbow trout, Salmo gairdnerii Richardson. Journal of fish biology, 24(2), pp.235-244.

Burrells, C., Williams, P.D. and Forno, P.F., 2001. Dietary nucleotides: a novel supplement in fish feeds: 1. Effects on resistance to disease in salmonids. Aquaculture, 199(1-2), pp.159-169.

Cahu, C.L., Infante, J.L.Z. and Barbosa, V., 2003. Effect of dietary phospholipid level and phospholipid: neutral lipid value on the development of sea bass (Dicentrarchus labrax) larvae fed a compound diet. British Journal of Nutrition, 90(1), pp.21-28.

Cai, Z., Feng, S., Xiang, X., Mai, K. and Ai, Q., 2016. Effects of dietary phospholipid on lipase activity, antioxidant capacity and lipid metabolism-related gene expression in large yellow croaker larvae (Larimichthys crocea). Comparative Biochemistry and Physiology Part B: Biochemistry and Molecular Biology, 201, pp.46-52.

Castex, M., Lemaire, P., Wabete, N. and Chim, L., 2010. Effect of probiotic Pediococcus acidilactici on antioxidant defences and oxidative stress of Litopenaeus stylirostris under Vibrio nigripulchritudo challenge. Fish & shellfish immunology, 28(4), pp.622-631.

Chadha, B.S., Rai, R. and Mahajan, C., 2019. Hemicellulases for lignocellulosics-based bioeconomy. In Biofuels: Alternative Feedstocks and Conversion Processes for the Production of Liquid and Gaseous Biofuels (pp. 427-445). Academic Press.

Chen, Y.J., Yuan, R.M., Liu, Y.J., Yang, H.J., Liang, G.Y. and Tian, L.X., 2015. Dietary vitamin C requirement and its effects on tissue antioxidant capacity of juvenile largemouth bass, Micropterus salmoides. Aquaculture, 435, pp.431-436.

Cheng C-H, Guo Z-X, Wang A-L (2018b) Growth performance and protective effect of vitamin E on oxidative stress pufferfish (Takifugu obscurus) following by ammonia stress. Fish Physiology and Biochemistry 44: 735–745.

Dawood, M.A., Koshio, S., Zaineldin, A.I., Van Doan, H., Ahmed, H.A., Elsabagh, M. and Abdel-Daim, M.M., 2019. An evaluation of dietary selenium nanoparticles for red sea bream (Pagrus major) aquaculture: growth, tissue bioaccumulation, and antioxidative responses. Environmental Science and Pollution Research, 26(30), pp.30876-30884.

de Castro Silva, T.S., Furuya, W.M., dos Santos, V.G., Botaro, D., Silva, L.C.R., Sales, P.J.P., Hayashi, C., dos Santos, L.D. and Furuya, V.R.B., 2005. Coeficientes de digestibilidade aparente da energia e nutrientes do farelo de soja integral sem e com fitase para a tilápia do Nilo (Oreochromis niloticus. Acta Scientiarum. Animal Sciences, 27(3), pp.371-376.

De Schryver, P., Sinha, A.K., Kunwar, P.S., Baruah, K., Verstraete, W., Boon, N., De Boeck, G. and Bossier, P., 2010. Poly-β-hydroxybutyrate (PHB) increases growth performance and intestinal bacterial range-weighted richness in juvenile European sea bass, Dicentrarchus labrax. Applied microbiology and biotechnology, 86(5), pp.1535-1541.

Debnath, D., Sahu, N.P., Pal, A.K., Jain, K.K., Yengkokpam, S. and Mukherjee, S.C., 2005. Mineral status of Pangasius pangasius (Hamilton) fingerlings in relation to supplemental phytase: absorption, whole-body and bone mineral content. Aquaculture Research, 36(4), pp.326-335.

Deguara, S., 1998. Effect of supplementary enzymes on the growth and feed utilisation of gilthead sea bream, Sparus aurata L.

Dias, J., Huelvan, C., Dinis, M.T. and Métailler, R., 1998. Influence of dietary bulk agents (silica, cellulose and a natural zeolite) on protein digestibility, growth, feed intake and feed transit time in European seabass (Dicentrarchus labrax) juveniles. Aquatic Living Resources, 11(4), pp.219-226.

El-Kader, A., Marwa, F., Fath El-Bab, A.F., Abd-Elghany, M.F., Abdel-Warith, A.W.A., Younis, E.M. and Dawood, M.A., 2021. Selenium nanoparticles act potentially on the growth performance, hemato-biochemical indices, antioxidative, and immune-related genes of European seabass (Dicentrarchus labrax). Biological Trace Element Research, 199(8), pp.3126-3134.

Eya, J.C. and Lovell, R.T., 1997. Net absorption of dietary phosphorus from various inorganic sources and effect of fungal phytase on net absorption of plant phosphorus by channel catfish Ictalurus punctatus. Journal of the World Aquaculture Society, 28(4), pp.386-391.

Fajardo, C., Martinez-Rodriguez, G., Blasco, J., Mancera, J.M., Thomas, B. and De Donato, M., 2022. Nanotechnology in aquaculture: Applications, perspectives and regulatory challenges. Aquaculture and Fisheries, 7(2), pp.185-200.

FAO, 2020. The State of World Fisheries and Aquaculture: Sustainability in action, The State of World Fisheries and Aquaculture (SOFIA). FAO, Rome, Italy. https://doi.org/10.4060/ca9229en

Ferguson, R.M.W., Merrifield, D.L., Harper, G.M., Rawling, M.D., Mustafa, S., Picchietti, S., Balcàzar, J.L. and Davies, S.J., 2010. The effect of Pediococcus acidilactici on the gut microbiota and immune status of on-growing red tilapia (Oreochromis niloticus). Journal of applied microbiology, 109(3), pp.851-862.

Fuchs, V.I., Schmidt, J., Slater, M.J., Zentek, J., Buck, B.H. and Steinhagen, D., 2015. The effect of supplementation with polysaccharides, nucleotides, acidifiers and Bacillus strains in fish meal and soy bean based diets on growth performance in juvenile turbot (Scophthalmus maximus). Aquaculture, 437, pp.243-251.

Fujiki, K., Matsuyama, H. and Yano, T., 1994. Protective effect of sodium alginates against bacterial infection in common carp, Cyprinus carpio L. Journal of Fish diseases, 17(4), pp.349-355.

Furuya, W.M., Michelato, M., Silva, L.C.R., dos Santos, L.D., Silva, T.S.C., Schamber, C.R., Vidal, L.V.O. and Furuya, V.R.B., 2008. Fitase em rações para juvenis de pacu (Piaractus mesopotamicus). Boletim do Instituto de Pesca, 34(4), pp.489-496.

Gao, J., Koshio, S., Ishikawa, M., Yokoyama, S., Nguyen, B.T. and Mamauag, R.E., 2013. Effect of dietary oxidized fish oil and vitamin C supplementation on growth performance and reduction of oxidative stress in Red Sea Bream Pagrus major. Aquaculture Nutrition, 19(1), pp.35-44.

Gao, J., Koshio, S., Wang, W., Li, Y., Huang, S. and Cao, X., 2014. Effects of dietary phospholipid levels on growth performance, fatty acid composition and antioxidant responses of Dojo loach Misgurnus anguillicaudatus larvae. Aquaculture, 426, pp.304-309.

García-Chavarría, M. and Lara-Flores, M., 2013. The use of carotenoid in aquaculture. Research Journal of Fisheries and Hydrobiology, 8(2), pp.38-49.

Gbadamosi, O.K., Fasakin, A.E. and Adebayo, O.T., 2016. Hepatoprotective and stress-reducing effects of dietary Moringa oleifera extract against Aeromonas hydrophila infections and transportation-induced stress in Nile tilapia, Oreochromis niloticus (Linnaeus 1757) fingerlings. International Journal of Environmental & Agriculture Research, 2, pp.121-128.

Gershwin, M.E. and Belay, A. eds., 2007. Spirulina in human nutrition and health. CRC Press.

Ghanei-Motlagh, R., Gharibi, D., Mohammadian, T., Khosravi, M., Mahmoudi, E., Zarea, M., Menanteau-Ledouble, S. and El-Matbouli, M., 2021. Feed supplementation with quorum quenching probiotics with anti-virulence potential improved innate immune responses, antioxidant capacity and disease resistance in Asian seabass (Lates calcarifer). Aquaculture, 535, p.736345.

Ghomi, M.R., Shahriari, R., Langroudi, H.F., Nikoo, M. and von Elert, E., 2012. Effects of exogenous dietary enzyme on growth, body composition, and fatty acid profiles of cultured great sturgeon Huso huso fingerlings. Aquaculture international, 20(2), pp.249-254.

Gonçalves, G.S., Pezzato, L.E., Barros, M.M., Kleeman, G.K. and Rocha, D.F., 2005. Efeitos da suplementação de fitase sobre a disponibilidade aparente de Mg, Ca, Zn, Cu, Mn e Fe em alimentos vegetais para a tilápia-do-nilo. Revista Brasileira de Zootecnia, 34, pp.2155-2163.

Gonzalez-Silvera, D., Herrera, M., Giráldez, I. and Esteban, M.Á., 2018. Effects of the dietary tryptophan and aspartate on the immune response of meagre (Argyrosomus regius) after stress. Fishes, 3(1), p.6.

Guerin, M., Huntley, M.E. and Olaizola, M., 2003. Haematococcus astaxanthin: applications for human health and nutrition. Trends in Biotechnology, 21(5), pp.210-216.

Guzmán, S., Gato, A., Lamela, M., Freire-Garabal, M. and Calleja, J.M., 2003. Anti-inflammatory and immunomodulatory activities of polysaccharide from Chlorella stigmatophora and Phaeodactylum tricornutum. Phytotherapy Research, 17(6), pp.665-670.

Hamdan, A.M., El-Sayed, A.F.M. and Mahmoud, M.M., 2016. Effects of a novel marine probiotic, Lactobacillus plantarum AH 78, on growth performance and immune response of Nile tilapia (Oreochromis niloticus). Journal of Applied Microbiology, 120(4), pp.1061-1073.

Hansen, J.Ø. and Storebakken, T., 2007. Effects of dietary cellulose level on pellet quality and nutrient digestibilities in rainbow trout (*Oncorhynchus mykiss*). Aquaculture, 272(1-4), pp.458-465.

Hassaan, M.S., Mohammady, E.Y., Soaudy, M.R., Elashry, M.A., Moustafa, M.M., Wassel, M.A., El-Garhy, H.A., El-Haroun, E.R. and Elsaied, H.E., 2021. Synergistic effects of Bacillus pumilus and exogenous protease on Nile tilapia (*Oreochromis niloticus*) growth, gut microbes, immune response and gene expression fed plant protein diet. Animal Feed Science and Technology, 275, p.114892.

He, S., Zhang, Y., Xu, L., Yang, Y., Marubashi, T., Zhou, Z. and Yao, B., 2013. Effects of dietary Bacillus subtilis C-3102 on the production, intestinal cytokine expression and autochthonous bacteria of hybrid tilapia Oreochromis niloticus♀× Oreochromis aureus♂. Aquaculture, 412, pp.125-130.

Herrera, M., Fernández-Alacid, L., Sanahuja, I., Ibarz, A., Salamanca, N., Morales, E. and Giráldez, I., 2020. Physiological and metabolic effects of a tryptophan-enriched diet to face up chronic stress in meagre (Argyrosomus regius). Aquaculture, 522, p.735102.

Hlophe-Ginindza, S.N., Moyo, N.A., Ngambi, J.W. and Ncube, I., 2016. The effect of exogenous enzyme supplementation on growth performance and digestive enzyme activities in O reochromis mossambicus fed kikuyu-based diets. Aquaculture research, 47(12), pp.3777-3787.

Hooshyar, Y., Abedian Kenari, A., Paknejad, H. and Gandomi, H., 2020. Effects of Lactobacillus rhamnosus ATCC 7469 on different parameters related to health status of rainbow trout (Oncorhynchus mykiss) and the protection against Yersinia ruckeri. Probiotics and Antimicrobial Proteins, 12(4), pp.1370-1384.

Hoseini, S.M., Ahmad Khan, M., Yousefi, M. and Costas, B., 2020. Roles of arginine in fish nutrition and health: insights for future researches. Reviews in Aquaculture, 12(4), pp.2091-2108.

Hoseini, S.M., Hosseini, S.A. and Soudagar, M., 2012. Dietary tryptophan changes serum stress markers, enzyme activity, and ions concentration of wild common carp Cyprinus carpio exposed to ambient copper. Fish Physiology and Biochemistry, 38(5), pp.1419-1426.

Hoseini, S.M., Mirghaed, A.T., Mazandarani, M. and Zoheiri, F., 2016. Serum cortisol, glucose, thyroid hormones' and non-specific immune responses of Persian sturgeon, Acipenser persicus to exogenous tryptophan and acute stress. Aquaculture, 462, pp.17-23.

Hossain, M.S., Koshio, S., Ishikawa, M., Yokoyama, S., Sony, N.M., Kader, M.A., Maekawa, M. and Fujieda, T., 2017. Effects of dietary administration of inosine on growth, immune response, oxidative stress and gut morphology of juvenile amberjack, Seriola dumerili. Aquaculture, 468, pp.534-544.

Hung, L.T., Thanh, N.T., Pham, M.A. and Browdy, C.L., 2015. A comparison of the effect of dietary fungal phytase and dicalcium phosphate supplementation on growth performances, feed and phosphorus utilization of tra catfish juveniles (P angasianodon hypophthalmus Sauvage, 1878). Aquaculture Nutrition, 21(1), pp.10-17.

Huynh, H.P. and Nugegoda, D., 2011. Effects of dietary supplements on growth performance and phosphorus waste production of australian catfish, tandanus tandanus, fed with diets containing soybean meal as fishmeal replacement. Journal of the World Aquaculture Society, 42(5), pp.645-656.

Iijima, N., Tanaka, S. and Ota, Y., 1998. Purification and characterization of bile salt-activated lipase from the hepatopancreas of red sea bream, Pagrus major. Fish Physiology and Biochemistry, 18(1), pp.59-69.

Ismail, M., Wahdan, A., Yusuf, M.S., Metwally, E. and Mabrok, M., 2019. Effect of dietary supplementation with a synbiotic (Lacto Forte) on growth performance, haematological and histological profiles, the innate immune response and resistance to bacterial disease in Oreochromis niloticus. Aquaculture Research, 50(9), pp.2545-2562.

Jacobsen, H.J., Samuelsen, T.A., Girons, A. and Kousoulaki, K., 2018. Different enzyme incorporation strategies in Atlantic salmon diet containing soybean meal: Effects on feed quality, fish performance, nutrient digestibility and distal intestinal morphology. Aquaculture, 491, pp.302-309.

Jami, M.J., Kenari, A.A., Paknejad, H. and Mohseni, M., 2019. Effects of dietary b-glucan, mannan oligosaccharide, Lactobacillus plantarum and their combinations on growth performance, immunity and immune related gene expression of Caspian trout, Salmo trutta caspius (Kessler, 1877). Fish & Shellfish Immunology, 91, pp.202-208.

Jiang, W.D., Hu, K., Zhang, J.X., Liu, Y., Jiang, J., Wu, P., Zhao, J., Kuang, S.Y., Tang, L., Tang, W.N. and Zhang, Y.A., 2015. Soyabean glycinin depresses intestinal growth and function in juvenile Jian carp (Cyprinus carpio var Jian): protective effects of glutamine. British Journal of Nutrition, 114(10), pp.1569-1583.

Kader, M.A., Bulbul, M., Abol-Munafi, A.B., Asaduzzaman, M., Mian, S., Noordin, N.B.M., Ali, M.E., Hossain, M.S. and Koshio, S., 2018. Modulation of growth performance, immunological responses and disease resistance of juvenile Nile tilapia (Oreochromis niloticus)(Linnaeus, 1758) by supplementing dietary inosine monophosphate. Aquaculture Reports, 10, pp.23-31.

Kanani, H.G., Ramezani, S. and Zoriezahra, S.J., 2014. Effects of two dietary exogenous multi-enzyme supplementation, Natuzyme® and beta-mannanase (Hemicell®), on growth and blood parameters of Caspian salmon (Salmo trutta caspius). Comparative Clinical Pathology, 23(1), pp.187-192.

Khajepour, F. and Hosseini, S.A., 2012. Citric acid improves growth performance and phosphorus digestibility in Beluga (Huso huso) fed diets where soybean meal partly replaced fish meal. Animal Feed Science and Technology, 171(1), pp.68-73.

Khan, M.I.R., Saha, R.K. and Saha, H., 2018. Muli bamboo (Melocanna baccifera) leaves ethanolic extract a non-toxic phyto-prophylactic against low pH stress and saprolegniasis in Labeo rohita fingerlings. Fish & Shellfish Immunology, 74, pp.609-619.

Kianersi, F., Salati, A.P. and Hosmand, H.H., 2020. Comparison of dietary sodium selenite and selenium nanoparticles on growth performance, liver enzymes and antioxidant status of yellowfin porgy, Acanthopagrus latus.

Kim, D., Beck, B.R., Heo, S.B., Kim, J., Kim, H.D., Lee, S.M., Kim, Y., Oh, S.Y., Lee, K., Do, H. and Lee, K., 2013. Lactococcus lactis BFE920 activates the innate immune system of olive flounder (Paralichthys olivaceus), resulting in protection against Streptococcus iniae infection and enhancing feed efficiency and weight gain in large-scale field studies. Fish & shellfish immunology, 35(5), pp.1585-1590.

Kim, Y.D., Kim, H.G., Kim, J.C., Sim, S.J., Min, J.Y., Hwang, J.G., Kang, S.M., Moon, H.S., Kim, J.K. and Choi, M.S., 2013. Effects of culture media on catechins and caffeine production in adventitious roots of tea tree (Camellia sinensis L.)., 47(1), pp.11-20.

Kolkovski, S., Tandler, A., Kissil, G.W. and Gertler, A., 1993. The effect of dietary exogenous digestive enzymes on ingestion, assimilation, growth and survival of gilthead seabream (Sparus aurata, Sparidae, Linnaeus) larvae. Fish Physiology and Biochemistry, 12(3), pp.203-209.

Kumar, P., Jain, K.K., Sardar, P., Sahu, N.P. and Gupta, S., 2017. Dietary supplementation of acidifier: effect on growth performance and haemato-biochemical parameters in the diet of Cirrhinus mrigala juvenile. Aquaculture International, 25(6), pp.2101-2116.

Kumar, S., Sahu, N.P., Pal, A.K., Subramanian, S., Priyadarshi, H. and Kumar, V., 2011. High dietary protein combats the stress of Labeo rohita fingerlings exposed to heat shock. Fish Physiology and Biochemistry, 37(4), pp.1005-1019.

Kuz'mina, V.V. and Golovanova, I.L., 2004. Contribution of prey proteinases and carbohydrases in fish digestion. Aquaculture, 234(1-4), pp.347-360.

Laining, A., Ishikawa, M., Koshio, S. and Yokoyama, S., 2012. Dietary inorganic phosphorus or microbial phytase supplementation improves growth, nutrient utilization and phosphorus mineralization of juvenile red sea bream, Pagrus major, fed soybean-based diets. Aquaculture Nutrition, 18(5), pp.502-511.

Lee, C.S., 2015. Dietary nutrients, additives and fish health.

Li, E., Arena, L., Lizama, G., Gaxiola, G., Cuzon, G., Rosas, C., Chen, L. and Van Wormhoudt, A., 2011. Glutamate dehydrogenase and Na+-K+ ATPase expression and growth response of Litopenaeus vannamei to different salinities and dietary protein levels. Chinese Journal of Oceanology and Limnology, 29(2), pp.343-349.

Li, J.S., Li, J.L. and Wu, T.T., 2009. Effects of non-starch polysaccharides enzyme, phytase and citric acid on activities of endogenous digestive enzymes of tilapia (Oreochromis niloticus× Oreochromis aureus). Aquaculture Nutrition, 15(4), pp.415-420.

Li, M.H. and Robinson, E.H., 1997. Microbial phytase can replace inorganic phosphorus supplements in channel catfish Ictalurus punctatus diets 1. Journal of the World Aquaculture Society, 28(4), pp.402-406.

Li, X., Ringø, E., Hoseinifar, S.H., Lauzon, H.L., Birkbeck, H. and Yang, D., 2019. The adherence and colonization of microorganisms in fish gastrointestinal tract. Reviews in Aquaculture, 11(3), pp.603-618.

Li, X.Q., Chai, X.Q., Liu, D.Y., Kabir Chowdhury, M.A. and Leng, X.J., 2016. Effects of temperature and feed processing on protease activity and dietary protease on growths of white shrimp, L itopenaeus vannamei, and tilapia, O reochromis niloticus× O. aureus. Aquaculture Nutrition, 22(6), pp.1283-1292.

Liebert, F. and Portz, L., 2005. Nutrient utilization of Nile tilapia Oreochromis niloticus fed plant based low phosphorus diets supplemented with graded levels of different sources of microbial phytase. Aquaculture, 248(1-4), pp.111-119.

Lin, Y.H., Wang, H. and Shiau, S.Y., 2009. Dietary nucleotide supplementation enhances growth and immune responses of grouper, Epinephelus malabaricus. Aquaculture Nutrition, 15(2), pp.117-122.

Lindsay, G.J.H. and Harris, J.E., 1980. Carboxymethylcellulase activity in the digestive tracts of fish. Journal of fish Biology, 16(3), pp.219-233.

Liu, B., Xu, P., Xie, J., Ge, X., Xia, S., Song, C., Zhou, Q., Miao, L., Ren, M., Pan, L. and Chen, R., 2014. Effects of emodin and vitamin E on the growth and crowding stress of Wuchang bream (Megalobrama amblycephala). Fish & Shellfish Immunology, 40(2), pp.595-602.

Liu, L., Zhou, Y., Wu, J., Zhang, W., Abbas, K., Xu-Fang, L. and Luo, Y., 2014. Supplemental graded levels of neutral phytase using pretreatment and spraying methods in the diet of grass carp, C tenopharyngodon idellus. Aquaculture Research, 45(12), pp.1932-1941.

Liu, L.W., Su, J. and Luo, Y., 2012. Effect of partial replacement of dietary monocalcium phosphate with neutral phytase on growth performance and phosphorus digestibility in gibel carp, C arassius auratus gibelio (B loch). Aquaculture Research, 43(9), pp.1404-1413.

Liu, L.W., Su, J.M., Zhang, T., Liang, X.F. and Luo, Y.L., 2013. Apparent digestibility of nutrients in grass carp (C tenopharyngodon idellus) diet supplemented with graded levels of neutral phytase using pretreatment and spraying methods. Aquaculture Nutrition, 19(1), pp.91-99.

Lorenz, R.T. and Cysewski, G.R., 2000. Commercial potential for Haematococcus microalgae as a natural source of astaxanthin. Trends in Biotechnology, 18(4), pp.160-167.

Maas, R.M., Verdegem, M.C., Dersjant-Li, Y. and Schrama, J.W., 2018. The effect of phytase, xylanase and their combination on growth performance and nutrient utilization in Nile tilapia. Aquaculture, 487, pp.7-14.

Maas, R.M., Verdegem, M.C., Stevens, T.L. and Schrama, J.W., 2020. Effect of exogenous enzymes (phytase and xylanase) supplementation on nutrient digestibility and growth performance of Nile tilapia (Oreochromis niloticus) fed different quality diets. Aquaculture, 529, p.735723.

Malicki, A., Zawadzki, W., Bruzewicz, S., Graczyk, S. and Czerski, A., 2004. Effect of formic and propionic acid mixture on Escherichia coli in fish meal stored at 12 C. Pakistan Journal of Nutrition, 3(6), pp.353-356.

Montero, D., Tort, L., Robaina, L., Vergara, J.M. and Izquierdo, M.S., 2001. Low vitamin E in diet reduces stress resistance of gilthead seabream (Sparus aurata) juveniles. Fish & Shellfish Immunology, 11(6), pp.473-490.

Mroz, Z., Moeser, A.J., Vreman, K., van Diepen, J.T.M., Van Kempen, T., Canh, T.T. and Jongbloed, A.W., 2000. Effects of dietary carbohydrates and buffering capacity on nutrient digestibility and manure characteristics in finishing pigs. Journal of Animal Science, 78(12), pp.3096-3106.

Munir, M.B., Hashim, R., Chai, Y.H., Marsh, T.L. and Nor, S.A.M., 2016. Dietary prebiotics and probiotics influence growth performance, nutrient digestibility and the expression of immune regulatory genes in snakehead (Channa striata) fingerlings. Aquaculture, 460, pp.59-68.

Murray, H.M., Gallant, J.W., Perez-Casanova, J.C., Johnson, S.C. and Douglas, S.E., 2003. Ontogeny of lipase expression in winter flounder. Journal of Fish Biology, 62(4), pp.816-833.

Murthy, K.C., Vanitha, A., Rajesha, J., Swamy, M.M., Sowmya, P.R. and Ravishankar, G.A., 2005. In vivo antioxidant activity of carotenoids from Dunaliella salina—a green microalga. Life sciences, 76(12), pp.1381-1390.

Nafisi Bahabadi, M., Torfi Mozanzadeh, M., Agh, N., Ahmadi, A. and Yaghoubi, M., 2018. Enriched Artemia with L-lysine and DL-methionine on growth performance, stress resistance, and fatty acid profile of Litopenaeus vannamei postlarvae. Journal of Applied Aquaculture, 30(4), pp.325-336.

Nasr-Eldahan, S., Nabil-Adam, A., Shreadah, M.A., Maher, A.M. and El-Sayed Ali, T., 2021. A review article on nanotechnology in aquaculture sustainability as a novel tool in fish disease control. Aquaculture International, 29(4), pp.1459-1480.

Neamat-Allah, A.N., Mahmoud, E.A. and Abd El Hakim, Y., 2019. Efficacy of dietary Nano-selenium on growth, immune response, antioxidant, transcriptomic profile and resistance of Nile tilapia, Oreochromis niloticus against Streptococcus iniae infection. Fish & Shellfish Immunology, 94, pp.280-287.

Nwanna, L.C. and Schwarz, F.J., 2007. Effect of supplemental phytase on growth, phosphorus digestibility and bone mineralization of common carp (Cyprinus carpio L). Aquaculture Research, 38(10), pp.1037-1044.

Owolabi, O.D. and Babarinsa, M.K., 2020, March. Assessment of growth performance, nutrient utilization and haematological profile of Clarias gariepinus fed with nanoselenium formulated diets. In IOP Conference Series: Materials Science and Engineering (Vol. 805, No. 1, p. 012014). IOP Publishing.

Paray, B.A., Hoseini, S.M., Hoseinifar, S.H. and Van Doan, H., 2020. Effects of dietary oak (Quercus castaneifolia) leaf extract on growth, antioxidant, and immune characteristics and responses to crowding stress in common carp (Cyprinus carpio). Aquaculture, 524, p.735276.

Portz, L. and Liebert, F., 2004. Growth, nutrient utilization and parameters of mineral metabolism in Nile tilapia Oreochromis niloticus (Linnaeus, 1758) fed plant-based diets with graded levels of microbial phytase. Journal of Animal Physiology and Animal Nutrition, 88(9-10), pp.311-320.

Qi, C., Han, F., Wang, X., Xu, C., Huang, Z., Li, E., Qin, J.G. and Chen, L., 2020. High protein diet alleviates the high pH stress in Chinese mitten crab Eriocheir sinensis. Aquaculture, 516, p.734523.

Qiu, X. and Davis, D.A., 2017. Effects of dietary carbohydrase supplementation on performance and apparent digestibility coefficients in Pacific white shrimp, Litopenaeus vannamei. Journal of the World Aquaculture Society, 48(2), pp.313-319.

Ramadan, A., Atef, M. and Afifi, N.A., 1991. Effect of the biogenic performance enhancer (ascogen" s") on growth rate of tilapia fish. Acta Veterinaria Scandinavica. Supplementum (Denmark).

Reda, R.M., Mahmoud, R., Selim, K.M. and El-Araby, I.E., 2016. Effects of dietary acidifiers on growth, hematology, immune response and disease resistance of Nile tilapia, Oreochromis niloticus. Fish & Shellfish Immunology, 50, pp.255-262.

Reda, R.M., Selim, K.M., Mahmoud, R. and El-Araby, I.E., 2018. Effect of dietary yeast nucleotide on antioxidant activity, non-specific immunity, intestinal cytokines, and disease resistance in Nile Tilapia. Fish & Shellfish Immunology, 80, pp.281-290.

Safari, O., Shahsavani, D., Paolucci, M. and Mehraban Sang Atash, M., 2015. The effects of dietary nucleotide content on the growth performance, digestibility and immune responses of juvenile narrow clawed crayfish, A stacus Leptodactylus leptodactylus Eschscholtz, 1823. Aquaculture Research, 46(11), pp.2685-2697.

Saha, A.K. and Ray, A.K., 1998. Cellulase activity in rohu fingerlings. Aquaculture International, 6(4), pp.281-291.

Sardar, P., Shamna, N. and Sahu, N.P., 2020. Acidifiers in aquafeed as an alternate growth promoter: A short review. Animal Nutrition and Feed Technology, 20(2), pp.353-366.

Sawanboonchun, J., Roy, W.J., Robertson, D.A. and Bell, J.G., 2008. The impact of dietary supplementation with astaxanthin on egg quality in Atlantic cod broodstock (Gadus morhua, L.). Aquaculture, 283(1-4), pp.97-101.

Sharma, S.A., Surveswaran, S., Arulraj, J. and Velayudhannair, K., 2021. Bromelain enhances digestibility of Spirulina-based fish feed. Journal of Applied Phycology, 33(2), pp.967-977.

Shi, Z., Li, X.Q., Chowdhury, M.K., Chen, J.N. and Leng, X.J., 2016. Effects of protease supplementation in low fish meal pelleted and extruded diets on growth, nutrient retention and digestibility of gibel carp, Carassius auratus gibelio. Aquaculture, 460, pp.37-44.

Stahl, W. and Sies, H., 2003. Antioxidant activity of carotenoids. Molecular Aspects of Medicine, 24(6), pp.345-351.

Stickney, R.R. and Shumway, S.E., 1974. Occurrence of cellulase activity in the stomachs of fishes. Journal of Fish Biology, 6(6), pp.779-790.

Stone, D.A., 2003. Dietary carbohydrate utilization by fish. Reviews in Fisheries Science, 11(4), pp.337-369.

Suprayudi, M.A., Harianto, D. and Jusadi, D., 2012. Kecernaan pakan dan pertumbuhan udang putih Litopenaeus vannamei diberi pakan mengandung enzim fitase berbeda The effect of phytase levels in the diet on the digestibility and growth performance of white shrimp Litopenaeus vannamei. J. Akuakultur Indones, 11(2), pp.103-108.

Tahmasebi-Kohyani, A., Keyvanshokooh, S., Nematollahi, A., Mahmoudi, N. and Pasha-Zanoosi, H., 2011. Dietary administration of nucleotides to enhance growth, humoral immune responses, and disease resistance of the rainbow trout (Oncorhynchus mykiss) fingerlings. Fish & Shellfish Immunology, 30(1), pp.189-193.

Tejpal, C.S., Pal, A.K., Sahu, N.P., Kumar, J.A., Muthappa, N.A., Vidya, S. and Rajan, M.G., 2009. Dietary supplementation of L-tryptophan mitigates crowding stress and augments the growth in Cirrhinus mrigala fingerlings. Aquaculture, 293(3-4), pp.272-277.

Tian, J., Zhang, Y., Zhu, R., Wu, Y., Liu, X. and Wang, X., 2021. Red elemental selenium (Se0) improves the immunoactivities of EPC cells, crucian carp and zebrafish against spring viraemia of carp virus. Journal of Fish Biology, 98(1), pp.208-218.

Venkat, H.K., Sahu, N.P. and Jain, K.K., 2004. Effect of feeding Lactobacillus-based probiotics on the gut microflora, growth and survival of postlarvae of Macrobrachium rosenbergii (de Man). Aquaculture Research, 35(5), pp.501-507.

Verlhac-Trichet, V., Vielma, J., Dias, J., Rema, P., Santigosa, E., Wahli, T. and Vogel, K., 2014. The Efficacy of a Novel Microbial 6-Phytase Expressed in Aspergillus oryzae on the Performance and Phosphorus Utilization of Cold-and Warm-Water Fish: Rainbow Trout, Oncorhynchus mykiss, and Nile Tilapia, Oreochromis niloticus. Journal of the World Aquaculture Society, 45(4), pp.367-379.

Wassef, E.A., Saleh, N.E., Abdel-Meguid, N.E., Barakat, K.M., Abdel-Mohsen, H.H. and El-bermawy, N.M., 2020. Sodium propionate as a dietary acidifier for European seabass (Dicentrarchus labrax) fry: immune competence, gut microbiome, and intestinal histology benefits. Aquaculture International, 28(1), pp.95-111.

Weerd, J.V., 1999. Balance trials with African catfish Clarias gariepinus fed phytase-treated soybean meal-based diets. Aquaculture Nutrition, 5(2), pp.135-142.

Wei, Z., Yi, L., Xu, W., Zhou, H., Zhang, Y., Zhang, W. and Mai, K., 2015. Effects of dietary nucleotides on growth, non-specific immune response and disease resistance of sea cucumber Apostichopus japonicas. Fish & Shellfish Immunology, 47(1), pp.1-6.

Weil, C., Lefèvre, F. and Bugeon, J., 2013. Characteristics and metabolism of different adipose tissues in fish. Reviews in Fish Biology and Fisheries, 23(2), pp.157-173.

Xavier, B., Sahu, N.P., Pal, A.K., Jain, K.K., Misra, S., Dalvi, R.S. and Baruah, K., 2012. Water soaking and exogenous enzyme treatment of plant-based diets: effect on growth performance, whole-body composition, and digestive enzyme activities of rohu, Labeo rohita (Hamilton), fingerlings. Fish Physiology and Biochemistry, 38(2), pp.341-353.

Yousefi, M., Hoseini, S.M., Vatnikov, Y.A., Kulikov, E.V. and Drukovsky, S.G., 2019. Rosemary leaf powder improved growth performance, immune and antioxidant parameters, and crowding stress responses in common carp (Cyprinus carpio) fingerlings. Aquaculture, 505, pp.473-480.

Yu, H., Zhang, C., Zhang, X., Wang, C., Li, P., Liu, G., Yan, X., Xiong, X., Zhang, L., Hou, J. and Liu, S., 2020. Dietary nano-selenium enhances antioxidant capacity and hypoxia tolerance of grass carp Ctenopharyngodon idella fed with high-fat diet. Aquaculture Nutrition, 26(2), pp.545-557.

Yúfera, M., Moyano, F.J., Astola, A., Pousao-Ferreira, P. and Martinez-Rodriguez, G., 2012. Acidic digestion in a teleost: postprandial and circadian pattern of gastric pH, pepsin activity, and pepsinogen and proton pump mRNAs expression. PLoS One, 7(3), p.e33687.

Zambonino-Infante, J.L. and Cahu, C., 2010. Effect of nutrition on marine fish development and quality.

Zhao, J., Ai, Q., Mai, K., Zuo, R. and Luo, Y., 2013. Effects of dietary phospholipids on survival, growth, digestive enzymes and stress resistance of large yellow croaker, Larmichthys crocea larvae. Aquaculture, 410, pp.122-128.

Zheng, C.C., Wu, J.W., Jin, Z.H., Ye, Z.F., Yang, S., Sun, Y.Q. and Fei, H., 2020. Exogenous enzymes as functional additives in finfish aquaculture. Aquaculture Nutrition, 26(2), pp.213-224.

Zhou, Y., Yuan, X., Liang, X.F., Fang, L., Li, J., Guo, X., Bai, X. and He, S., 2013. Enhancement of growth and intestinal flora in grass carp: The effect of exogenous cellulase. Aquaculture, 416, pp.1-7.

Section 2
Fisheries Biology, Resources and Post-Harvest Management

5

Jellyfish Apocalypse- An Opportunity?

[1]Meenatchi. S, [2]Abuthagir Iburahim.S, [2]Vidhya.V, [3]Kamalii A., [4]Ulaganathan Arisekar

[1]TANII project, Department of Fish Biology and Fisheries Resource Management Fisheries College and Research Institute, Thoothukudi, Tamil Nadu

[2]Fisheries Resource Harvest and Post Harvest Management Division ICAR-Central Institute of Fisheries Education (ICAR-CIFE), Mumbai Maharashtra

[3]Department of Aquaculture, Dr. M.G.R. Fisheries College and Research Institute Ponneri, Tamil Nadu

[4]Department of Fish Quality Assurance and Management, Fisheries College and Research Institute, Thoothukudi, Tamil Nadu

Abstract

Jellyfishes are beautiful, larger zooplankton, that take the form of umbrellas drift with the current. Jellyfish comes under the phylum Cnidaria and the sub phylum, Meduzoa. There are about 10347 species of jellyfish exist in the world include the classes namely Scyphozoa, Cubozoa, Hydrozoa, Anthozoa and Staurozoa. Around 13% (1396 species) of the total jellyfish species were reported from India. Jellyfish have no circulatory system, nervous system, skeletal system, or blood supply; they digest internally; and they have no heart, brain, feet, ears, head, bones, or blood. Jellyfish are not only edible, but also extremely nutritious due to their low calorie, fat, and protein/collagen content. Jellyfish are known for their refined taste, which can lean toward the salty side. Apart from the food uses, jellyfishes are used in plenty of ways such as marvelous medicine, fertilizers, and pets. Potential benefits aside, nowadays they are portrayed as a potential danger because of increasing events of jellyfish blooms in ocean. It's true that twenty to forty people per year died from box jellyfish stings in the Philippines but this victim list mainly includes child, elderly person and immune suppressed people. Systematic classification, Diversity, potential uses, potential uses, potential threats, bloom forming potential, marine capture, culture and processing of jellyfish are detailed in this chapter. Aim of this chapter, is to highlight the better utilization of augmenting jellyfish fishery such as the increasing blooms in a variety of beneficial ways. Today's danger (Jellyfish bloom) will be on your delicious plate soon, with increasing demand, this can be achieved by proper processing aids.

Keywords: Jellyfish, Diversity, Smack, Processing, Commercial uses, Threats, Threat to treat

Introduction

A few decades back, no one would ever consider the potential danger posed by an ocean overrun with jellyfish? But many researchers around the world assure us that we can turn that danger into an asset with the help of scientific ideas and careful management. Gelatinous zooplankton like jellyfish are thought to have existed millions of years before the dinosaurs did. There are cases where these emotionless, mindless, bloodless organisms kill an adult human (Wilson, 1947). Despite being an annoyance, they could greatly benefit human society in many ways. The unfortunate truth is that even though they have 500 million years of history, they still appear to be an untapped opportunity. Jellyfish are not only edible, but also extremely nutritious due to their low calorie, fat, and protein/collagen content. Jellyfish are known for their refined taste, which can lean toward the salty side. The texture is more of a factor; it's not as gelatinous as one might think, falling somewhere between a summer squash and a glass noodle. According to the Intergovernmental Panel on Climate Change (IPCC) report published in 2019, human activities have had an impact on virtually all marine organisms, including corals, oysters, and other marine organisms. However, jellyfish are more resistant to the acidic effects of the world's oceans. This does not necessarily imply that these organisms are highly immune, but they are doing better than average. There are about hundred known venomous jellyfish species out of an estimated 10,000 worldwide. These box jellyfish, or *Chironex flecker*, are among the deadliest in the ocean. The length of a lion's mane jellyfish ranges from about 1.5 feet (40 centimetres) to 6.5 feet (2.0 m), making it the largest species of jellyfish (200cm). Although jellyfish can be found in both the pelagic and benthic regions of the ocean, many of the species that pose a threat to human mariners are found in the former. In addition, their presence on the Indian coast is only observed during certain times of the year (Revelles et al., 2007). Even though they have been present earlier than dinosaurs, we seem to be having more trouble with them now than we did in the past. The question then becomes, "Is it a human or a jellyfish who did this?" The simple answer is that the human race is to blame for creating the conditions that have resulted in a global proliferation of jellyfish. Research suggests that the jellyfish population has grown dramatically over the past ten year, though the trend likely began in the 1980s. According to Primo et al., 2012, jellyfishes have the potential to play a keystone species role because they feed on the eggs and larvae of previous winners of a community, thereby reducing or controlling their population and helping to flourish other nektons. We can extrapolate from this the fact that two-thirds of fish species form partnerships with jellyfish, which protect them from larger predators and provide food for them in various instances (Morgan & Hodgkinson, 1999) and thus continue to serve both human needs and greed.

Table 1 : Systematic classification of jellyfish

Kingdom	**Animalia**	
Phylum	Cnidaria	Over 11,000 different species of marine and freshwater organisms make up the phylum Cnidaria, which is part of the kingdom Animalia. Cnidocytes, specialised cells used primarily in prey capture, are a defining characteristic of these animals.
Subphylum	Medusozoa	Within the subphylum Medusozoa, nearly 4,055 different invertebrate species can be found in the four classes Hydrozoa, Cubozoa, Staurozoa, and Scyphozoa.
Class	Scyphozoa	Scyphozoa, also known as "true jellyfish," are a marine-only subphylum of the phylum Cnidaria (or "true jellies").
	Cubozoa	Cnidarian invertebrates known as box jellyfish have a distinctive boxy shape. Tentacle contact with certain species of box jellyfish can result in a painful reaction due to the venom they secrete.
	Hydrozoa	Generally found in salt water, the phylum Hydrozoa is comprised of very tiny predatory animals, both solitary and colonial. Some colonial species have extremely large colonies, and the individuals are so specialised that they simply cannot survive if they were to be separated from the group.
	Anthozoa	The sea anemone, stony coral, and soft coral are all examples of the class Anthozoa, which consists of marine invertebrates. Nearly all anthozoan adults are anchored to the bottom of the ocean, but their larvae float freely among the plankton.
	Staurozoa	Staurozoa, also called as stalked jellyfishes (benthic cnidarians)

Systematic Classification

Uniqueness of Jellyfish

Jellyfish that take the form of umbrellas drift with the current. They are zooplankton, so their bodies are made up of the umbrella, oral arms near the mouth, and stinky tentacles. Their bodies are primarily water, accounting for 96% of their total mass (Hsieh et al., 1996). By blending in with their surroundings, these opportunistic carnivorous jellyfish are able to ambush their prey and ingest it whole with the aid of their tentacles and oral arms. Jellyfish have no circulatory system, nervous system, skeletal system, or blood supply; they digest internally; and they have no heart, brain, feet, ears, head, bones, or blood. Jellyfish are exhibiting gonochoristic behaviour when, following fertilisation, the zygote releases polyps that must settle down and attach to the seafloor or another substrate in order to undergo asexual reproduction and produce new free-swimming planulae, larvae.

No of jellyfish species reported

Table 2 : Number of jellyfish reported throughout the world and India

Class	World	India
Scyphozoa	192	37
Hydrozoa	3673	212
Anthozoa	6395	1134
Cubozoa	38	5
Staurozoa	49	8
Total	10347	1396

Table 3 : Jellyfishes with bloom forming potential

Jellyfishes reported in India	Bloom forming potential
Atolla wyvillei	yes
Periphylla periphylla	yes
Cyanea nazakii	yes
Chrysaora melanaster	yes
Aurelia aurita	yes
Crambionella orsini	yes
Rhoplema hispidum	yes

Jellyfish bloom occurrence throughout the world

Table 4 : Jellyfish bloom problems created in different years throughout the world

Species	Area of bloom occurrence	Year
Aurelia aurita	Seto Inland sea,Japan	1983, 2003
Nemopilema nomurai	Japan sea	1995
Mastigias sp	Jellyfish lake, Palau	1980
Pelagia noctiluca	Mediterranean sea	1985, 1993
Aurelia aurita, Cyanea capillata, C. lamarckii	North sea	1971
Chrysaora quinquecirrha	Chesapeake Bay, USA	1995

Species	**Area of bloom occurrence**	**Year**
Chrysaora melanaster	Eastern Bering Sea	2005
Chrysaora fuscescens	Oregon, USA	2000-2002
Liriope tetraphylla	Mediterranean sea	1993
Rhopalonema velatum	Mediterranean sea	1993
Solmundella bitentaculata	Mediterranean sea	1993
Aglantha digitale	Greenland	1968
Aglantha digitale	North sea	1994
Mitrocoma cellularia	Monterey Bay, USA	1998
Colobonema sericeum	Monterey Bay, USA	1998
Aequorea victoria	Vancouver Island, Canada	1987
Aurelia aurita	Lake Hachirogata, Akita prefecture	1950
Porpita porpita	Kyoto, Fukui prefectures	2000
Pelagia noctiluca	Ehime prefecture	2004
Aurelia sp	Tasmania	2001
Rhizostome scyphozoan	Goa, India	2006
Aurelia aurita	Northwest Ireland	2010
Pelagia noctiluca	Brittany, France	1994
Pelagia noctiluca	Western Ireland	2007
Cyanea capillata	Loch Fyne, Scotland	1996
Solmaris crona	Western Ireland	2009
Muggiacea atlantica	Norway	2002
Phialella quadrata	Scotland	1984
Porpita porpita	Astranga coast, India	2016
Porpita porpita	Puri coast, India	2018
Pelagia noctiluca	Rushikulya estuary, India	2012
Crambiionella stulmanni	East vishakapatnam coast, India	2015
Physalia physalis	Goa,India	2015
Physalia physalis	Mumbai (west), India	2018
Purple striped jellyfish	Goa (West), India	2019
Porpita porpita	Chennai coast, India	2019
Crambiolella orisini	Kavvayi Island, Kerala, India	2021

Life-Cycle of Jellyfish

They seem to do best in saltier and warmer water, 20–40 miles out from the coast, and do not prefer the colder, less saline coastal waters. However, during times of drought, when coastal waters receive less riverine water or no rainwater at all, the salinity of the coastal water approaches that of the ocean, allowing us to find jellyfish in coastal regions as well. When washed up on a beach, they tend to perish rather quickly. Australia, the Philippines, the Indian Ocean, and the middle of the Pacific Ocean are home to the most lethal species of jellyfish (Calder & Hester, 1978). Climate change, which raises sea levels and, in turn, increases the prevalence of toxic jellyfish, is a major problem today. Their tentacles are their main line of defence against predators, and humans are not on their prey list. It's true that the odour they leave behind can be painful, but most people only experience minimal discomfort, redness, tingling, itching, or numbness at most. It is more likely that a child, an elderly person, or someone whose immune system is compromised will die (Firth, 1969). According to recent reports, each summer in the Mediterranean, an estimated 150,000 people seek medical attention after being stung by jellyfish. Twenty to forty people per year died from box jellyfish stings in the Philippines.

Possible Causes of Jellyfish Apocalypse

The proliferation of jellyfish poses an unfathomable threat to humanity. Beautiful schools of jellyfish are called swarms or smacks (Barzansky et al., 1975). It is possible that nutrient upwelling, climate change, overfishing, aquaculture, anthropogenic interventions, and the introduction of exotic species are all to blame for the rapid increase in their population, which is known as a bloom. The abundance of jellyfish varies with climate, often on a centennial scale, as shown by analyses of long-term (8-100 year) changes in population. Many independent pieces of evidence point to ongoing increases; however, the length of the most recent time series is insufficient for ruling out naturally occurring, cyclical variations in the climate (Kramp, 1961). The factors that contribute to a jellyfish bloom and their effects are outlined below.

Major Causes of Jellyfish Bloom

Eutrophication

Eutrophication is a leading form of pollution on a global scale, and it is caused by both an increase in nutrients and a shift in the ratio of nutrients in the environment (Mayer, 1910). Increases in biomass can be observed at nearly every trophic level after nutrient enrichment was introduced. The primary role that jellyfish play as carnivores allowed their population to explode. Enrichment with nutrients has the potential to shift the lower trophic structure toward a food web dominated by microplankton. Since most fish are visual predators that prefer larger zooplankton, this shrinking of the lower trophic levels is thought to be detrimental to them, while benefiting jellyfish, which are not visual and can filter small as well as large prey. Phytoplankton with a high N:P ratio consists more of flagellated organisms and jellyfish rather than diatoms (MacMillen, 1994).

Overfishing

Jellyfish are being devoured by more than hundreds of fishes and other marine animals.

Table 4 : Major predators of Jellyfish

Major Predators of jellyfish		
Common name	**Scientific name**	**Photographs**
Ocean sunfish	*Mola mola*	
Leatherback turtle	*Adermocheyelys coriacea*	
Grey triggerfish	*Balistes caprisas*	

Whale shark	*Rhincodon typus*	
Hermit crab	*Pagurus.sp*	
Seabird	*Fulmarus glacialis*	

People all over the world are concerned about overfishing. Neither a lack of food nor the presence of predators threatens jellyfish populations. The dramatic decline of leatherbacks in the Pacific has led to an increase in the number of jellyfish there. As a result, overfishing has a beneficial effect on this cnidarian society. Food competition comes from predator depletion as well as the global demand for fish meal made from planktivorous fishes like sardines and anchovies. With such an abundance of food and no natural enemies to worry about, jellyfish thrive (Leong, 1995). Then, without a doubt, they'll proliferate in the seawater and move into the void left by other fish communities.

Aquaculture

There are 30.8 million tonnes of mariculture produced globally (SOFIA- 2020). There are multiple ways in which aquaculture may contribute to rising jellyfish populations, both knowingly and unknowingly (Mianzan & Corneliu, 1999). These lucky jellyfish are able to mature slowly on the leftovers of mariculture operations. Jellyfish production is helped along by the nitrogenous and organic waste products of farmed fish. Mariculture uses cage and raft structures because polyps attach to them during reproduction; this increases polyp production.

Aquaculture techniques are used to artificially increase jellyfish production in China and Malaysia.

Alien Invasion

Releasing the ballast water from ship is a significant contributor to biological pollution (Richardson et al., 2009). However, the groundwork has been laid for subsequent large blooms when favourable circumstances encourage, and for the expansion of communities into alien ecosystems. It is estimated that 21 of the 45 major marine ecosystems have been invaded by non-native species of jellyfish (Roohi et al., 2010).

Table 5 : Invasion of jellyfish throughout the world

Invading species	Areas of invasion
Aurelia . sp	Multiple invasion
Rhopilema nomadica	Eastern Mediterranean sea
Phyllorhiza punctata	Western Australia, Hawaii & San Diego, USA, Caribbean and Mediterranean seas, Gulf of Mexico
Cassiopea andromeda	Hawaii and Mediterranean seas
Sanderia malayensis	Yangtze River estuary
Craspidacusta sowerbii	All continents except Antarctica
Moerisia lyonsi	Chesapeake and san francisco bays, and Lake ponchatrain,, USA
Blackfordia virginica	San Francisco Bay
Maeotias inexspectata	San Francisco Bay
Mnemiopsis leidyi	Black, Azov, Caspian, Mediterranean and North seas
Beroe ovata	Black sea

Climate Change

We are now in the most worrisome stage of the climate crisis, and we can all feel those cnidarian shifts. Elevated levels of both temperature and salinity in the water have no effect on jellyfish. Ocean acidification, the gradual decrease in ocean pH that results primarily from the ocean's absorption of carbon dioxide from the atmosphere, has had a devastating effect on virtually every marine organism (Rippingale & Kelly, 1995). However, jellyfish are less vulnerable to the effects of ocean acidification than other marine organisms. So, climate change is not having a major impact on the development and population of jellyfish.

Jellyfish Culture

Jellyfish culture is important, but first we need to know why. Nowadays, jellyfish are popular among the processing industry as a tasty dinner, hobbyists as an entertaining addition to public aquariums, and scientists for its adaptable potential. A deeper understanding of the biology, physiology, life cycle, and environmental needs of jellyfish breeding is crucial for the culture activities (You et al., 2007). Collecting fertile broodstock using a drift net, caring for the broodstock in well-designed tanks, allowing the polyps to mature, feeding the polyps debris, releasing the planulae actively, and rearing the medusae larvae under strict supervision are all part of their culture practises. Aquaculture of jellyfish began in China in 1983. In response to a severe decline in *Rhopilema esculentum*, a popular edible jellyfish, Chinese researchers were compelled to launch a jellyfish aquaculture programme, which ultimately proved successful. By 2004, Malaysia had also begun cultivating jellyfish.

Ideal water conditions for jellyfish culture

Table 6 : Optimum water quality parameters required for the culture of jellyfish

Parameters	**Limit**
DO	75-99%
pH	8.0-8.4
Salinity	32-35 ppt
Nitrate	< 20 ppm
Nitrire	<2 ppm
Copper	< 0.01 ppm
Ammonia	< 2ppm

Marine Capture of Jellyfish (*Source:* SOFIA-2020)

Over 842 different species of jellyfish have been documented in Indian waters, with 212 belonging to the hydrozoa class and 25 to the scyphozoa class. Therefore, there are plenty of means to advance jellyfish aquaculture in India. Jellyfish fisheries along the coasts of Andhra Pradesh, Gujarat, and Kerala are quite profitable due to the presence of several species, including Crambionella stuhlmanni, Crambionella Orsini, Catostylus perezi, and Rhopilema hispidum (Martinussen and Bamstedt, 1999). Markets for export of these four species are robust as well. Over the past two decades, Kerala has shipped 1,092 metric tonnes of jellyfish to the countries of Southeast Asia. Consequently, there is obviously a massive demand for aquacultured jellyfish all over the world. With this goal in mind, the University of Kerala has announced the plans to host the seventh International jellyfish bloom symposium in Thiruvananthapuram, India, in 2022. This event will serve as an attempt to fill the knowledge gap regarding jellyfish in the region.

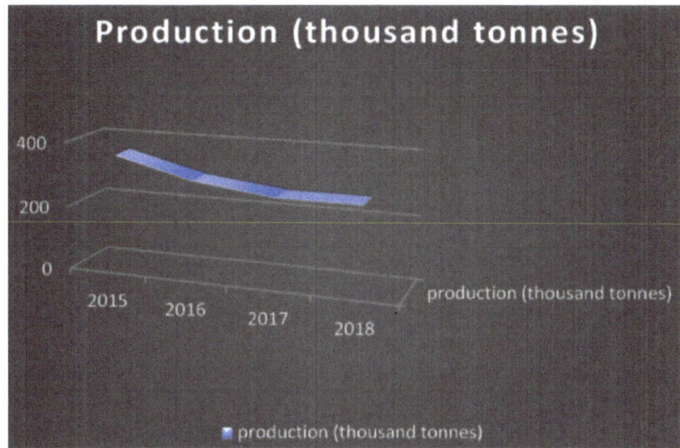

Figure 1 : Marine capture of jellyfish (source: SOFIA-2020)

Edible Jellyfishes from India

In China, eating jellyfish is not just a novelty but also a cultural rite. Like many vegetables, it is a natural diet. Restaurants, ceremonies, and banquets across many Asian countries have featured delectable dishes prepared with jellyfish. Most of these nations' economies depend on the export of processed jellyfish to Asia. Countries like China, Japan, Malaysia, South Korea, Taiwan, Singapore, Thailand, and Hong Kong are among the top markets for jellyfish exports (Rapoza et al., 2005). Although eating jellyfish is still a novel experience for the vast majority of people in the western hemisphere, increasing multiculturalism in major cities has led to an increase in jellyfish imports.

Table 7 : List of edible jellyfishes reported from India

Family	Species
Cepheidae	*Cephea cephea*
Catostylidae	*Catostylus mosaicus, Crambione mastigophora, Crambionella orsini*
Lobonematidae	*Lobonema smithi, Lobonemoides gracilis*
Rhizostomatidae	*Rhopilema esculentum, Rhopilema hispidum, Rhopilema plumo*
Stomolophidae	*Stomolophus meleagris, Stomolophus nomurai*

Proximate composition of jellyfish

Table 8 : Proximate composition of commercially important jellyfish

Proximate composition	*Cyanea capillata*	*Rhizostoma octopus*
Protein (%DM)	16.5	12.8
Lipid (%DM)	0.50	0.32
CHO (%DM)	0.88	0.83
Ash (%DM)	67.8	73.6
G.E (KJg/drymass)	4.22	2.80

Jellyfish Processing

Jellyfish that can be consumed by humans are distinguished by having more oral arms than tentacles (Zeller et al., 2015). They're more perishable because of the water they contain, so processing them quickly is essential.

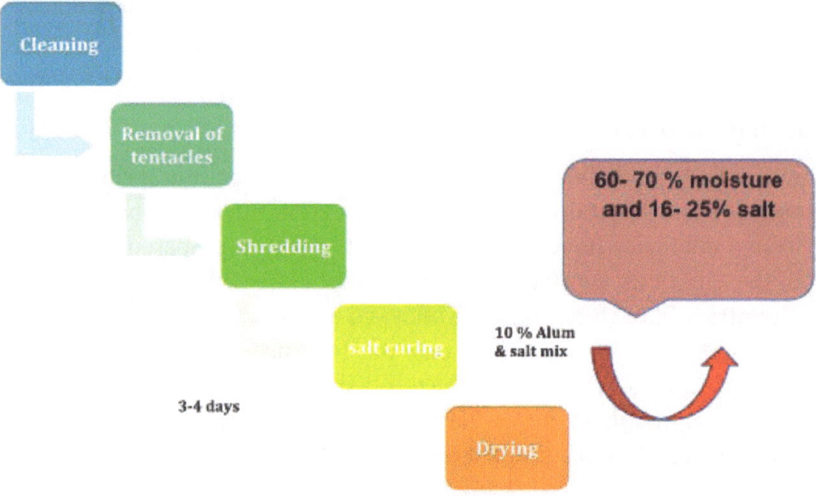

Figure 2 : Flow chart of jellyfish processing

Most of the time, it takes 20 to 40 days. Mechanical dryers are a good way to cut down on the time it takes to do something. Processing jellyfish could be expensive and not very good because it goes bad quickly, takes a long time to process, and needs a lot of work.

Ready to Eat Products from Jellyfish

Some flavours, oil, sauces, and condiments are added to jellyfish before it is packed and sold in Chinese and Japanese markets as a ready-to-eat product. In China, people like to eat sliced jellyfish salads. In Thailand, jellyfish noodles are considered a delicacy, and in Japan, people like to buy shredded jellyfish and mustard in small packages. Students at Obama Fisheries High School in Japan use Nomura's jellyfish to make candy, which is pretty cool (Zhong et al., 2004). Some of the best candies are made of sugar, starch syrup, and dried jellyfish powder to make a sweet and salty caramel. Lick Me I'm Delicious, a well-known ice cream company owned by Mr. Charlie Francis, made an ice cream that glows in the dark. The glow comes from jellyfish fluorescent proteins. Jellyfish fluorescent proteins are also found in glow-in-the-dark beer, which was made by a former NASA biologist named Josiah Zaynera.

Table 9 : Non food uses of jellyfish

Category	Uses
Agriculture	Livestock feeds, fertilizers, insecticides
Aquaculture	Finfish and shellfish feeds
Cosmetics	Gelatin, emulsifier
Environmental monitoring	Pollution detection
Fishing	Bait
Industrial product	Absorbent polymers, cement additives, copolymer films, nanoparticle filtering
Parmaceuticals	Antihypertensive peptides, anticoagulants, antimicrobiotics, antioxidants, collagen, bioactive peptides

Jellyfish as Marvelous Medicine

Television shows and magazines aimed at Korean women promoted jellyfish as a means of weight loss. It is well-known that jellyfish is rich in collagen and has significant medicinal value (Joseph, 1979). According to traditional Chinese medicine, consuming jellyfish can alleviate symptoms of arthritis, hypertension, and back pain, as well as aid in skin softening, digestion, and circulation. There is preliminary evidence that collagen peptides isolated from jellyfish can boost immune functions in mice (Zhang et al., 2008, 2014, 2016).

Jellyfish as Fertilizer

Studies show that jellyfish can contribute to organic farming of agricultural produce as organic fertilizer. In Japan and Korea, they use jellyfish in rice cultivation and also to improve the soil nutrient content.

Sea Jellies in Space Research

Space-based jellyfish reproduction was pioneered in 1991. Their ability to sense gravity was accomplished through the use of calcium crystals, just as calcium crystals in the inner ears of humans serve the same purpose (Hooper

& Ackman, 1973). Jellyfish are able to reproduce in space, but their offspring have a hard time adjusting to Earth's gravity after being born there. This suggests that human babies born in space will have similar difficulties.

Pet Jellyfish

The most common species of jellyfish kept as a pet is the moon jellyfish (Aurelia aurita). Historically, the incredible exhibit at the Monterey Bay Aquarium that opened in 1992 has been credited with sparking a renewed interest in jellyfish breeding in recent decades. Keeping jellyfish as pets is becoming increasingly popular, and jellyfish exhibits can now be found in many public aquariums worldwide (Hsieh, 1994).

Other Benefits

There is currently no standard technology for treating microplastic particles, but researchers have found that jellyfish can trap them. Jellyfish have been linked to a long-chain protein similar to the mucins used in artificial tears (Burke, 1976). Jellyfish use this protein to clean their bodies and maintain hydration. Therefore, this protein can be incorporated into synthetic tears. Jellyfish were utilised by an Israeli firm to create biodegradable diapers. In 30 days, it will be completely biodegraded. Medical bandages and sanitary napkins made from dried jellyfish are a real possibility. These can also be used as edible packaging (Huang, 1988).

Problems with Jellyfish

Tourism, fish mortality, and major fisheries are all negatively impacted by jellyfish. Overfishing in the 1960s and 1970s killed off the sardine population in the waters off the southwest coast of Namibia, making way for the many-ribbed jellyfish and the sea nettle (Larson, 1976). Most fish would flee from swarms of stinging jellyfish, but not the resilient little African fish known as the barbed goby. Jellyfish in coastal cages can be lethal to fish for reasons other than their sting, including gill irritation, which can cause bleeding and suffocation. Jellyfish strandings on beaches can have a negative impact on coastal economies. Large amounts of water used by power plants and desalinization facilities are contaminated by jellyfish blooms (Mills, 2001). Further, organic carbon is transported downward thanks to the swarming behaviour of jellyfish during their decline phase. Adding insult to injury, the natural ecosystem of the ocean could be negatively impacted by the disposal of waste products from jellyfish processing. In 2013, four jellyfish processing factories in Quang Minh province dumped their wastewater into Ha Long Bay, which is on the list of World Heritage Sites maintained by UNESCO (Kraeuter & Setzler, 1975).

Conclusion

Always exploring the unexplored resources can be a better management tool to reduce the pressure on the resources which are over exploited. Jelly fishes are potential resources to explore more though they cause some menaces. With smart processing approach, marketing strategies, jellyfish apocalypse can be converted to a better opportunity. This article suggests the importance and the way of Converting the threat (Jellyfish bloom) into a wonderful treat. Today's danger (Jellyfish bloom) will be on your delicious plate soon, with increasing demand, this can be achieved by proper processing aids as discussed in this article.

References

Barzansky, B., H. M. Lenhoff & H. Bode,(1975). Hydra mesoglea:similarity of its amino acid and neutral sugar composition to that of vertebrate basal lamina. Comp. Biochem. Physiol. 50B, 419–424.

Burke, W. D., (Biology and distribution of the macrocoelenterates of Mississippi Sound and adjacent waters. Gulf Res. Rep.5, 17–28.

Calder, D. R. & B. S. Hester, (1978). Phylum Cnidaria. In Zingmark,R. G. (ed.), An Annotated Checklist of the Biota of the Coastal Zone of South Carolina. Univ. South Carolina Press, Columbia, pp. 87–93.

Firth, F. E., (1969). The Encyclopedia of Marine Resources. Van Nostrand Reinhold Co., New York, pp. 324–325.

Hooper, S. N. & R. G. Ackman, (1973). Distribution of trans-6- hexadecenoic acid, 7-methyl-7-hexadecenoic acid and common fatty acids in lipids of the ocean sunfish Mola mola. Lipids 8, 509–516.

Hsieh, Y-H. P. & J. Rudloe, (1994). Potential of utilizing jellyfish as food in Western countries. Trends Food Sci. Tech. 5, 225–229.

Hsieh, Y-H. P., F-M. Leong & K.W. Barnes, (1996). Inorganic constituents in fresh and processed cannonball jellyfish (Stomolophus meleagris). J. Agric. Food Chem. 44, 3117–3119.

Huang, Y. W., (1988). Cannonball jellyfish, Stomolophus Meleagris as a food resource. J. Food Sci. 53, 341–343.

Joseph, J. D., (1979). Lipid composition of marine and estuarine invertebrates. Porifera and Cnidaria. Prog. Lipid Res. 18, 1–30.

Kimura, S., S. Miura & Y. H. Park, (1983). Collagen as the major edible component of jellyfish (Stomolophus nomurai). J. Food Sci. 48, 1758–1760.

Kraeuter, J. N. & E. M. Setzler, (1975). The seasonal cycle of Scyphozoa and Cubozoa in Georgia estuaries. Bull. mar. Sci. 25, 66–74.

Kramp, P. L., (1961). Synopsis of the medusae of the world. J. mar. biol. Ass. U. K. 40, 1–469.

Larson, R. J., (1976). Marine flora and fauna of the northeastern United States. Cnidaria: Scyphozoa. NOAA Tech. Rep. NMFS Circ. 397, 17.

Leong, F-M., (1995). Processing, chemical composition, and quality evaluation of cannonball jellyfish. Diss:Auburn University, Alabama, U.S.A.

MacMillen, O. (1994). Zoomobile effectiveness: sixth graders learning vertebrate classification. Annual Proceedings of the American Association of Zoological Parks and Aquariums. 181–183.

Martinussen, M.B. and Bamstedt, U. (1999). Nutritional ecology of gelatinous planktonic predators. Digestion rate in relation to type and amount of prey. Journal of Experimental Marine Biology and Ecology. 232. 61-84.

Mayer, A. G., (1910). Medusae of the World. Volume 3. 'The Scyphomedusae'. Carnegie Institution of Washington, Washington. pp. 711.

Mianzan, H. W., & Cornelius, P. F. S. (1999). Cubomedusae and Scyphomedusae. In D. Boltovskay (Ed.), South Atlantic Zooplankton. Leiden: Backhuys Publishers. pp. 513–559.

Mills, C. E. (2001). Jellyfish blooms: are populations increasing globally in response to changing ocean conditions? Hydrobiologia, 451, 55–68.

Morgan, J. M., & Hodgkinson, M. (1999). The motivation and social orientation of visitors attending a contemporary zoological park. Environment and Behavior. 31. 227–239.

Ortiz-Corp's, E., Cutress C. E., & Cutress, B. M. (1987). Life historyof the coronate scyphozoan Linuche unguiculata (Swatz, 1788). Caribbean Journal of Science. 23. 432–443.

Primo, A. L., Marques, S. C., Falcão, J., Crespo, D., Pardal, M. A., & Azeiteiro, U. M. (2012). Environmental forcing on jellyfish communities in a small temperate estuary. Marine environmental research, 79, 152-159.

Rapoza R., Novak D. & Costello J.H. (2005). Life-stage dependent, in situ dietary patterns of the lobate ctenophore Mnemiopsis leidyi Agassiz 1865. Journal of Plankton Research. 27. 951–956.

Revelles M., Cardona L., Aguilar A. & Fernàndez G. (2007). The diet of pelagic loggerhead sea turtles (Caretta caretta) off the Balearic archipelago (western Mediterranean): relevance of ling-line baits. Journal of the Marine Biological Association of the United Kindgom. 87, 805–813.

Richardson A.J., Bakun A., Hays G. & Gibbons M.J. (2009). The jellyfish joyride: causes, consequences and management responses to a more gelatinous future. Trends in Ecology & Evolution. 24, 312–322.

Rippingale R.J. & Kelly S.J. (1995). Reproduction and survival of Phyllorhiza punctate (Cnidaria: Rhizostomeae) in a seasonally fluctuating salinity regime in western Australia. Marine and Freshwater Research. 46, 1145–1151.

Roohi A., Kideys A., Sajjadi A., Hashemian A., Pourgholam R., Fazli H., Khanari A.G. & Eker-Develi E. (2010). Changes in biodiversity of phytoplankton, zooplankton, fishes and macrobenthos in the southern Caspian Sea after the invasion of the ctenophore Mnemiopsis leidyi. Biological Invasions. 12, 2343–2361.

UNEP. (1991). Jellyfish blooms in the Mediterranean. Proceedings of the II Workshop on Jellyfish in the Mediterranean Sea. MAP Technical Reports Series, No.47; UNEP, Athens

Wilson D.P. (1947). The Portuguese Man-of-War, Physalia physalis L., in British and adjacent seas. Journal of the Marine Biological Association of the United Kingdom. 27, 139–172.

You K., Ma C.0, Gao H., Li F., Zhang M., Qui Y. & Wang B. (2007). Research on the jellyfish (Rhopilema esculentum Kishinouye) and associated aquaculture techniques in China: current status. Aquaculture International. 15, 479–488.

Zeller, D., S. Harper, K. Zylich, & D. Pauly (2015). Synthesis of underreported small-scale fisheries catch in Pacific island waters, Coral Reefs. 34, 25-39.

Zhang, H., J.Y. Zhang, H.L. Wang, P.J. Luo, & J.B. Zhang (2016). The revision of aluminum-containing food addidive provisions in China, Biomedical and Environmental Sciences. 29, 461-466.

Zhang, J., R. Duan, L. Huang, Y. Song, & J.M. Regenstein (2014). Characterisation of acid-soluble and pepsin-soubilised collagen from jellyfish (Cyanea nozakii Kishinouye), Food Chemistry. 150, 22-26.

Zhang, W.-T., L.-H. Tang, W. Chen, D.-Q. Chen, & C. Deng (2008). Therapeutic effect of collagen from Cyanea nozakii on adjuvant arthritis in rats, Chinese Journal of Marine Drugs. 27, 35-38.

Zhong, X.M., J.H. Tang, & P.T. Liu (2004). A study on the relationship between Cyanea nozakii breaking out and ocean ecosystem, Modern Fisheries Information. 19, 15-17.

6

Cybertaxonomy A New Biodiversity Science Tool

[1]Sanjay Chandravanshi, [2]Sudhan Chandran, [3]Hari Prasad Mohale, [4]Sahil Rai and [5]H.S. Mogalekar

[1]Department of Fisheries Biology and Resource Management, Fisheries Resources, Fisheries College and Research Institute,TNJFU, Tamil Nadu, India
[2]Department of Fisheries Resource Management, Fisheries Resources Harvest and Post-Harvest Management Division, ICAR-CIFE, Mumbai Maharashtra, India
[3]Department of Fisheries Biology and Resource Management Fisheries College and Research Institute, TNJFU, Tamil Nadu India
[4]Department of Fisheries Resource Management Fisheries Resources, College of Fisheries, GADVASU, Ludhiana, India
[5]Department of Aquatic Environment Management, College of Fisheries Dholi Muzaffarpur, RPCAU, Bihar, India

---

Intoduction

Cybertaxonomy is an abbreviation for "cyber-enabled taxonomy." It shares the traditional goals of taxonomy: to explore, discover, characterize, name, and classify species; to study their phylogenetic relationships; and to map their ecological associations and geographic distributions. To improve the speed with which we can help in providing taxonomic products and information, there is undoubtedly a requirement to move taxonomic endeavour into the digital era. Currently, the rate at which we record and maintain morphological information that enables species discovery is the primary constraint on taxonomic productivity (Lasalle, et al., 2009). Cybertaxonomy is one attempt to accomplish this complex goal.

Cybertaxonomy employs a collection of electronic tools to assist and equip the science of taxonomy in accelerating species invention and the implementation of taxonomic expertise in biodiversity investigation.Cybertaxomy can yield results more quickly and efficiently than ever before thanks to adoption of digital technologies and cybertaxonomy. It shares the traditional goals of taxonomy but employs software, instrumentation, non-traditional hardware, communication tools, and work practices, bringing information science and technologies to carry on the data and information derived by the study of organisms, their genes, and their interactions. The new internet resources facilitate and simplify taxonomic work, as well as aid in the filling of biodiversity knowledge gaps (Smith, 2013).

Like traditional taxonomy, cybertaxonomy is integrative. That is, taxonomists collect, synthesize, and analyze all evidence available that is relevant at the taxonomic level(s) under consideration. Data sources that are commonly used include molecular, morphological, ontogenetic and fossil data, together with physiological, biochemical, ethological, other data sources, as appropriate. Cybertaxonomy can be represented as a GIS-like environment that contains several data layers, including molecular, distributional, morphological, image, and sound recordings (to name just a few). There are also layers with applications and algorithms for processing data based on spatial, temporal, ecological or phylogenetic relationships.In a multi-layered "mesh," users can activate any perfect blend of "layers" to retrieve required information. There are numerous and diverse possibilities, such as dichotomous or interactive diagnostic keys, checklists of species in specific areas plotted over seasonal occurrences, distribution maps (point data or predicted ranges based on climate and environment), three-dimensional visualizations of phenotypic variation, and so on. Independent taxonomists have traditionally synthesized all knowledge available into periodically published monographs, but international teams of specialists working collaboratively in cyberspace will be able to gather and maintain all data and information pertaining to a taxon's species with up-to-date accuracy. Ecologists will even be able to combine these taxon knowledge bases to compare and difference species that coexist in a particular environment.

Tools in Cybertaxonomy

A diverse set of hardware, software, instrumentation, and communication tools contribute to the creation of biodiversity information as well as its efficient and expedited management and delivery. Electronic publications, digital image archives, freely accessible electronic databases, taxon factsheets, and interactive identification keys help and speed up the identification process. Several kinds of software now allow for the creation of online taxonomic databases and identification keys. Thus, morphological nomenclatural, distributional and bibliographical data, as well as illustrations/digital images, are organized a solitary database, providing end users with all necessary baseline information about a taxon. These innovative work practices enable taxonomists to perform more effectively while ensuring the highest excellence levels (Wheeler and Valdecasas, 2004).

Digital Imaging, an Essential Component in Cybertaxonomy

Recent advancements in digital imaging have profited many areas of study in biology, particularly entomology. By prefixing a good image/good illustration to the verbal part, a character state is best described and information is imparted more meaningfully in taxonomy. Until recently, light microscopy photographs might barely reproduce the requisite taxonomic characters in detail. This has been a constraint in taxonomic studies of small insects. A taxonomist, who is typically very busy, had to spend hours drawing line diagrams. The SEM (Scanning Electron Microscopy) was the sole option for capturing high-quality images until recently. However, using light microscopy it is now possible to produce high resolution images of taxonomic characters even in the case of minute specimens.In comparison to SEM, the procedure is significantly less expensive and simpler. There are additional benefits, such as the specimens remaining unchanged after the imaging process. While the procedure is less expensive and simpler than SEM, it has the added benefit of leaving the specimens unchanged after the imaging procedure and reproducing the original colors in the images.

State of Cybertaxonomy

Cybertaxonomy will have an effect on almost every aspect of taxonomic information creation and use when implemented fully. Taxonomists need original sources that date back to 1753 as well as thousands number of specimen from all of a species' known geographic ranges, form specimens to ensure nomenclatural stability, and specimens and data from dozens of museums or herbaria in many countries. The promise of cybertaxonomy is incredible access to such resources as archives, open-access databases, digital image electronic publishing tools remotely, electronic publishing tools and operable instrumentation.

Printed traditional sources of information about taxonomy are nearly always out of date, and frequently by the time of publication. There might have been several new species, distribution records, name changes or natural history assumptions added since publication, all of which are extremely valuable to the ecologist. Cybertaxonomy will provide access to complete and up-to-date information derived from a taxon base of knowledge that is updated regularly by the taxonomic community.

Cybertaxonomy will expand the field ecologist's arsenal of species identification tools, such as interactive diagnostic keys, automated identifications from photographs browser-based field guides, direct access to a specialist, or the collection of samples of tissues for molecular identifications — the latter being especially useful for associating very dissimilar life stages, such as those observed between larval and adult insects.

Interactive Identification Keys

Accurate identification is of fundamental importance in biodiversity studies and relies on efficient identification tools. Identification is the process of finding the taxon to which a specimen actually belongs. An identification key, also known as a taxonomic key, aids in tracing the identity of unknown biological entities. The age old tradition in taxonomy has been the use of conventional paper based keys. But the past two decades witnessed the development and advancement of many computer based interactive keys. Compared to paper-based keys, desktop identification keys offer more alternatives, frequently permit backtracking, and also include images of both eliminated and still-under-consideration species. This greatly speeds up the identifying procedure. A set of advantages of interactive keys over conventional keys can be accessed at http://delta-intkey.com/www/interactivekeys.htm.

Taxonomic Names

Accurate identification is critical in biodiversity studies and is dependent on effective identification tools. Identification is the method of determining which taxon a specimen belongs to. A taxonomic key, also known as identification key, aids in determining the identity of unidentified biological entities. The use of traditional paper-based keys has been a long-standing tradition in taxonomy. However, many computer-based interactive keys have been developed and improved during the previous two decades. Computer-based identification keys have even more options than paper-based keys and generally support back-tracking, as well as providing images of both excluded and still-under-consideration species. This significantly speeds up the identification process (http://www.fishbase.org/manual/english/FishBaseIdentification keys.htm).

Species Identifications

Cybertaxonomy will offer a variety of tools to aid in the immediate and precise identification of species, in addition to training field ecologists to recognize the target taxa and video-mediated consultation.

Checklists

Cybertaxonomy will make complete and up to the minute checklists available for specific localities and ecosystems.

Names

Scientific names are unique identifiers that are used to keep and retrieve data in publications and databases. Taxonomists will be able to use cybertaxonomy to enhance the stability, informativeness and reliability of names, as well as retrieve all citations and relevant data stored under older names.

Phylogenetic Classification

Ecologists could indeed make predictions about species significant contribution to ecological functions by viewing them in their phylogenetic context. Sources of food, physiologies, reproductive strategies, and behaviors of closely related species are frequently similar.

Geographic Distributions

Cybertaxonomy is mobilizing the information content of the natural history collections of the world, whose estimated three billion specimens provide a wealth of information about the distributions of species, including irreplaceable historical records.

Virtual Ecology Assemblages

Specimens are chosen for taxonomy research based on their evolutionary connections. Using cybertaxonomy tools, the ecologist can virtually reassemble all the specimens, regardless of taxon, collected in any one place at the same time or over a sequence of times.

Conservation-Relevant Information

Cybertaxonomy is mobilizing the factual content of the world's natural history collections, which contain an expected three billion specimens and contain a wealth of information about species distributions, which include irreplaceable historical records.

Morphologically Structured Information

As image archives expand to include digitized publications as well as images of specimens, such visual information can be harvested and analyzed to better understand phenotypic diversity in connection with population structure, morphoclines, environmental conditions, population structure and other factors.

Online Access to Museum Specimens

Pathologists and histologists has been using telemicroscopy for decades and will soon have the ability to network global collections so that actual specimens, as well as stored images, can be made accessible, compared measured, manipulated and studied in real - time basis.

Conclusion

Cybertaxonomy is changing not only the way taxonomists work, but also how ecologists could perhaps access and use information, knowledge and taxonomic data. Ecologists will be able to better understand complex ecosystems, detect and supervise environmental change, and achieve the goals for self sustaining ecological services as new methods for harvesting, structuring, and analyzing taxonomic and associated natural history information emerge.

References

http://cybertaxonomy.eu/blog/2007/04/11/what-is-cybert axonomy.html

http://darwin.zoology.gla.ac.uk/~vsmith/cybertaxonomy/ http://www.biocorder.org/

http://onlinelibrary.wiley.com/doi/10.1111/1744-7917.1 208 8/abstract

http://taxonomytraining.eu/content/basics-taxonomydescribing-illustrating-and-communicating-biodiversity#overlay-context=content/modern-taxonomy-2014-2015%3Fdestination%3Dnode/233

http://www.e-taxonomy.eu/

http://www.fao.org/docrep/019/i3354e/i3354e.pdf

La Salle, J., Wheeler Q. D., Jackway, P., Winterton, S., Hobern, D., Lovell, D., 2009. Accelerating taxonomic discovery through automated character extraction Zootaxa 2217: 43–55.

Louette, M. and Mergen, P., 2008. Taxonomy training and cybertaxonomy activities at the Royal Museum for Central Africa, Tervuren, Belgium. In Origin and Evolution of Natural Diversity: Proceedings of the International Symposium, The Origin and Evolution of Natural Diversity, held from 1-5 October 2007 in Sapporo, Japan, pp. 241-243.

Rajmohana, K. and Bijoy, C., 2012. Cybertaxonomy:Anovel tool in biodiversity science. Biodiversity: Utilization, Threats and Cultural Linkages, Pages 55–64.

Smith, V. (2013). Cybertaxonomy. In D. Shorley and M. Jubb (Eds.), The Future of Scholarly Communication, pp 63-74.

Sudhan, C. and Moulitharan, N., 2017. A novel approach in fish cyber taxonomy. National Journal of Advanced Research, 3(2): 47-50.

Wheeler Q.D., Raven, P.H. & Wilson, E.O. (2004) Taxonomy: impediment or expedient? Science, 303, 285.

Wheeler QD, ed. 2008. The New Taxonomy. The Systematics Association Special Volumes Series 76. Boca Raton, FL: CRC Press, 2008.

Winterton S.L. (2009) Revision of the stiletto fly genus Neodialineura Mann (Diptera: Therevidae): an empirical example of cybertaxonomy. Zootaxa, 2157, 1–33.

Zhang, Z.-Q. (2006) The making of a mega-journal in taxonomy. Zootaxa, 1385, 67–68.

Zhang, Z.-Q. (2008) Zoological taxonomy at 250: showcasing species descriptions in the cyber era. Zootaxa, 1671, 1–2.

7

Delineation of Stocks through Integrated Approaches for Management and Conservation of Fishes

[1]Dhanlakshmi. M, [2]Shivkumar ,[3]Rinkesh N Wanjari, [4]Harshavarthini. M

[1]Department of Fisheries Resource Management, Fisheries Resources, Harvest and Post-Harvest Management Division, ICAR-Central Institute of Fisheries Education (ICAR-CIFE), Mumbai, Maharashtra

[2]Department of Fish Nutrition and Feed Technology, Fish Nutrition Biochemistry and Physiology and Division ICAR-CIFE, Mumbai, Maharashtra

[3]Fisheries Resource Management Division, Faculty of Fisheries, Rangil Ganderbal, Jammu & Kashmir

[4]Department of Fish Genetics and Breeding, Fish Genetics and Biotechnology Division, ICAR-Central Institute of Fisheries Education (ICAR-CIFE) Mumbai, Maharashtra

Abstract

Stock is the fundamental unit used for the management of fishery resources on which dynamic population models are applied to understand the status and to adopt appropriate management measures to ensure long term sustainability. The stock identification is an integral component of modern fisheries used for effective fisheries management. For the management of the fishery resources there is a requirement of specific stock level tactic corresponding to different regions. So for the assessment of stock and its status, stock identification and differentiation which are evaluated through compilation of various methods such as truss morphometry, morphometric and meristic count study and otolith shape analysis give better understanding of stocks compared to reliable only on few methods for stock differentiation.

Keywords: Stock, morphometric, truss, otolith shape

Introduction

The fish is an important source of protein as it constitute about 17 of the total animal protein (SOFIA, 2018) and the rising demand for fish has led to unsustainable exploitation of marine and inland water resources and fish stocks are getting overexploited, due to which the output from capture fisheries is uncertain. This has put difficulty for the management of our open water resources. It is now urgent to develop a management strategy for conservation of natural resources for future generation as our responsibility.

Hilborn and Walters (1992) define stocks are arbitrary groups of fish large enough to be essentially self-reproducing, with members of each group having similar life history characteristics. Stock is defined as "a part of a fish population usually with a particular migration pattern, specific spawning grounds, and subject to a distinct fishery" by the International Council for the Exploration of the Seas. Understanding the importance of delineating the stocks and their boundaries which have become an essential part of fishery management and conservation. The stock concept contains two basic principles that fish are subdivided into local populations, and they are having genetic differences between local populations that are adaptive (MacLean & Evans, 1981). According to Cushing (1968), stock as a population in which important parameters of growth, recruitment, and mortality are homogeneous. Ihssen et al., (1981) defined stock as an intraspecific group of randomly mating individuals with temporal and spatial integrity. According to Larkin (1981), a stock as "a population of organisms which share a common gene pool, is sufficiently discrete to warrant consideration as a self-perpetuating system which can be managed.

Variation in the growth, development, and maturation in different individuals of the same species results in a variety of body shapes. (Cadrin,2000). In fact, a fish population can be composed of more subpopulations, each member having its own demographic properties and capabilities for recovery after overexploitation. Such subpopulations can be relatively discrete and self-replenishing while still maintain sufficient exchange to preserve genetic homogeneity through the dispersal of individuals (Hanski & Simberloff, 1997; Hanski, 1999). Therefore, population identity and the underlying metapopulation structure should be together taken within resource assessments, which help in fishery management (Reiss et al., 2009). Otherwise, it could lead to the local depletion of the stocks and the associated loss of genetic diversity (Stephenson, 1999; Hutchinson, 2008; Ying et al., 2011). The stock structure of a species and the way in which fishing effort and mortality are dispersed for the optimum yield is considered important for effective fishery management (Grimes et al., 1987). An understanding of stock structure is also important in designing appropriate management regulations in fisheries where multiple stocks are exploited differently (Ricker, 1981). The delineation of stocks through morphometrics, meristics, truss and otolith shape indices are discussed below.

Conventional Morphometrics and Meristics

Anatomical features such as morphometrics and meristics have been traditionally used in fisheries biology to describe various categories of fishes by fishery biologists (Utter, 1981). Both these characters form a backbone in stock identification of fishes (Ihssen et al., 1981). These characters are most commonly employed in the stock anatomical features such as morphometrics, and meristics have been traditionally used fisheries biology to describe various categories of fishes by fishery biologists (Utter, 1981). These characters are most commonly employed in the stock delineation of fish (Reist, 1985; Silva, 2003). Morphometric includes the analysis of body shape or the size of particular morphological features of various body dimensions or parts. Morphometrics is used for measuring discreteness and relation among stocks (Ihssen et al., 1981; Melvin et al., 1992). Bookstein (1991) showed that morphometrics is elegant assimilation of geometry and biology bonded together empirically. Morphometric variables for stock identification were analysed earlier by univariate comparisons, then bivariate analysis, and then by multivariate morphometrics (Cadrin, 2000).

Morphometric variations have been identified as an important basis for examining the population structure (Ihssen et al., 1981). Begg et al., (1999) opined that morphometric variations between stocks could provide a basis for stock structure, and maybe more appropriate for studying short-term, environmentally induced variation for the fisheries management.

According to Cadrin (2000), spatial isolation can also result in the development of different morphological features between stocks because the interactive effects of the environment, selection, and genetics on individual ontogenies produce morphometric differences within a species. The traditionally used morphometrics which is still used widely in identification approaches (Marcus, 1990; Rohlf & Marcus, 1993), which typically apply multivariate statistical methods such as principal component analysis, canonical variate analysis, discriminant function analysis, multivariate analysis of variance, factor analysis to a set of variables measured on each individual. Now numerous morphometric tools for stock differentiation like morphometric landmark methods, truss network analysis, and outline methods are developed. For stock characterization of fishery resources, these techniques may be used alone or in combinations (Pazhayamadom, 2006). For enhancing the utility of morphometric research in fish stock identification image analysis systems played a major role in the development of morphometric techniques (Cadrin & Friedland, 1999)

Meristic counts are measurable traits such as counts of fin rays, gill rakers, scales in Lateral lines / Transverse scales, or vertebrae that are countable in a serial fashion. Meristic characters are generally set early in ontogeny and remain stable throughout life; thus, reflecting environmental factors like temperature, salinity, dissolved oxygen, and light over a relatively brief time of larval development (Begg & Waldman, 1999). These characters are determined partially by genetics and environmental conditions during egg and larval development (Swain et al., 2005).

Truss Network System

A "truss" is a line connecting two morphological landmarks that can be linked in a network of polygons to quantify the longitudinal, vertical, and oblique dimensions of the body. It is a system widely used for morphometric measurements for stock differentiation of fishes which are physically looking similar. Traditional measurements have the disadvantage of biased coverage over the body (Strauss & Bookstein, 1982), which was overcome by the truss network system in which fish in a uniform network and increases the probability of extracting morphometric differences within species (Turan,1999). When the differences are not specific to a few structures, the truss network system augments the discrimination between the species compared to most of the traditional methods used (Strauss & Bookstein, 1982). Truss network systems were to evaluate the potential impact of farmed fish escaped to the wild environment (Arechavala-Lopez et al., 2012).

From the perspective of interdisciplinary stock differentiation, the morphometric analysis provides information on phenotypic stocks, groups of individuals with similar growth, mortality, and reproductive rates. Landmarks refer to some arbitrarily selected points on a body of fish, and with these points, the individual fish shape can be investigated. Landmarks are a point of correspondence on an object that matches between and within populations (Barlow, 1961; Swain & Foote, 1999; Hossain et al., 2010). Landmarks selected for the truss network should be homologous, signifying the same developmental feature among specimens, and should be easily located (Bookstein, 1990).

The points that can be easily found are the most efficient landmarks like the tip of snout, insertion point of different fins. The number of landmarks used by different researchers on different species ranged from 6 to 17, and the truss distances range from 6 to 38. Comparing the performances of traditionally measured finfish dimensions to box-truss distances and the number of investigators reported that the latter provided a more accurate classification of individuals (Strauss & Bookstein, 1982; Schweigert & Withler, 1990; Roby et al., 1991).The landmark-based method of geometric morphometrics has no restriction on the directions of variation and localization of shape changes and is highly effective in capturing information about the shape of an organism (Cavalcanti et al., 1999). Computer-based image-capture and analysis of fish with truss networks allow for the simple and precise measurement of truss lines and boxes (Fitzgerald et al., 2002).

Table 1: Studies on truss network performed by different authors

S.No.	Species	Locality	Truss		Statistical Methods used for analysis	Author
			Landmarks	Distances		
1	*Oncorhynchus kisutch.*	Wild and Hatchery stocks	15	34	PCA	Swaine et al., 1991
2	*Salmo salar*	Wild and Farmed stocks	19	21	ANOVA, PCA	Fleming and Einum, 1997
3	*Sardina pilchardus*	North eastern Atlantic and the Western Mediterranean stocks	10	19	DFA,PCA	Silva, 2003
4	*Rastrelliger kanagurta*	Indian coast	10	21	PCA	Jayasankar et al.,2004
5.	*Liza abu*	Three different rivers	13	24	DFA,PCA	Turan et al., 2004
6.	*Engraulis encrasicolus*	Black, Marmara, Aegean Sea	12	25	ANOVA, DFA	Turan et al., 2004
7.	*Pomatomus saltatrix*	Four different seas	13	27	ANOVA,PCA,DFA	Turan et al., 2006
8.	*Lates calcarifer*	East and West coast of India	10	21	PCA	Gopikrishna et al., 2006
9.	*Ocyurus chrysurus*	Tropical West Atlantic	10	21	CVA	Vasconcellos et al., 2008
10.	*Chalcalburnus Chalcoides*	Haraz and Shiruz rivers	15	16	MANOVA,CDA,PCA	Bagherian and Rahmani, 2009
11	*Labeo calbasu*	Hatchery and two different rivers	11	22	DFA	Hossain et al., 2010
12.	*Megalopsis cordyla*	Indian coast	15	33	CFA	Sajina et al.,2011
13.	*Decapterus russelli*	East and West coast of India	11	23	FA, DFA	Sen et al., 2011
14.	*Sparus aurata & Dicentrarchus labrax*	Farmed and Wild population	16 & 17	31 & 30	PCA,ANOVA,DFA	Arechavala-Lopez et al., 2011
15.	*Channa punctatus*	Three different rivers	10	27	DFA,PCA	Khan et al., 2013
16.	*Rasbora group*	Lake Laut Tawar	8	14	ANOVA, DFA	Muchlisin, 2013
17.	*Alburnus chalcoides*	Southern Caspian Sea	13	25	MANOVA, ANOVA, PCA	Mohaddasi et al., 2013

S.No.	Species	Locality	Truss		Statistical Methods used for analysis	Author
			Landmarks	Distances		
18.	*Liza aurata*	Caspian Sea	13	35	DFA,PCA	Kohestan - Eskandari et al., 2014
19.	*Alosa brashnicowi*	Southern Caspian Sea coasts	15	72	ANOVA,PCA	Paknejad et al., 2014
20.	*Schizothorax esocinus*	Kashmir Himalaya	14	31	PCA,DFA	Mir et al., 2014
21.	*Rastrelliger kanagurta*	Both Southern coast	10	21	FA	Remya et al., 2014
22.	*Sparus aurata*	Wild and farmed from Mali Ston Bay	12	27	PCA	Šegvić - Bubić et al., 2014
23.	*Harpadon nehereus*	Indian coast	11	24	FA,DFA,ANOVA	Pazhayamadom et al., 2015
24.	*Clarias batrachus*	Gangetic river system	11	23	ANOVA,DFA	Miyan et al., 2016
25.	*Channa gachua*	North-Eastern Thailand.	12	26	PCA,DFA	Jearranaiprepame, 2017
26	*Channa punctatus*	Northern and Eastern Regions of India	11	23	PCA,DFA	Kashyap et al., 2016
27	*Aristeus alcocki*	Indian coast	18	39	PCA, DFA	Purushothaman et al., 2018
28	*Anodontostoma chacunda*	Four rivers of Bangladesh	12	28	DFA,PCA,ANOVA	Hanif et al., 2019
29	*Chanos chanos*	Indian waters	10	21	PCA	Hari et al., 2019
30	*Clupisoma garua*	Ganga riverine system	13	27	PCA,DFA	Biswas et al., 2019
31	*Scomberomorus niphonius*	China Coast	8	16	PCA,DFA	Jiang et al., 2020
32	*Rastrelliger kanagurta*	Indonesian waters	16	34	DFA	Hakim et al., 2020
33	*Labeo rohita*	Fish pond in Nepal	11	22	PCA	Amatya .,2021
34	*Botia dario*	Bangladesh	12	23	FA,CVA,CA,PCA	Mahfuj et al.,2022

For phenotypic characters, multivariate analysis has a greater probability of detecting complex changes than univariate methods. Multivariate techniques simultaneously consider the dissimilarity in several characters and thereby assess the similarities between samples (Turan, 1999). Statistical analysis like Principal Component Analysis (PCA), Common Factor Analysis (CVA), Canonical Variate Analysis (CVA), and Discriminant Function Analysis (DFA) is used most for morphometric analysis. In all landmark-based morphometric methods faces the difficulty of removing the size variation in extracted measurements (Parsons et al., 2003). The standard method used to reduce the size variation includes ratios (Humphries et al., 1981), allometric methods, regression models, and factor or component analyses (Bookstein et al., 1985). The regression equation given by Elliott et al., (1995) and Reist (1985) was found to be a common method followed by several researchers to eliminate any variation resulting from allometric growth.

Otolith Shape Analysis

Otoliths, comprise of calcium carbonate are crystallized structures located in the inner ear of fish, are an indirect means for examining fish populations and evaluating the relationship between the environment and the organisms (Lord et al., 2012; Zengin et al., 2015). In different stocks or populations, otolith shapes are varying (Campana, 2004; Schulz-Mirbach et al., 2008). Sagittal otoliths are mostly used in studies of otolith morphology (Campana & Casselman, 1993). Fish sagittal otoliths are formed during the egg stage itself and begin developing and grow continuously and forming ring structures around the core. Otoliths are an important tool for fishery biologists in the study of fish populations (Gillanders & Kingsford 2000; Longmore et al., 2010). Otolith shapes were species-specific (Stransky et al., 2008), and also spatial variation in otolith shapes could be related to stock differences (Campana & Casselman, 1993; Stransky et al., 2008). They have been most widely used in aging and growth studies and information about growth patterns from individual to population level (Campana & Thorrold 2001; Longmore et al., 2010).

Analysis of otolith shape is a successful tool in differentiating between and within fish stocks (Campana & Casselman,1993; Longmore et al., 2010). For analysis of otolith shape, two main morphometric methods are used: analysis of landmark (Libungan et al., 2015) and analysis of outline (Bookstein et al., 1985; Libungan et al., 2015). By the outline analysis of otolith, boundary shapes are done so that patterns of variation of shape within and among stocks can be evaluated (Libungan et al., 2015). Otolith morphology is impacted by biotic and abiotic factors (Mahe et al., 2016). Otolith size changes with body growth, temperature, and food quantity (Libungan et al., 2015). The growth rate of fish has a direct relation to the size and shape of the otoliths (Gauldie & Nelson, 1990; Hari et al., 2019). Fluctuations of the diet in quantity and frequency also determine the shape of the otolith over a short period of time. (Gagliano and McCormick, 2004; Hari et al., 2019).

Otoliths are recorders of development of growth, and their structure and development are influenced by external environmental conditions as well as the physiological state of individual fish (Campana & Neilson, 1985). Environmental characteristics of the area inhabited by the fish throughout its life sea such as temperature, salinity, and food availability have an impact on fish growth rates (Longmore et al., 2010), which impact the growth pattern and final shape of the otolith (Campana & Neilson, 1985; Longmore et al., 2010). Otolith shape is not affected by short-term changes in fish conditions (Campana & Casselman, 1993; Mahe et al., 2016).

Otolith shape is one of these techniques and is more reliable than other ones because it is less to short-term variability caused by changes in feeding or spawning condition (Castonguay et al., 1991) The shape analysis of otoliths is the most effective alternative method to differentiate the population over genetic analysis as these are time-consuming and more expensive (Deepa et al., 2019). Cadrin & Friedland (1999) reported that image-processing techniques have more significantly enhanced morphometric analysis, such as shape analysis of scales and otoliths, and can complement other methods of stock identification (DeVries et al., 2002). The advantage of these methods is that they are cost-effective (Libungan et al., 2015).

When the relationship between otolith length and total length in a species is examined, the total length or standard length of a fish from its otolith length can be estimated, or vice versa (Zengin et al., 2015) Otolith length was defined as the greatest distance between anterior and posterior edges, and Otolith breadth was defined as the greatest distance from dorsal to ventral edges (Zengin et al., 2015). The utilization of shape indices may be considered a good technique to separate stocks. From a fisheries management perspective, populations delineated based on phenotypic otolith traits can be considered as separate management units (Cadrin & Friedland ,1999; Swain & Foote ,1999).

Table 2: Studies on otolith shape analysis by different authors

S.No	Species	Location	Statistical Method	Author
1	*Melanogrammus Aeglefinus*	Georges bank waters	ANCOVA, MANOVA, CDA	Begg and Brown, 2000
2	*Scomberomorus cavalla*	Eastern Gulfof Mexico and Atlantic Ocean waters	ANOVA, DFA	DeVries et al.,2002
3	*Serranus cabrilla*	Atlantic and Mediterranean waters	ANOVA , CDA	Tuset et al., 2003
4	*Sebastes maritnus S. mentella*	North Atlantic waters	MANCOVA, LDA	Stransky, 2005
5	*Solea solea*	North West Mediterranean waters	ANOVA, PCA, CDA	Mérigot et al., 2007
6	*Clupea harengus*	Irish Sea	ANCOVA , DFA	Burke et al., 2008
7	*Trachurus trachurus*	Northeast Atlantic and Mediterranean waters	ANCOVA , DFA	Stransky et al., 2008
8	*Coryphaenoides rupestris*	North Atlantic waters	ANCOVA, ANOVA, DFA	Longmore et al., 2010
9	*Helicolenus Dactylopterus*	Portuguese waters	MANOVA, CDA	Neves, 2011
10	*Scomberesox saurus saurus*	Northeast Atlantic and Western Mediterranean waters	ANCOVA , PCA, DFA, MANOVA	Agüera and Brophy, 2011
11	*Lophius piscatorius*	Northeast Atlantic waters	LDA	Caňás et al., 2012
12	*Encrasicholina punctifer*	Persian Gulf and Oman sea	ANOVA	Daryaei et al., 2013
13	*Micromesistius australis*	Atlantic waters	ANOVA , CDA	Leguá et al., 2013
14	*Gadus morhua*	Baltic waters	CDA	Paul et al., 2013
15	*Phycis phycis*	Northeast Atlantic waters	MANOVA, CDA	Vieira et al., 2014
16	*Lutjanus johnii*	Persian Gulf and Oman Sea	PCA, CDA, ANCOVA	Sadighzadeh et al., 2014
17	*Engraulis encrasicolus*	Black and Marmara Seas	ANOVA	Zengin et al., 2015
18	*Clupea harengus*	Canada, the Faroe Islands, Iceland, Ireland, Norway and Scotland, U.K.	ANCOVA , LDA	Libungan et al., 2015

S.No	Species	Location	Statistical Method	Author
19	*Micromesistius poutassou,*	Northeast Atlantic	ANCOVA, HSD , PCA	Mahe et al., 2016
20	*Sardinops sagax*	Australian waters	EFC, PERMANOVA	Izzo et al., 2017
21	*Trachurus picturatus*	Northeast Atlantic waters	MANOVA, DFA	Vasconcelos et al., 2018
22	*Thunnus alalunga*	North Atlantic waters	ANCOVA DFA	Duncan et al., 2018
23	*Chanos chanos*	Indian waters	MANOVA , PC1, PC2	Hari et al., 2019
24	*Bembrops caudimacula*	Arabian Sea and Andaman Sea	PCA and MANOVA	Deepa et al., 2019
25	*Genypterus blacodes*	Chilean patagonia	DA	Wiff et al.,2020
26	*Channa Striata*	Major rivers in Northern India	PERMANOVA and MGLM	Khan et al.,2021
27	*Upeneus vittatus*	Indian waters	LDA and CAP	Nama et al.,2022

Conclusion

In this chapter we have dealt with different stock identification methods used for delineating the stocks.It is prime most important to know about the stock ,so every stock can be conserved efficiently to standardize their yield and for their substainability for future generations. An operational fisheries management can top to intense variations in the biological aspects and productivity rates of a species.Though, not able to correctly regulate a species' stock structure might top to the overfishing and reduction of less productive stocks while discrepancy re-establishment among unidentified stock components may top to a absence of anticipation of future requirements for stocks under re-establishment programmes.In future we can reduce overfishing and maintain the sustainable management of resources.

References

Agüera, A. and Brophy, D., 2011. Use of saggital otolith shape analysis to discriminate Northeast Atlantic and Western Mediterranean stocks of Atlantic saury, Scomberesox saurus saurus (Walbaum). Fisheries Research, 110(3), pp.465-471.

Amatya, B., 2021. Truss-based morphometrics of pond cultured Indian major carp, Labeo rohita (Hamilton, 1822) from Chitwan District, Nepal.

Arechavala-Lopez, P., Sanchez-Jerez, P., Bayle-Sempere, J. T., Sfakianakis, D. G. and Somarakis, S., 2012. Morphological differences between wild and farmed Mediterranean fish. Hydrobiologia, 679(1):217-231.

Bagherian, A. and Rahmani, H., 2009. Morphological discrimination between two populations of Shemaya, (Actinopterygii, Cyprinidae) using a truss network, Animal Biodiversity and Conservation., 32 (1):1-8.

Barlow, G.W., 1961. Causes and significance of morphological variation in fishes. Sys. Zool., 10(3):105-117.

Begg, G.A. and Brown, R.W., 2000. Stock identification of haddock Melanogrammus aeglefinus on Georges Bank based on otolith shape analysis. Transactions of the American Fisheries Society, 129(4), pp.935-945.

Begg, G.A. and Waldman, J.R., 1999. An holistic approach to fish stock identification. Fish. Res., 43(1):35-44.

Begg, G.A., Friedland, K.D. and Pearce, J.B., 1999. Stock identification and its role in stock assessment and fisheries management: an overview. Fish. Res., 43(1):1-8.

Biswas, I., Nagesh, T. S. and Sajina, A. M., 2019. Stock delineation in Clupisoma garua (Hamilton, 1822) populations of Ganga riverine system using truss network analysis. Indian J. Fish, 66(2), pp.1-7.

Booke, H.E., 1981. The conundrum of the stock concept—are nature and nurture definable in fishery science?. Canadian Journal of Fisheries and Aquatic Sciences, 38(12):1479-1480.

Bookstein, F. L., 1990. Introduction to methods for landmark data. In Proceedings of the Michigan Morphomertic Workshop. University of Michigan Museum of Zoology Special Publication., 2: 215-226.

Bookstein, F. L., 1991. Morphometric tools for landmark data: Geometry and biology. Cambridge Univ. Press, New York, pp 435.

Bookstein, F.L., Chernoff, B., Elder, R.L., Humphries, J.M., Smith, G.R. and Strauss, R.E., 1985. Morphometrics in evolutionary biology: the geometry of size and shape change, with examples from fishes.

Burke, N., Brophy, D. and King, P.A., 2008. Otolith shape analysis: its application for discriminating between stocks of Irish Sea and Celtic Sea herring (Clupea harengus) in the Irish Sea. ICES Journal of Marine Science: Journal du Conseil, 65(9):1670-1675.

Cadrin, S.X. and Friedland, K.D., 1999. The utility of image processing techniques for morphometric analysis and stock identification. Fish. Res., 43(1):129-139.

Cadrin, S.X., 2000. Advances in morphometric identification of fishery stocks. Rev. F. Bio. Fish., 10(1):91-112.

Campana, S.E. and Casselman, J.M., 1993. Stock discrimination using otolith shape analysis. Canadian Journal of Fisheries and Aquatic Sciences, 50(5), pp.1062-1083.

Campana, S.E. and Neilson, J.D., 1985. Microstructure of fish otoliths. Canadian Journal of Fisheries and Aquatic Sciences, 42(5):1014-1032.

Campana, S.E. and Thorrold, S.R., 2001. Otoliths, increments, and elements: keys to a comprehensive understanding of fish populations?. Canadian Journal of Fisheries and Aquatic Sciences, 58(1), pp.30-38.

Campana, S.E., 2004. Photographic atlas of fish otoliths of the Northwest Atlantic Ocean Canadian special publication of fisheries and aquatic sciences No. 133. NRC Research press.

Cañás, L., Stransky, C., Schlickeisen, J., Sampedro, M.P. and Fariña, A.C., 2012. Use of the otolith shape analysis in stock identification of anglerfish (Lophius piscatorius) in the Northeast Atlantic. ICES Journal of Marine Science: Journal du Conseil, p.fss006.

Castonguay, M., Simard, P. and Gagnon, P., 1991. Usefulness of Fourier analysis of otolith shape for Atlantic mackerel (Scomber scombrus) stock discrimination. Canadian Journal of Fisheries and Aquatic Sciences, 48(2):296-302.

Cavalcanti, M.J., Monteiro, L.R. and Lopes, P.R., 1999. Landmark-based morphometric analysis in selected species of serranid fishes (Perciformes: Teleostei). Zoological Studies-TaipeiI-, 38(3):287-294.

Cushing, D.H., 1968. Fisheries Biology: A Study in Population Dynamics. The University of Wisconsin Press, Madison, WI, 200 pp.

Daryaei, N.A., Kamrani, E., Salarzadeh, A.R. and Salaripour, A., 2013. Identification of Anchovy, Encrasicholina punctifer Stocks on Persian Gulf and Oman Sea Using Analysis of Otolith Morphometrics. Environmental Sciences, 1(2):53-63.

Deepa, K.P., Kumar, K.A., Kottnis, O., Nikki, R., Bineesh, K.K., Hashim, M., Saravanane, N. and Sudhakar, M., 2019. Population variations of Opal fish, Bembrops caudimacula Steindachner, 1876 from Arabian Sea and Andaman Sea: Evidence from otolith morphometry. Regional studies in marine science, 25, p.100466.

DeVries, D.A., Grimes, C.B. and Prager, M.H., 2002. Using otolith shape analysis to distinguish eastern Gulf of Mexico and Atlantic Ocean stocks of king mackerel. Fisheries Research, 57(1), pp.51-62.

Duncan, R., Brophy, D. and Arrizabalaga, H., 2018. Otolith shape analysis as a tool for stock separation of albacore tuna feeding in the Northeast Atlantic. Fisheries Research, 200, pp.68-74.

Elliott, N.G., Haskard, K. and Koslow, J.A., 1995. Morphometric analysis of orange roughly (Hoplostethus atlanticus) off the continental slope of southern Australia. Oceanographic Literature Review, 9(42):790.

Fitzgerald, D.G., Nanson, J.W., Todd, T.N. and Davis, B.M., 2002. Application of truss analysis for the quantification of changes in fish condition. Journal of Aquatic Ecosystem Stress and Recovery (Formerly Journal of Aquatic Ecosystem Health), 9(2):115-125.

Fleming, I. A. and Einum, S., 1997. Experimental tests of genetic divergence of farmed from wild Atlantic salmon due to domestication. ICES Journal of Marine Science: Journal du Conseil, 54(6):1051-1063.

Gagliano, M. and McCormick, M.I., 2004. Feeding history influences otolith shape in tropical fish. Marine Ecology Progress Series, 278:291-296.

Gauldie, R.W. and Nelson, D.G.A., 1990. Otolith growth in fishes. Comparative Biochemistry and Physiology Part A: Physiology, 97(2):119-135.

Gillanders, B.M. and Kingsford, M.J., 2000. Elemental fingerprints of otoliths of fish may distinguish estuarine'nursery'habitats. Marine Ecology Progress Series, 201, pp.273-286.

Gopikrishna, G., Sarada, C. and Sathianandan, T. V., 2006. Truss morphometry in the Asian seabass-Lates calcarifer. Journal of the Marine Biological Association of India, 48(2):220-223.

Grimes, C. B., Johnson, A. G. and Fable Jr., W. A., 1987. Delineation of king mackerel (Scomberomorus cavalla) stocks along the US east coast and in the Gulf of Mexico. In: Kumpf, H.E., Vaught, R.N., Grimes, C.B., Johnson, A.G., Nakamura, E.L. (Eds.), Proceedings of the Stock Identification Workshop, Panama City Beach, 1985., pp. 186-187.

Hakim, A. A., Kurniavandi, D.F., Mashar, A., Butet, N.A., Maduppa, H. and Wardiatno, Y., 2020, January. Study on stock structure of Indian mackerel (Rastrelliger kanagurta Cuvier, 1816) in Fisheries Management Area 712 of Indonesia using morphological characters with Truss Network Analysis approach. In IOP Conference Series: Earth and Environmental Science (Vol. 414, No. 1, p. 012006). IOP Publishing.

Hanif, M. A., Chaklader, M. R., Siddik, M. A., Nahar, A., Foysal, M. J. and Kleindienst, R., 2019. Phenotypic variation of gizzard shad, Anodontostoma chacunda (Hamilton, 1822) based on truss network model. Regional Studies in Marine Science, 25, p.100442.

Hanski, I., 1999. Habitat connectivity, habitat continuity, and metapopulations in dynamic landscapes. Oikos, 87: 209-219, DOI: 10.1143/JJAP.45.6974.

Hanski, I., and Simberloff, D., 1997. The metapopulation approach, its history, conceptual domain, and application to conservation. In: Metapopulation Biology: Ecology, Genetics, and Evolution. Hanski, I., and Gilpin, M. E., eds., Academic Press, San Digego, 5-26, DOI: 10.1016/B978-012323445-2/50003-1.

Hari, M. S., Kathrivelpandian, A., Bhavan, S. G., Sajina, A. M., Gangan, S. S. and Abidi, Z. J., 2019. Deciphering the stock structure of Chanos chanos (Forsskål, 1775) in Indian 91 waters by truss network and otolith shape analysis. Turkish Journal Fisheries and Aquatic Science, 20(2), pp.103-111.

Hilborn, R. and Walters, C.J., 1992. Quantitative fisheries stock assessment: choice, dynamics and uncertainty. Reviews in Fish Biology and Fisheries, 2(2):177-178.

Hossain, M. A., Nahiduzzaman, M., Saha, D., Khanam, M. U. H. and Alam, M. S., 2010. Landmark-based morphometric and meristic variations of the endangered Carp, Kalibaus Labeo calbasu, from stocks of two isolated rivers, the Jamuna and Halda, and a hatchery. Zoological Studies, 49(4):556-563.

Hossain, M.A., Nahiduzzaman, M., Saha, D., Khanam, M.U.H. and Alam, M.S., 2010. Landmark-based morphometric and meristic variations of the endangered Carp, Kalibaus Labeo calbasu, from stocks of two isolated rivers, the Jamuna and Halda, and a hatchery. Zoological Studies, 49(4):556-563.

Humphries, J.M., Bookstein, F.L., Chernoff, B., Smith, G.R., Elder, R.L. and Poss, S.G., 1981. Multivariate discrimination by shape in relation to size. Sys. Bio., 30(3):291308.

Hutchinson, W. F., 2008. The dangers of ignoring stock complexity in fishery management: The case of the North Sea cod. Biology Letters, 4: 693-695, DOI: 10.1098/rsbl.2008.0443

Ihssen, P.E., Booke, H.E., Casselman, J.M., McGlade, J.M., Payne, N.R. and Utter, F.M., 1981. Stock identification: materials and methods. Canadian Journal of Fisheries and Aquatic Sciences, 38(12):1838-1855.

Izzo, C., Ward, T.M., Ivey, A.R., Suthers, I.M., Stewart, J., Sexton, S.C. and Gillanders, B.M., 2017. Integrated approach to determining stock structure: implications for fisheries management of sardine, Sardinops sagax, in Australian waters. Reviews in fish biology and fisheries, 27(1), pp.267-284.

Jayasankar, P., Thomas, P. C., Paulton, M. P. and Mathew, J., 2004. Morphometric and genetic analyzes of Indian mackerel (Rastrelliger kanagurta) from peninsular India. Asian Fisheries Science, 17, pp.201-215.

Jearranaiprepame, P., 2017. Morphological differentiation among isolated populations of dwarf snakehead fish, Channa gachua (Hamilton, 1822) using truss network analysis. Acta Biologica Szegediensis, 61(2), pp.119-128.

Jiang, Y., Zhang, C., Ye, Z., Xu, B., Tian, Y. and Watanabe, Y., 2020. Stock Structure Analysis of the Japanese Spanish Mackerel Scomberomorus niphonius (Cuvier, 1832) Along the China Coast Based on Truss Network. Journal of Ocean University of China, 19(2), pp.446-452.

Kashyap, A., Awasthi, M. and Serajuddin, M., 2016. Phenotypic variation in freshwater murrel, Channa punctatus (Bloch, 1793) from Northern and Eastern Regions of India using truss analysis. International Journal of Zoology, 2016.

Khan, M. A., Miyan, K. and Khan, S., 2013. Morphometric variation of snakehead fish, Channa punctatus, populations from three Indian rivers. Journal of Applied Ichthyology, 29(3):637-642.

Khan, S., Schilling, H.T., Khan, M.A., Patel, D.K., Maslen, B. and Miyan, K., 2021. Stock delineation of striped snakehead, Channa striata using multivariate generalised linear models with otolith shape and chemistry data. Scientific reports, 11(1), pp.1-11.

Kohestan-Eskandari, S., AnvariFar, H. and Mousavi-Sabet, H., 2014. Detection of morphometric differentiation of Liza aurata (Pisces: Mugilidae) in southeastern of the Caspian Sea, Iran. Our Nature, 11(2):126-137.

Larkin, P.A., 1981. A perspective on population genetics and salmon management. Canadian Journal of Fisheries and Aquatic Sciences, 38(12):1469-1475.

Leguá, J., Plaza, G., Pérez, D. and Arkhipkin, A., 2013. Otolith shape analysis as a tool for stock identification of the southern blue whiting, Micromesistius australis. Latin American Journal of Aquatic Research, 41(3):479.

Libungan, L.A., Óskarsson, G.J., Slotte, A., Jacobsen, J.A. and Pálsson, S., 2015. Otolith shape: a population marker for Atlantic herring Clupea harengus. Journal of fish biology, 86(4), pp.1377-1395.

Longmore, C., Fogarty, K., Neat, F., Brophy, D., Trueman, C., Milton, A. and Mariani, S., 2010. A comparison of otolith microchemistry and otolith shape analysis for the study of spatial variation in a deep-sea teleost, Coryphaenoides rupestris. Environmental biology of fishes, 89(3-4), pp.591-605.

Lord, C., Morat, F., Lecomte-Finiger, R. and Keith, P., 2012. Otolith shape analysis for three Sicyopterus (Teleostei: Gobioidei: Sicydiinae) species from New Caledonia and Vanuatu. Environmental Biology of Fishes, 93(2), pp.209-222.

MacLean, J.A. and Evans, D.O., 1981. The stock concept, discreteness of fish stocks, and fisheries management. Canadian Journal of Fisheries and Aquatic Sciences, 38(12):1889-1898.

Mahe, K., Oudard, C., Mille, T., Keating, J., Gonçalves, P., Clausen, L.W., Petursdottir, G., Rasmussen, H., Meland, E., Mullins, E. and Pinnegar, J.K., 2016. Identifying blue whiting (Micromesistius poutassou) stock structure in the Northeast Atlantic by otolith shape analysis. Canadian Journal of Fisheries and Aquatic Sciences, 73(9), pp.1363-1371.

Mahfuj, M.S., Ahmed, F.F., Hossain, M.F., Islam, S.I., Islam, M.J., Alam, M.A., Hoshan, I. and Nadia, Z.M., 2022. Stock Structure Analysis of the Endangered Queen Loach, Botia dario (Hamilton 1822) from Five Rivers of Northern Bangladesh by Using Morphometrics: Implications for Conservation. Fishes, 7(1), p.41.

Marcus, L. 1990. Traditional morphometrics. In: F.J. Rohlf & F. L. Bookstein (Ed.) Proceedings of the Michigan Morphometrics Workshop. Special Publication No. 2, University of Michigan Museum of Zoology, Ann Arbor, pp. 77–123.

Melvin, G.D., Dadswell, M.J. and McKenzie, J.A., 1992. Usefulness of meristic and morphometric characters in discriminating populations of American shad (Alosa sapidissima)(Ostreichthyes: Clupeidae) inhabiting a marine environment. Canadian J. Fish. Aq. Sci., 49(2):266-280.

Mérigot, B., Letourneur, Y. and Lecomte-Finiger, R., 2007. Characterization of local populations of the common sole Solea solea (Pisces, Soleidae) in the NW Mediterranean through otolith morphometrics and shape analysis. Marine Biology, 151(3):997-1008.

Mir, J. I., Mir, F. A. and Patiyal, R. S., 2014. Phenotypic Variations Among Three Populations of Chirruh Snowtrout, Schizothorax esocinus (Heckel, 1838) in Kashmir Himalaya with Insights from Truss Network System. Proceedings of the National Academy of Sciences, India Section B: Biological Sciences, 84(1):105111.

Miyan, K., Khan, M. A., Patel, D. K., Khan, S. and Ansari, N. G., 2016. Truss morphometry and otolith microchemistry reveal stock discrimination in Clarias batrachus (Linnaeus, 1758) inhabiting the Gangetic river system. Fisheries Research, 173:294-302.

Mohaddasi, M., Shabanipour, N. and Abdolmaleki, S., 2013. Morphometric variation among four populations of Shemaya (Alburnus chalcoides) in the south of Caspian Sea using truss network. The Journal of Basic & Applied Zoology, 66(2):87-92.

Muchlisin, Z. A., 2013. Morphometric Variations of Rasbora Group (Pisces: Cyprinidae) in Lake Laut Tawar, Aceh Province, Indonesia, Based on Truss Character Analysis. Journal of Biosciences, 20(3):138-143.

Nama, S., Bhushan, S., Ramteke, K.K., Jaiswar, A.K., Nayak, B.B., Pathak, V. and Akter, S., 2022. Stock structure analysis of Upeneus vittatus based on morphometric, meristic and otolith shape analysis along the Indian coast.

Neves, A., Sequeira, V., Farias, I., Vieira, A.R., Paiva, R. and Gordo, L.S., 2011. Discriminating bluemouth, Helicolenus dactylopterus (Pisces: Sebastidae), stocks in Portuguese waters by means of otolith shape analysis. Journal of the Marine Biological Association of the United Kingdom, 91(6), pp.1237-1242.

Paknejad, S., Heidari, A. and Mousavi-Sabet, H., 2014. Morphological variation of shad fish Alosa brashnicowi (Teleostei, Clupeidae) populations along the southern Caspian Sea coasts, using a truss system. International Journal of Aquatic Biology, 2(6):330- 336.

Parsons, K. J., Robinson, B. W. and Hrbek, T., 2003. Getting into shape: An empirical comparison of traditional truss-based morphometric methods with a newer geometric method applied to New World cichlids. Environmental Biology of Fishes, 67: 417–431.

Paul, K., Oeberst, R. and Hammer, C., 2013. Evaluation of otolith shape analysis as a tool for discriminating adults of Baltic cod stocks. Journal of Applied Ichthyology, 29(4), pp.743-750.

Pazhayamadom, D. G., 2006. A study on the variation in Harpadon nehereus (Hamilton, 1822) stocks from east and west coast of India. M. F. Sc. Thesis, Central Institute of Fisheries Education, Mumbai.

Pazhayamadom, D. G., Chakraborty, S. K., Jaiswar, A.K., Sudheesan, D., Sajina, A. M. and Jahageerdar, S., 2015. Stock structure analysis of 'Bombay duck'(Harpadon nehereus Hamilton, 1822) along the Indian coast using truss network morphometrics. Journal of Applied Ichthyology, 31(1):37-44.

Purushothaman, P., Chakraborty, R. D., Kuberan, G., Maheswarudu, G., Baby, P.K., Sreesanth, L., Ragesh, N. and Pazhayamadom, G., 2018. Stock structure analysis of 'Aristeus alcocki Ramadan, 1938 (Decapoda: Aristeidae)'in the Indian coast with truss network morphometrics. Canadian Journal of Zoology, 96(5), pp.411-424.

Reiss, H., Hoarau, G., Dickey-Collas, M., and Wolff, W. J., 2009. Genetic population structure of marine fish: Mismatch between biological and fisheries management units. Fish and Fisheries, 10 (4): 361-395, DOI: 10.1111/j.1467-2979.2008.00324.x.

Reist, J.D., 1985. An empirical evaluation of several univariate methods that adjust for size variation in morphometric data. Canadian Journal of Zoology, 63(6):14291439.

Remya, R., Vivekanandan, E., Sreekanth, G. B., Ambrose, T. V., Nair, P. G., Manjusha, U., Thomas, S. and Mohamed, K. S., 2014. Stock structure analysis of Indian mackerel Rastrelliger kanagurta (Cuvier, 1817) from south-east and south-west coasts of India using truss network system. Indian Journal of Fisheries, 61(3):16-19.

Ricker, W.E., 1981. Changes in the average size and average age of Pacific salmon. Canadian J. Aq. Sci., 38(12):1636-1656.

Roby, D., Lambert, J.D. and Sevigny, J.M., 1991. Morphometric and electrophoretic approaches to discrimination of capelin (Mallotus villosus) populations in the estuary and Gulf of St. Lawrence. Canadian J. Fish. Aq Sci., 48(11):2040-2050.

Rohlf, F.J. and Marcus, L.F., 1993. A revolution morphometrics. Trends in Eco. Evol., 8(4):129-132.

Sadighzadeh, Z., Valinassab, T., Vosugi, G., Motallebi, A.A., Fatemi, M.R., Lombarte, A. and Tuset, V.M., 2014. Use of otolith shape for stock identification of John's snapper, Lutjanus johnii (Pisces: Lutjanidae), from the Persian Gulf and the Oman Sea. Fisheries Research, 155, pp.59-63.

Sajina, A. M., Chakraborty, S. K., Jaiswar, A. K., Pazhayamadom, D. G., & Sudheesan, D., 2011. Stock structure analysis of Megalaspis cordyla (Linnaeus, 1758) along the Indian coast based on truss network analysis. Fisheries Research, 108(1), 100-105. https://doi.org/10.1016/j.fishres.2010.12.006

Schulz-Mirbach, T., Stransky, C., Schlickeisen, J. and Reichenbacher, B., 2008. Differences in otolith morphologies between surface-and cave-dwelling populations of Poecilia mexicana (Teleostei, Poeciliidae) reflect adaptations to life in an extreme habitat. Evolutionary Ecology Research, 10(4), pp.537-558.

Schweigert, J.F. and Withler, R.E., 1990. Genetic differentiation of Pacific herring based on enzyme electrophoresis and mitochondrial DNA analysis. In Am. Fish. Soc. Symp., 7:459-469.

Šegvić-Bubić, T., Talijančić, I., Grubišić, L., Izquierdo-Gomez, D. and Katavić, I., 2014. Morphological and molecular differentiation of wild and farmed gilthead sea bream Sparus aurata: implications for management. Aquaculture Environment Interactions, 6(1):43-54.

Sen, S., Jahageerdar, S., Jaiswar, A. K., Chakraborty, S. K., Sajina, A. M. and Dash, G. R., 2011. Stock structure analysis of Decapterus russelli (Ruppell, 1830) from east and west coast of India using truss network analysis. Fisheries research, 112(1):38-43.

Silva, A., 2003. Morphometric variation among sardine (Sardina pilchardus) populations from the northeastern Atlantic and the western Mediterranean. ICES Journal of Marine Science: Journal du Conseil, 60(6):1352-1360.

Silva, A., 2003. Morphometric variation among sardine (Sardina pilchardus) populations from the northeastern Atlantic and the western Mediterranean. ICES Journal of Marine Science: Journal du Conseil, 60(6):1352-1360.

SOFIA, 2018. The State of World Fisheries and Aquaculture. FAO Fisheries and Aquaculture Department.FAO, Rome,194pp.

Stephenson, R. L., 1999. Stock complexity in fisheries management: A perspective of emerging issues related to population sub-units. Fisheries Research, 43 (1-3): 247- 249, DOI: 10.1016/ S0165-7836(99)00076-4

Stransky, C., 2005. Geographic variation of golden redfish (Sebastes marinus) and deep-sea redfish (S. mentella) in the North Atlantic based on otolith shape analysis. ICES Journal of Marine Science, 62(8), pp.1691-1698.

Stransky, C., Baumann, H., Fevolden, S.E., Harbitz, A., Høie, H., Nedreaas, K.H., Salberg, A.B. and Skarstein, T.H., 2008. Separation of Norwegian coastal cod and Northeast Arctic cod by outer otolith shape analysis. Fisheries Research, 90(1-3), pp.26-35.

Strauss, R.E. and Bookstein, F.L., 1982. The truss: body form reconstructions in morphometrics. Systematic Biology, 31(2):113-135.

Swain, D.P. and Foote, C.J., 1999. Stocks and chameleons: the use of phenotypic variation in stock identification. Fisheries Research, 43(1):113-128.

Swain, D.P. and Foote, C.J., 1999. Stocks and chameleons: the use of phenotypic variation in stock identification. Fisheries Research, 43(1):113-128.

Swaine, D. P., Ridell, B. E. and Murray, C. B., 1991. Morphological differences between hatchery and wild populations of coho salmon (Oncorhynchus kisutch): Environmental versus genetic origin. Canadian Journal of Fisheries and Aquatic Sciences., 48: 1783-1791.

Turan, C., 1999. A note on the examination of morphometric differentiation among fish populations: the truss system. Turkish Journal of Zoology, 23(3):259-264.

Turan, C., Ergüden, D., Gurlek, M. and Turan, F., 2004. Genetic and morphologic structure of Liza abu (Heckel, 1843) populations from the rivers Orontes, Euphrates and Tigris. Turkish Journal of Veterinary and Animal Sciences, 28(4):729-734.

Turan, C., Ergüden, D., Gürlek, M., BAŞUSTA, N. and Turan, F., 2004. Morphometric structuring of the anchovy (Engraulis encrasicolus L.) in the Black, Aegean and Northeastern Mediterranean Seas. Turkish Journal of Veterinary and Animal Sciences, 28(5):865-871.

Turan, C., Oral, M., Öztürk, B. and Düzgüneş, E., 2006. Morphometric and meristic variation between stocks of Bluefish (Pomatomus saltatrix) in the Black, Marmara, Aegean and northeastern Mediterranean Seas. Fisheries Research, 79(1):139-147.

Tuset, V.M., Lozano, I.J., Gonzalez, J.A., Pertusa, J.F. and García-Díaz, M.M., 2003. Shape indices to identify regional differences in otolith morphology of comber, Serranus cabrilla (L., 1758). Journal of Applied Ichthyology, 19(2):88-93.

Utter, F.M., 1981. Biological criteria for definition of species and distinct intraspecific populations of anadromous salmonids under the US Endangered Species Act of 1973. Canadian Journal of Fisheries and Aquatic Sciences, 38(12):1626-1635.

Vasconcellos, A. V., Vianna, P., Paiva, P. C., Schama, R. and Solé-Cava, A., 2008. Genetic and morphometric differences between yellowtail snapper (Ocyurus chrysurus, Lutjanidae) populations of the tropical West Atlantic. Genetics and Molecular Biology, 31(1):308-316.

Vasconcelos, J., Vieira, A.R., Sequeira, V., González, J.A., Kaufmann, M. and Gordo, L.S., 2018. Identifying populations of the blue jack mackerel (Trachurus picturatus) in the Northeast Atlantic by using geometric morphometrics and otolith shape analysis. Fishery Bulletin.

Vieira, A.R., Neves, A., Sequeira, V., Paiva, R.B. and Gordo, L.S., 2014. Otolith shape analysis as a tool for stock discrimination of forkbeard (Phycis phycis) in the Northeast Atlantic. Hydrobiologia, 728(1), pp.103-110.

Wiff, R., Flores, A., Segura, A.M., Barrientos, M.A. and Ojeda, V., 2020. Otolith shape as a stock discrimination tool for ling (Genypterus blacodes) in the fjords of Chilean Patagonia. New Zealand Journal of Marine and Freshwater Research, 54(2), pp.218-232.

Ying, Y., Chen, Y., Lin, L., and Gao, T., 2011. Risks of ignoring fish population spatial structure in fisheries management. Canadian Journal of Fisheries and Aquatic Sciences, 68 (12): 2101-2120, DOI: 10.1139/f2011-116.

Zengin, M., Saygin, S. and Polat, N.A.Z.M.I., 2015. Otolith shape analyses and dimensions of the anchovy Engraulis encrasicolus L. in the Black and Marmara Seas. Sains Malaysiana, 44(5), pp.657-662.

8

Current Biodiversity Status of the Chilika Lake and Their Sustainable Management

[1]Sanjay Chandravanshi, [2]Rishikesh Venkatrao Kadam, [3]Koshalya Rohidas, [4]Sudhan Chandran and [5]Abinaya R.

[1]Department of Fisheries Biology and Resource Management, Fisheries Resources, Fisheries College and Research Institute,TNJFU, Tamil Nadu, India

[2]Department of Aquatic Environment Management, Fisheries College and Research Institute,TNJFU, Tamil Nadu, India

[3]Department of Fisheries Extension, Economics and Statistics, Fisheries College and Research Institute,TNJFU, Tamil Nadu, India

[4]Department of Fisheries Resource Management, Fisheries Resources, Harvest and Post-Harvest Management Division, ICAR-CIFE, Mumbai, Maharashtra India

[5]Department of Fisheries Resource Management, Kerala University of Fisheries and Ocean Studies Panangad Cochin, India

---

Abstract

In the eastern Indian state of Odisha, there is a brackish water lake called Chilika Lake that is also a shallow lagoon with estuarine characteristics. It is distributed throughout the districts of Puri, Khurda, and Ganjam. The Ramsar Convention has recognized Lake Chilika as a Wetland of International Importance (IUCN). Chilika Lakes fish fauna includes 336 species which spread across 217 genera, 92 families and 23 orders, 66 species of shellfish, 155 to 165 species of Irrawaddy dolphin, 103 species of birds, 66 species of macrophytes, including five species of seagrasses and 14 species of different seaweed. A distinctive combination of freshwater, brackish water and marine life is supported by the lagoon. Significant economic contributions are made by marine products and tourism in the area of the lagoon in Odisha.

Keywords: Chilika Lake, Biodiversity, Flora and fauna, Sustainable management

Introduction

The Chilika lagoon situated in the eastern coast of India within 85°05' and 85°38' east longitude and 19°54' North latitude near Bay of Bengal and one of the biggest estuarine-style brackish water lagoon in Asia and the greatest migratory waterfowl wintering area on the Indian subcontinent (Suresh et al., 2018). In the monsoon & summer time, the lagoon's size varies between 1165 and 906 sq. km. The total catchment area of the lagoon is 4,406 sq km, of which 32% is made up of the Mahanadi delta and 68% is western catchment (Bengtsson et al., 2012). The lake has a distinctive combination of freshwater, brackish, and marine ecosystems with estuary characteristics. In 1981, Chilika Lake was chosen as the India's first "Ramsar site" and it considered as one of the biodiversity hotspots of the nation (Kumar et al., 2012). The lake is very productive and has abundant fisheries resources. Over 0.15 million fishermen live around the lagoon and rely on its productive fishing grounds for their livelihood (Suresh et al., 2018). The lake area is home to certain uncommon, endangered and IUCN threatened species.

Chilika Lake Biodiversity (according to Chilika Development Authority)

Faunal Diversity

Finfishes

There is a wide variety of ichthyofauna diversity in Chilika Lake. The recently updated checklist of Chilika lake's fish fauna includes 336 species which spread across 217 genera, 92 families, and 23 orders, which include one new order (Torpediniformes), four new families (Lethrinidae, Apogonidae, Narcinidae and Synanceiidae), and 17 newly documented species (Suresh et al., 2018). *Ompok bimaculatus, Chanos chanos, Wallago attu,Lates calcarifer, Calamus bajonda,* and other fish species are frequently encountered. Over time, the abundant brackish water ichthyofauna has seen a significant drop in the number of native species caught, including some of the most common fishes like sea bass, mullets, and pearl spot species observed in Chilika Lake (Madhusmita, 2012).

Shellfishes

The Chilika Lake has 66 species of shellfish (shrimp/prawns, brachyuran crabs, and lobsters, 29 of which are prawn species, 35 of which are species of brachyuran crabs, and two of which are lobster species). These species are divided into 26 families. The upgraded checklist of shrimp and prawns includes 29 species that are under the order of Decapoda and belong to 17 genera and

10 families, whereas the brachyuran crab checklist includes 35 species that are belonging to 27 genera and 15 families. Suresh et al. (2018) recorded two species of lobster belonging genus *Panulirus*.

Dolphins

The Irrawaddy dolphin, which is prevalent in the Satapada region of Chilika Lake, has drawn tourists from all around the world in addition to those from India. The most important dolphin species in Chilika Lake is the *Orcaella brevirostris*, also known as the Irrawaddy Dolphin. There are only 2 lagoons worldwide where Irrawaddy dolphins may be found, and Chilika is one of those two. In five of the remaining six locations where it is known to exist, it is categorized as severely endangered. Also migrating from the sea into the lagoon is a low number of bottlenose dolphins.

The population of dolphins in Chilika Lake is 155 to 165 in 2022. According to sources, Chilika Development Authority (CDA) monitored the area for an entire year and directly observed 156 dolphins. In Chilika, there were 162 dolphins according to a survey conducted the previous year. Table 1 below provides statistics on the population of dolphins in Chilika Lake during the past few years.

Table 1 : Dolphin's population in Chilika Lake

Year	The dolphin population in Chilika lake
2013	152
2014	158
2015	160
2016	145
2017	121
2018	155
2019	150
2020	156
2021	162
2022	155-165

Source: Chilika Development Authority

Figure 1 : Faunal biodiversity of Chilika Lake

Migratory Birds

During the winter, Chilika lagoon welcomes hundreds of bird visitors from distant locations like Siberia. The abundance of the incredible coastal ecosystems in Orissa, which draw millions of tourists to see this rare natural occurrence, is enhanced by the presence of such exquisite birds. The results are encouraging because the Chilika Wildlife Division's water bird census from the first week of January had estimated that there were 10, 36,220 birds of 103 species in total. The greatest migratory bird wintering area on the Indian subcontinent is Chilika Lake which is the most important nation's biodiversity hotspots.

Zooplankton

Current investigations have revealed a large regional and temporally fluctuation in the distribution and abundance, the density of population, and species diversity of zooplankton. The most prevalent groups found in the lake waters include cladocerans, copepods, mysids, lucifers, euphausids, chaetognaths, siphonophores and sergestids. The contribution of copepods to total density is almost 70% (Bengtsson et al., 2012). It is discovered that changes in lake salinity have a major impact on zooplankton distributions.

Floral Diversity

Phytoplankton

A variety of freshwater, brackish-water and marine water taxa make up the community of phytoplankton of Chilika lagoon. Example- Dinoflagellates (Pyrrophyceae and Dinophyceae), diatoms (Bacillariophyceae), blue-green algae (Cyanobacteria) and green algae. From the lake, a total of 128 species of phytoplankton have been recorded (Panigrahi et al., 2009). Diatoms are the predominant groups which consist of 79 species and 13 species of dinoflagellates are present in the community. There are 18 species of blue-green algae and green algae representing 9 families and 15 genera (Bengtsson et al., 2012).

Macro Algae and Higher Plants

In Chilika Lake, there is a very high macrophytes and macroalgae diversity that shows spatiotemporal variability in four ecological regions. Seagrass beds provide vital ecological support in the form of nitrogen cycles, boosting coral reef fish production for millions of fish, avian and invertebrate species. 66 species of macrophytes, including five species of seagrass, have distribution records (Bhatta and Patra, 2018).

Even though the shoreline lacks any rocky intertidal areas, some green, brown, and red algae flourish on the rock islands of Chilika Lake. There were 14 different species of seaweed found in the lake, including 8 Chlorophyceae species and 6 Rhodophyceae species (Sahoo et al., 2003).The agarophyte *Gracillaria verucosa* is the most noticeable of them all. There are plentiful plants species that are associated with mangroves, including, *Avicinnia spp., Pongamia pinnata, Colubrina asiatica, Capparis spp., Aegiceras corniculatus, Excoecaria agalloch, Salvadora persica, Macrotyloma ciliatum*, and many more.

Five types of seagrass were identified in Chilika Lake in 2022 when it was being monitored. Several seagrasses species were identified, including *Holodule pinifolia, Halophila ovalis, Holodule uninervis, Halophila ovata,* and *Halophila beccarii*. These ecosystems have raised awareness of the global trend toward decline, and as a result, Chilika today boasts 33% of India's seagrass region (Bhatta and Patra, 2018). This helps the Chilika eco-system be more climates resilient and serves as a carbon sink.

Figure 2 : Floral diversity of Chilika Lake

Tourism

A well-liked ecotourism attraction is Chilika Lake. The Chilika Wildlife Refuge, often known as the famed bird sanctuary, is located on Nalabana Island. Chilika Lake is a well-liked birding location because of its diversity of bird species. Visitors can take a boat tour around the lake, and fishing and angling are two other popular activities there. It can be delightful to travel to the tourist attractions near the lake. The majority of tourist destination in the lake is Satapada, where the renowned Chilika dolphins are frequently spotted in quite considerable numbers (Bengtsson et al., 2012).

Measures for Sustainable Management

The following key actions are recommended for the management and conservation of the lake's finfish and shellfish biodiversity in light of the aforementioned facts. The adoption of proper legislation with sanctions against lake-damaging activities and illicit fishing. If the law is strictly enforced in Chilika, the majority of human-induced factors causing the decline of fish biodiversity in the lake might be efficiently managed or monitored. The recommended management actions are listed below:

Gheries and other forms of aquaculture may be used to rid Chilika Lake of illicit prawn culture enterprises. To enable unfettered movement and migration of brooder fishes & juveniles from sea to lagoon and vice versa, khanda fishing in the important fish migration channels, especially at Magarmukh, needs to be regulated. The total prohibition of unauthorized and illicit shrimp farming will boost the level of Chilika's fishermen's income. Most economically significant species reproduce between June to August and November to January, thus any fishing regulations for brooder protection must be designed during these months. Developing the fishermen's awareness and skills for practicing and encouraging responsible fisheries.The lakes recognized breeding grounds for inhabitant species need to be safeguarded with the help of the nearby fishing community and an effective awareness campaign. To promote free migratory brooders of fish, prawns, and crabs, the 14 km long Palur canal that connects the southern part of the lagoon with connected (BOB) The Bay of Bengal through to the mouth of the Rushikulya river should be dredged, deepened for natural movement of tidal water, and rendered free of any physical barriers (Suresh et al., 2018). Declare a "Prohibited Fishing Zone" including all lakeward and seaward locations within one kilometers radius of the new lagoon outlet (Pattnaik et al., 2007).A complete ban on harvesting wild shrimp seedlings at the lake's mouth since doing so depletes the lake's fish stocks. Chilika places restrictions on mechanical boats because they contaminate the water and stunt the growth of surface water fish. 'Zero net' should be banned since it depletes stocks of juvenile fish, shrimp, and crab by destroying their seedlings. Trawler fishing is prohibited in the Bay of Bengal near Chilika because it prevents fish and shrimp from entering the lake. Elimination of floating aquatic weeds, especially in Chilika's north and west sources where worsen siltation and impede boat operation.

Along with improving the socio and economic circumstances of the underprivileged fisherman, all these elements have the potential to significantly increase fish productivity. Regulations are applied consistently and effectively, and oversight is ongoing. Chilika Lake requires the implementation of a comprehensive conservation and management strategy that protects the lake's biodiversity, sustainable fisheries, and fishing communities' means of subsistence. Dredging of significant river mouths should continue in the northern part of the lake along with routine maintenance procedures for hydrological intervention.

Conclusion

One of India's biodiversity hotspots, the Chilika lagoon has a specific ecological state with both marine and fresh water characteristics, which results in an extraordinarily productive ecosystem thanks to effective cycling of nutrients. Due to its biodiversity, the lagoon has established itself as a top research location and tourism destination. A rich range of animals and plants can be found in the fresh water, brackish and marine water ecosystems that makeup the Chilika lagoon. The area around the lagoon is home to more over two lakh people, who are sustained by this incredibly productive ecosystem. Poor and primarily of the low caste, the local fishing community depends on Chilika for subsistence and a living.

References

Bengtsson, L., Herschy, R.W. and Fairbridge, R.W., 2012. Encyclopedia of lakes and reservoirs. Monographiae Biologicae, 53: pp.10-26.

Bhatta, K. and Patra, H.K., 2018. Distribution of macrophytes in Chilika. Int J Innov Sci Res Technol, 3, pp.95-99.

Chilika Health 2017-18. Ecosystem Health Report card.

https://kalingatv.com/features/latest-flora-and-fauna-survey-in-odishas-chilika-lake-out check-details/

https://economictimes.indiatimes.com/news/india/chilika-lake-a-hotspot-ofbiodiversity/articleshow/89342023.cms?from=mdr.

Madhusmita, T., 2012. Biodiversity of Chilika and its conservation, Odisha, India. Int. Res. J. Environment Sci, 1(5): pp.54-57.

Panda, P.C., Pattnaik, A.K., Rath, J. and Patnaik, S.N., 2002. Flora of Chilika lake and its immediate neighbourhood: a checklist. Journal of Economic and Taxonomic Botany, 26(1), pp.1-20.

Panigrahi, S., Wikner, J., Panigrahy, R.C., Satapathy, K.K. and Acharya, B.C., 2009. Variability of nutrients and phytoplankton biomass in a shallow brackish water ecosystem (Chilika Lagoon, India). Limnology, 10(2), pp.73-85.

Pattanaik, S., 2007. Conservation of environment and protection of marginalized fishing communities of Lake Chilika in Orissa, India. Journal of human ecology, 22(4), 291-302.

Sahoo, D., Sahu, N. and Sahoo, D., 2003. A critical survey of seaweed diversity of Chilika Lake, India. Algae, 18(1): pp.1-12.

Suresh, V.R., Mohanty, S.K., Manna, R,k., Bhatta, K.S., Mukherjee, M., S. K. Karna, S.K., Sharma, A.P., Das, B.K., Pattnaik, A.K., Nanda Susanta & Lenka, S., 2018. Fish and shellfish diversity and its sustainable management in Chilika lake, ICAR- Central Inland Fisheries Research Institute, Barrackpore, Kolkata and Chilika Development Authority, Bhubaneswar. 376p.

Wisa (2012). Chilika: An Integrated Management Planning Framework for Conservation and Wise Use. Wetlands International - South Asia & Chilika Development Authority, 162 pp.

9

Integrative Taxonomical Approach for Portunid Crab Species Identification

[1]Preetysh Nanda Patnaik, [1]Adyasha Sahu, [4]Sudhan C, [2]Venkataramani V.K., [3]Jayakumar N., [4]R. Durairaja

[1]Department of Fisheries Biology and Resource Management, Fisheries College & Research Institute, Thoothukudi, Tamil Nadu

[2]Fisheries College & Research Institute, Department of Fisheries Biology and Resource Management, Thoothukudi, TNJFU, Tamil Nadu

[3]Associate Professor and Head, Dept. of Fisheries Biology and Resource Management, Fisheries College & Research Institute, Thoothukudi, TNJFU Tamil Nadu

[4]Assistant Professor, Department of Fisheries Biology and Resource Management, Fisheries College & Research Institute, Thoothukudi, TNJFU Tamil Nadu

Abstract

The focus of evolutionary biology investigations is lineage divergence, and a taxonomist's purpose is to establish where along the continuum species classifications should be applied, and reaching a consensus on this creates a lot of disputes among taxonomists. Integrative taxonomy has evolved as a result of the recent incorporation of population biology research, sexual mating patterns, phylogeographical data and molecular phylogenetic information, and other evolutionary disciplines for species delimitation. Two of the most important and fast expanding areas are molecular phylogenetics and improved morphological taxonomy. In this review paper, we focused on the methodology of Portunid crab Species delineation in general by combining the two disciplines of morphological and molecular taxonomy. In the morphological technique, a sophisticated method called Truss morphometry is used to identify species, while DNA Barcoding is used in the molecular approach. Following the acquisition of procedural findings from both fields, the data sets are combined, and species can be distinguished as a result.

Keywords: Crab morphometry, Integrated approach, Meta analysis, Stock characterisation and delineation.

Background

The Enormous diversity of our planet harbors individuals of the sexually reproducing population that are unique and non-identical in nature. This unique nature is the source of this vast diversity. Taxonomy is critical for identification and classification. The oldest of the scientific disciplines, taxonomy, is described as the theory and practice of classifying organisms (Mayr, 1982; Mayr and Ashlock, 1991). Taxonomy is a branch of systematics that deals with the idea and practice of classifying and describing diversity (Nelson, 2006). According to the global estimates of species (Mora et al., 2011; Costello et al., 2013), 1.2-1.5 million species are considered to be described and valid to date (Costello et al., 2013; Mora et al., 2011; Zhang et al., 2011). The enormous and intriguing variation among the species, as well as anatomical differences between species, developmental stages, and geographical ranges, make identification difficult. Rapid documenting of biodiversity should be a top priority before it is lost. This necessitates quick and accurate taxonomic expertise as well as enhanced identification methods, paving the way for the emergence of Integrative taxonomy (Dayrat, 2005), which combines all available data, whether morphological, behavioral, ecological, or molecular.

Global Status and Trends of Portunid Crab

Global total marine catches increased from 81.2 million tonnes in 2017 to 84.4 million tonnes in 2018, but were still below the peak catches of 86.4 million tonnes in 1996. Crustacean catches declined to about 5.9 million tonnes in 2018 from 6.02 million tonnes in 2017. But it has been almost constant over the last 20 years. Blue swimming crabs, *Portunus pelagicus*, constitutes 5% of the total crustacean catch, i.e., 0.29 million tonnes in 2018 which has decreased from 0.3 million tonnes in 2017, and their catches have remained at relatively high levels that has marked their almost continuous growth over the last 20 years (SOFIA, 2020).

Indian Status and Trends of Portunid Crab

Crabs constitute an important resource in the marine fishery in India and contributed an overall average of 9.6% to the total crustacean landings during 1975-2020. Many species of crabs are exploited along the east and west coasts of India, mainly in trawls as a by-catch and as a targeted resource in gillnet in some regions. While making a comparison between east and west coasts of India, east coast was found more productive contributing 56.7% to the marine crab landings. The overall trend of the fishery showed increase at national level, recording a maximum landing of 57354 tonnes during 2018. Otherwise, the scenario varied between maritime states, for instance earlier years (2008-2012), state of Gujarat was leading in crab production, later the position was taken over by Tamil Nadu with very clear dominance. It is very evident from the national data on species wise production of marine crabs (2007-2020) and at present in overall production Gujarat is in second position followed by Andhra Pradesh & Kerala (Josileen, 2021).

Edible crabs landed in India belong to the family Portunidae and around 61% of the landings were recorded by three species of marine crabs *Portunus sanguinolentus* (28.2%), *Portunus pelagicus* (25%) and *Charybdis feriata* (7.7%). In Gujarat, *P. sanguinolentus* has recently emerged as the dominant species followed by *P. pelagicus*, pushing down *C. feriata* to the third position. Similarly in Tamil Nadu, the hitherto dominating species *Portunus pelagicus* (30.7%) has been overtaken by *P. sanguinolentus* (33.3%) with a marginal increase registered in the landings. The other important edible species included in the fishery in appreciable quantities were *Charybdis lucifera, Charybdis natator, Charybdis smithii, Charybdis annulata, Portunus gladiator* (revised as *Monomia gladiator*), *Podophthalmus vigil, Scylla serrata* and *Scylla olivacea* (Josileen, 2021).

Introduction

Taxonomists around the world have been facing crisis about the future of taxonomy and to find a way out, the problems relating to taxonomy must be clearly distinguished. Some of the main tasks are to delineate the species and to classify them, for which appropriate identification tools must be provided. All the phylogeneticists, population biologists, molecular ecologists or phylogeographers share a common interest that is delineating species to measure the extent of life's diversity and eventually to classify these. Another important task for taxonomists is the classification of species which need various identification tools to accurately identify the species. For many years, the conventional ways of identification were used, which to certain extent has caused this 'taxonomy crisis' (Gewin, 2002; Godfray, 2002; Mallett and Willmott, 2003; Wilson, 2003; Wheeler, Raven and Wilson, 2004). Alternative and complementary techniques, such as molecular taxonomy (Tautz et al., 2003; Hebert et al., 2003), the formation of investment funds (Wheeler, 2007), expanded use of cybertools (Pyle et al., 2008; La Salle et al., 2009), have been proposed in order to rejuvenate traditional taxonomy and assist it rise above the taxonomic crisis. To make the identification process easier and more efficient for users from various fields such as physiologists, conservation biologists, and ecologists, simple and dependable yet modern methods of identification must be employed. Modern identification techniques combine a variety of methods, such as morphological identification, molecular taxonomy, and phylogenetics, to provide an accurate and reliable way of species delimitation.

Crabs are decapod crustaceans with short stalked eyes and short, broad, and more or less flattened bodies (carapace), as well as small abdomens folded under the thorax. Brachyuran crabs (the subject of the present study) are one of the most diverse animal groups at the infra-order level (Stevcic', 2005), with approximately 1,271 genera and 6793 species worldwide (Stevcic', 2005). (Ng et al., 2008; De Grave et al., 2009). Brachyuran crabs play an important role in marine benthic communities, according to Boudreau and Worm (2012), ranging from intertidal to deep waters. They are prey for a wide range of invertebrates and vertebrates that are successful and flexible predators that feed on a variety of trophic levels. Crabs engage in a range of interactions with their surroundings and occupants, including providing refuge for smaller invertebrates and competing for food. The morphological diagnostic characteristics of Brachyuran crabs are used to classify them. Various innovative approaches, however, have been utilized to improve the accuracy of their identification.

We are emphasizing on the Methodology of Delineation of Portunid crab species, in this study by combining two fields, Morphological and Molecular Taxonomy. For species identification, an advanced method called Truss Morphometry (Strauss and Bookstein, 1982) is utilized in the morphological approach, while DNA Barcoding (Hebert et al., 2003) is employed in the molecular approach. Following the acquisition of procedural findings from both fields, the sets of data are combined utilizing integration methods such as congruence and cumulation, and species are discovered and categorized accordingly.

En route to Integrative Taxonomy

Integrative or Multidisciplinary Taxonomy is a collaborative or combined strategy for accurate species identification that utilizes a variety of complementary disciplines or fields of study. For the delineation of species, numerous disciplines must be studied in an integrated manner for a knowledge of taxonomy beyond the naming of species, that is, the fundamental processes bringing them about. Morphology, Molecular, Ecology, Behavior, Life History, Cytogenetics, Phylogenetic, and Evolutionary biology are examples of such disciplines. Integrative Taxonomy does not replace traditional taxonomy; rather, it refines the traditional slow method to a rigorous one by combining the findings of multiple disciplines into a single technique, resulting in a powerful tool for identification.

Truss Morphometry, An Advanced Morphological Approach

The empirical integration of geometry and biology is known as morphometrics. Its approaches must explicitly account for two completely different types of data: geographic location and biological homology. The goal of morphometrics is to investigate the relationships, causes, and effects between phenotypic features. Morphometric analysis is the study of landmarks, which is the analysis of data obtained from discrete morphometric points, as well as geometric relationships between them. Because of the substantial diversity associated with age, routine morphometric features have limitations in their use for stock identification. However, using the truss approach to examine landmarks can help avoid age reliance and aid in unbiased stock identification.

Methods for comparing biological shapes and forms range from classic verbal and pictorial representations (Thompson, 1917), methods involving the tracing of outlines (Zelditch et al., 1995; Loy et al., 2000), listing of measured linear distances among pairs of distinguishable landmarks on an organism (Robinson et al., 1993), and newer methods involving the geometric locations of landmarks (Walker and Bell, 2000). To quantify shape nowadays, multivariate analysis of numerous linear metrics ('trusses') throughout the body form is used. TRUSS Morphometry (Strauss and Bookstein, 1982) is a landmark analysis technique that examines data obtained from discrete morphometric points. Truss network systems constructed with the help of landmark points are powerful tools for stock identification of fish species. A sufficient degree of isolation may result in notable morphological, meristic, and shape differentiation among stocks of a species which may be recognizable as a basis for identifying the stocks. The characteristics may be more applicable for studying short – term, environmentally induced disparities, and the findings can be effectively used for improved fisheries management.

Studies on Truss Works Done by Different Authors

Using truss network approach in determining the stock structure of aquatic species has been initiated from the 20th century onwards. This method is applied in several crustacean studies, including shrimp (Paramasivam et al., 2017; Marini et al., 2017; Rebello et al., 2013), and in Blue Swimming crab (Bhosale et al., 2018). Truss based morphometric approach for the analysis of body shape in Portunid crabs (*Charybdis feriatus, Portunus pelagicus* and *P. sanguinolentus*) has been performed along Ratnagiri coast, India (Bhosale et al., 2018). Truss morphometric work in Portunid crabs has been in desuetude or not much work has been done. Different works carried out on Truss morphometric studies of finfishes and shellfishes are given in the following Table 1.

Table 1 : Review on Truss morphometry on various fauna

S. No.	Species	Aim	Author	Year
Finfishes:				
1.	*Salvelinus namaycush* Order: Salmoniformes	Shape difference among different morphotypes	Moore and Bronte	2001
2.	Family: Sparidae Order: Perciformes	Morphological study	Palma *et al*	2002
3.	Family: Cichlidae Order: Perciformes	Correlation of head shape and ecological variables	Bouton *et al*	2002
4.	*Oncorhynchus sp.* Order: Salmoniformes	Morphometric variation	Ruiz-Campos *et al*	2003
5.	*Sardina pilchardus* Order: Clupeiformes	Morphometric variation	Silva	2003
6.	*Rastrelliger kanagurta* Order: Perciformes	Population variation analysis	Jayasankar *et al*	2004
7.	*Lates calcarifer* Order: Perciformes	Stock differences in juveniles	Gopikrishna *et al*	2006
8.	*Megalaspis cordyla* Order: Carangiformes	Stock structure analysis	Sajna *et al*	2010
9.	*Labeo calbasu* Order: Cypriniformes	Morphometric variation	Hossain *et al*	2010
Shellfishes:				
10.	*Penaeus setiferus, P. vannamei, P. Stylirostris* Order: Decapoda	Species discrimination for selective breeding	Lester *et al*	1990
11.	*Homarus americanus* Order: Decapoda	Discrimination of Stock structure	Cadrin, S. X.	1995
12.	*Panulirus homarus* Order: Decapoda	Length-weight relationship	Senevirathna, J. D. M., *et al*	2012
13.	Family: Penaeidae Order: Decapoda	Conformation of Phylogenetic Relationship	Rajakumaran, P., *et al*	2014
14.	*Penaeus semisulcatus* Order: Decapoda	Phenotypic plasticity and genetic variation	Munasinghe, D. H. N., *et al*	2015
15.	*Charybdis feriatus, Portunus pelagicus* and *P. sanguinolentus* Order: Decapoda	Multivariate techniques to differentiate species	Bhosale, Mangesh M., *et al*	2016

16.	*Charybdis feriatus, Portunus pelagicus* and *P. sanguinolentus* Order: Decapoda	Species differentiation	Bhosale, Mangesh M., *et al*	2017
17.	*Penaeus merguiensis* Order: Decapoda	Identifying Stock structure	Marini, Melfa, *et al*	2017
18.	*Charybdis feriatus, Portunus pelagicus* and *P. sanguinolentus* Order: Decapoda	Analysis of Body Shape	Bhosale *et al*	2018
19.	*Aristeus alcocki* Order: Decapoda	Stock structure analysis	Purushothaman, P., *et al*	2018
20.	*Portunus pelagicus* Order: Decapoda	Identifying Stock structure	Afifah, Nurhaya, *et al*	2020
21.	*Penaeus monodon* Order: Decapoda	Morphometric Variation Analysis	Anjani, P. A. D. L., *et al*	2020
22.	*Heterocarpus chani* Order: Decapoda	Morphological differentiation	Kuberan, G., *et al*	2020
23.	*Portunus pelagicus* Order: Decapoda	Morphometric character variation	Riani, E., *et al*	2020
24.	Mantis shrimp (Stomatopods) Order: Stromatopoda	Morphometric differentiation	Pardhini, V., *et al*	2021
25.	Chinese mitten crabs *(Eriocheir sinensis)* Order: Decapoda	Geographical Origin Authentication	Zheng, Chaochen, *et al*	2021

The literature analysis reveals the global works on crustacean population species in multi-dimensional aspects (Fig I-III) which includes taxonomy, population biology, stock assessment and characterization (Table II), and conservation biology.

Table 2 : Global review on truss networking in shellfishes

Functional group	Countries	Species	Author	Year
Shrimp	USA	*Penaeus setiferus*	Lester *et al*	1990
Shrimp	USA	*Penaeus vannamei*	Lester *et al*	1990
Shrimp	USA	*Penaeus stylirostris*	Lester *et al*	1990
Lobster	England	*Homarus americanus*	Cadrin	1995
Lobster	Sri Lanka	*Panulirus homarus*	Senevirathna *et al*	2012
Shrimp	India	*Penaeus* sp.	Rajakumaran *et al*	2014
Shrimp	Sri Lanka	*Penaeus semisulcatus*	Munasinghe *et al*	2015
Crab	India	*Charybdis feriatus*	Bhosale *et al*	2016
Crab	India	*Portunus pelagicus*	Bhosale *et al*	2016
Crab	India	*Portunus sanguinolentus*	Bhosale *et al*	2016

Crab	India	*Charybdis feriatus*	Bhosale *et al*	2017
Crab	India	*Portunus pelagicus*	Bhosale *et al*	2017
Crab	India	*Portunus sanguinolentus*	Bhosale *et al*	2017
Shrimp	Indonesia	*Penaeus merguiensis*	Marini *et al*	2017
Crab	India	*Charybdis feriatus*	Bhosale *et al*	2018
Crab	India	*Portunus pelagicus*	Bhosale *et al*	2018
Crab	India	*Portunus sanguinolentus*	Bhosale *et al*	2018
Shrimp	India	*Aristeus alcocki*	Purushothaman *et al*	2018
Crab	Indonesia	*Portunus pelagicus*	Riani *et al*	2020
Crab	Indonesia	*Portunus pelagicus*	Afifah et al	2020
Caridean Shrimp	India	*Heterocarpus chani*	Kuberan et al	2020
Shrimp	Sri Lanka	*Penaeus monodon*	Anjani *et al*	2020
Crab	China	*Eriocheir sinensis*	Zheng *et al*	2021
Mantis Shrimp	Indonesia	*Stomatopoda*	Pardhini et al	2021

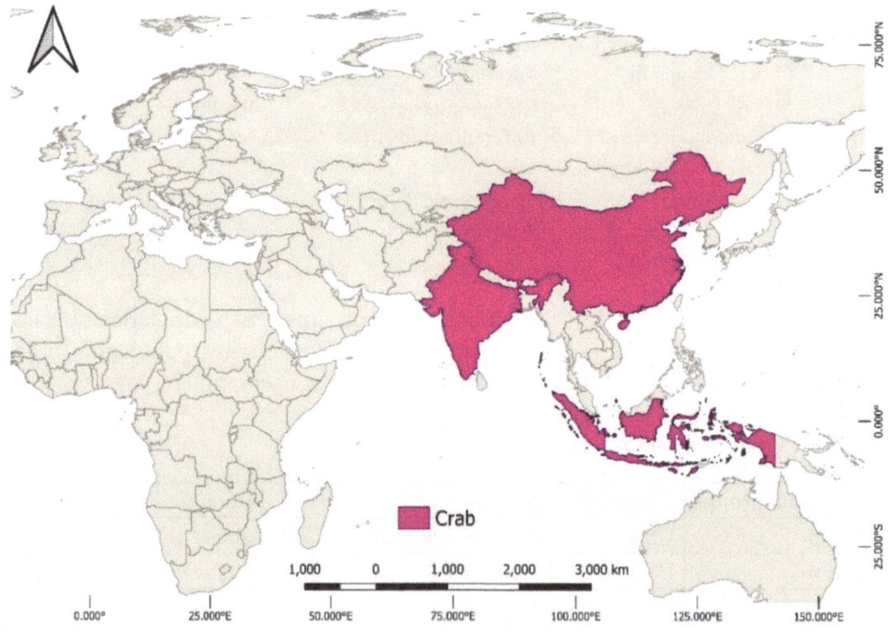

Figure 1 : Global scale representation of meta-analysis on Portunid crab stock characterization, morphometric differentiation and phenotypic plasticity works

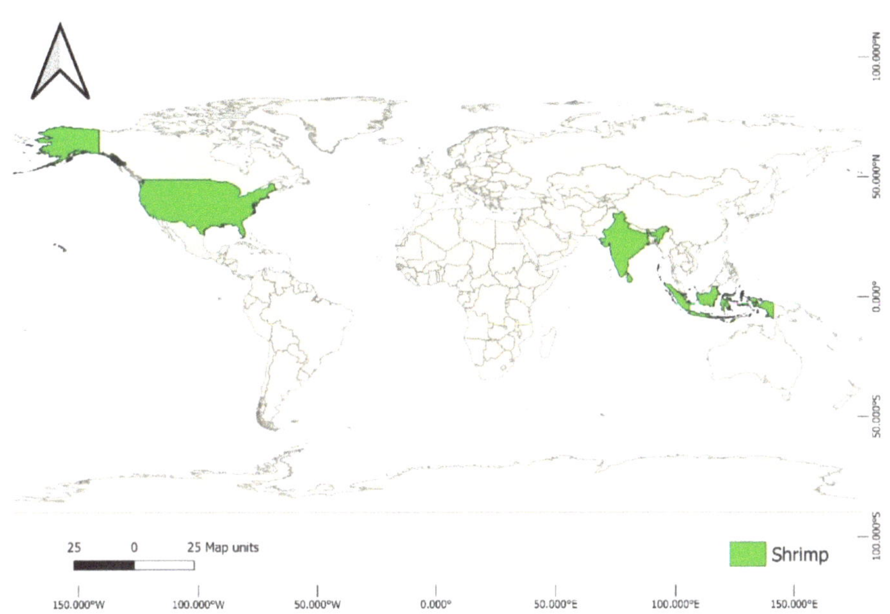

Figure 2 : Global scale representation of meta-analysis on shrimp stock characterization, morphometric differentiation and phenotypic plasticity works

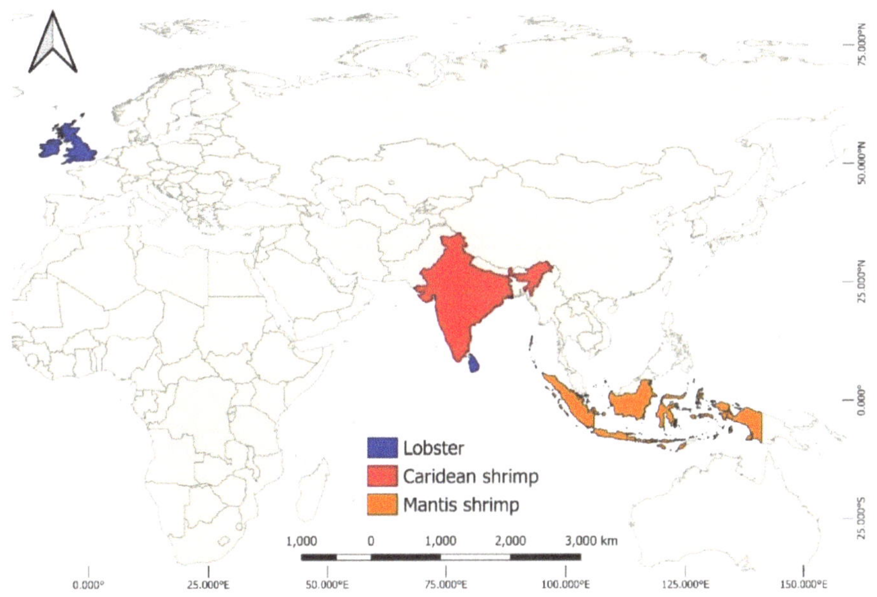

Figure 3 : Global scale representation of meta-analysis on other crustacean's stock characterization, morphometric differentiation and phenotypic plasticity works

Truss measures are a powerful tool for analyzing shape, and they're designed to cover the entire body, or at least the majority of it. In comparison to the classic morphometrics method, the truss network is a more useful and effective strategy for form descriptions; it has superior data collecting and a wider range of analytical tools. As a result, it can distinguish between phenotypic stocks because the created landmarks cover the complete fish body with no loss of information.

This review focuses on the Truss morphometrics methodology for delineating crab species, for which a set of landmarks is supplied (Table III & Figure IV). Landmarks are places that are chosen to split the body into functional units based on local morphological factors. Following the selection of landmarks, the Truss approach is applied.

Table 3: Landmarks, codes and descriptions used for Crab Truss morphometry

Landmarks	Codes	Description
1-8	UP1	Carapace Height (CH)
3-13	UP2	Carapace Width (CW)
4-12	UP3	Upper Carapace Width (UCW3)
5-11	UP4	Upper Carapace Width (UCW4)
6-10	UP5	Upper Carapace Width (UCW5)
1-7	UP6	Midpoint of Abdomen to Left Antennule
1-9	UP7	Midpoint of Abdomen to Right Antennule
7-9	UP8	Between Antennule
1-3	D7	Lateral Carapace Width Lower Left (LCWLL)
1-13	D8	Lateral Carapace Width Lower Right (LCWLR)
3-8	D9	Lateral Carapace Width Upper Left (LCWUL)
8-13	D10	Lateral Carapace Width Upper Right (LCWUR)

Figure 4: Schematic depiction oaf landmarks of carapace of crab species (*Portunus pelagicus*)

Sample collection	The Crab specimens are collected and identified using FAO species identification catalogue. Before being digitized, the specimens are cleaned, tagged and stored in 5% formalinized seawater.
Digitization of samples	Samples to be digitized are placed on a flat platform with laminated graph paper which is used in calibrating the coordinates of digital images. The pereiopods and the abdominal flaps are placed on the platform in such a way that makes their origin and insertion points, visible and then pinning is done. Digitization is done with a digital camera by mounting on a levelling tripod with a bubble level as an indicator of the inclination. Images from both dorsal and ventral aspects of the cephalothorax of the species are obtained and after digitization, the specimens are preserved in formalin for further analysis. Based on the tags attached, all digital images are identified.
Obtaining morphometric measurements	A series of software TPSUtil, TPSDig2 and Paleontological Statistics (PAST) is used to extract morphometric data from each individual image. TPSUtil is a utility program basically used to convert all images from JPEG (*.jpeg) format to TPS (*.tps). TPSDig2 is a window-based programme for digitizing landmarks (Table-II) and outlines in the images of objects for geometric morphometric analyses. The landmarks are digitized on each image using the 'Digitize landmarks' mode of the software and the landmark data are encrypted into the TPS files as X-Y coordinates. PAST is a multivariate analysis software package which is designed particularly for statistical analysis and plotting. The data encrypted TPS format image files are then used as input source in PAST and the data on distances between the landmarks is extracted using the 'All distances from landmarks' and '2 dimensional' options under the 'Geomet' menu of PAST.
Statistical Analysis	The applications of statistical analyses such as multivariate analyses namely Principal components analysis (PCA) and Factor analysis (FA) are correlated to differentiate among taxonomic groups and geometrical interpretations of studied Crab specimens.

Figure 5: Crab TRUSS Morphometry

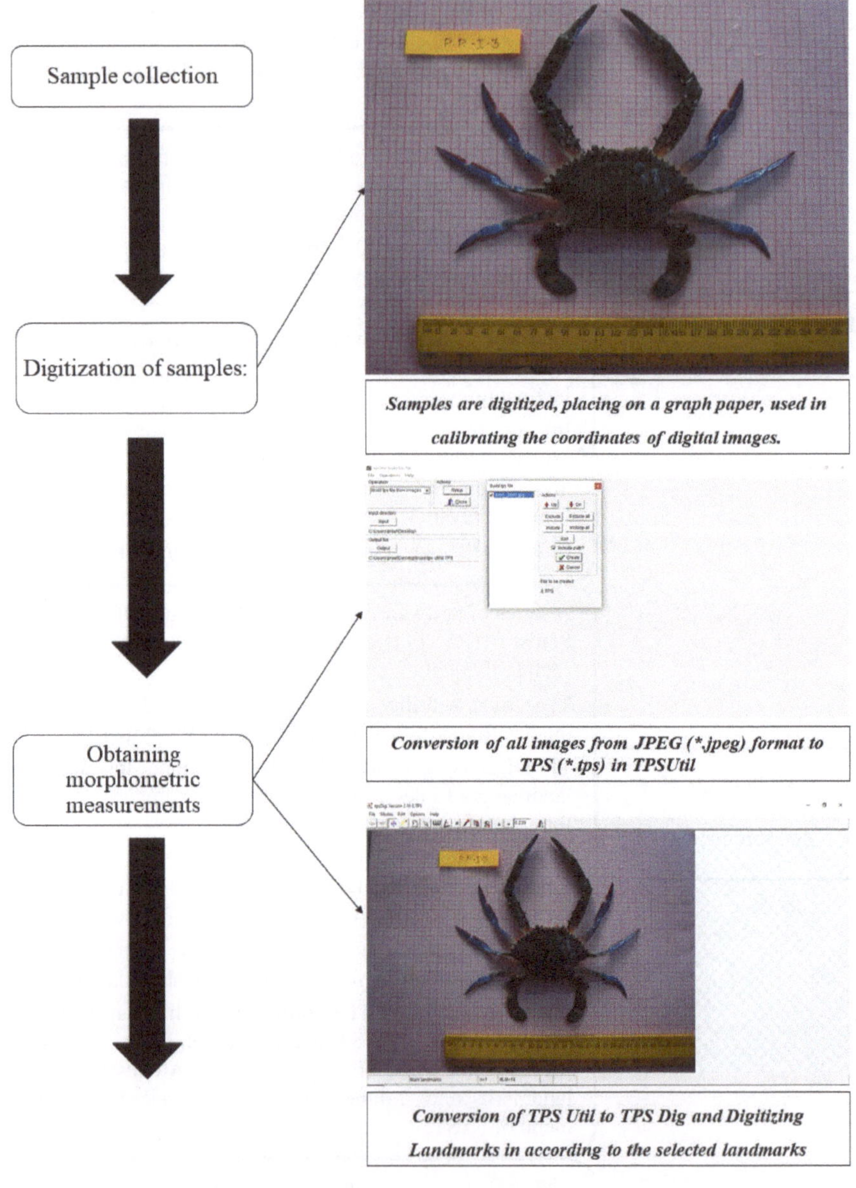
Sample collection
Digitization of samples:
Obtaining morphometric measurements
Samples are digitized, placing on a graph paper, used in calibrating the coordinates of digital images.
Conversion of all images from JPEG (*.jpeg) format to TPS (*.tps) in TPSUtil
Conversion of TPS Util to TPS Dig and Digitizing Landmarks in according to the selected landmarks

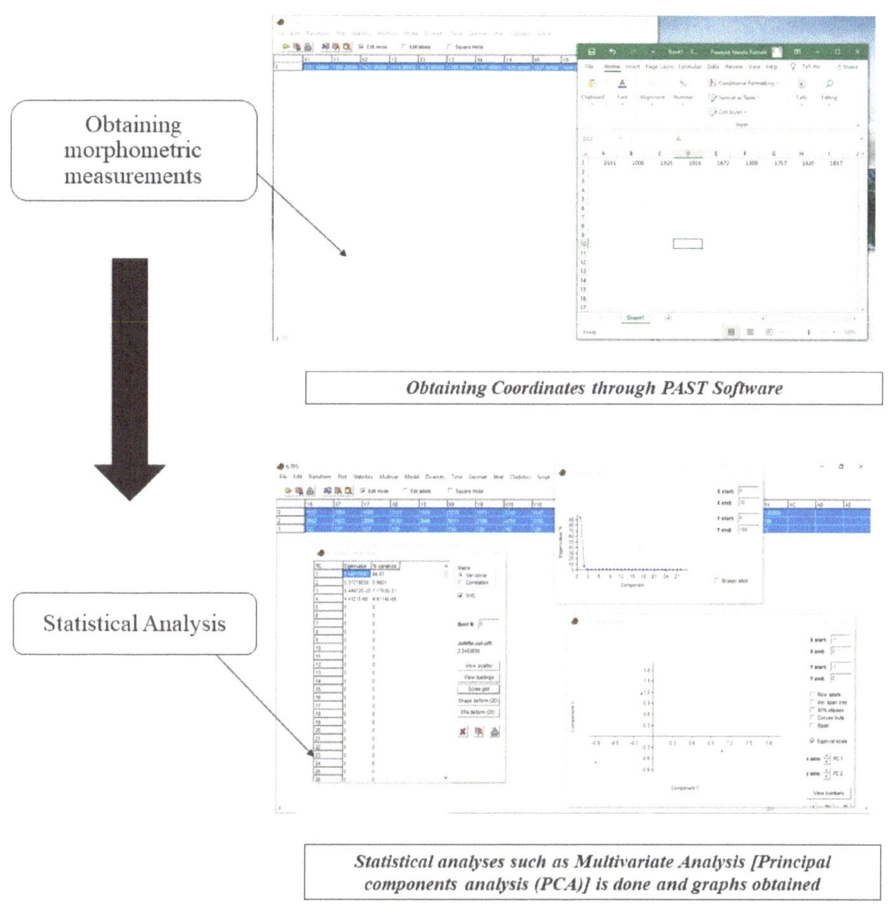

Figure 6. Portunid crab TRUSS morphometry using software

Reinvigorating taxonomy through DNA barcoding

DNA Barcoding (Hebert et al., 2003) is a reliable taxonomic tool for confirmation of species. Despite the fact that DNA sequences of closely related species are often extremely similar, there are still variances in the sequence. The section of the DNA sequence that is distinct to a given organism's species and generates a unique and specific DNA barcode. DNA barcoding (Hebert et al., 2003) has been shown to be effective in identifying and separating novel species from various groupings (Hebert et al., 2004; Ward et al., 2005; Cywinska et al., 2006; Hajibabaei et al., 2006a, b; Smith et al., 2007; Borisenko et al., 2008; Kerr et al., 2009). This method has received a high level of acceptance because it is simple and affordable (Padial and De La Riva, 2007). DNA barcoding has also focused on the creation of a global barcoding database as a species identification tool (Hebert et al., 2003). The technique of DNA barcoding identification is quite simple, and it relies on the quantifiable matching of COI sequences from unknown specimens to previously recorded specimens. DNA barcoding, which uses short sequences from the mitochondrial gene cytochrome oxidase subunit1 (COI), has been proposed as a method for detecting and identifying species quickly and accurately (Hebert et al., 2003; Marshall, 2005; Hajibabaei et al., 2007). Costa et al. (2007) confirmed the COI gene's effectiveness in assigning decapod species in their proper taxonomic order by studying the divergence of the COI gene's barcode region in 150 crustacean families. Congruent DNA barcoding and morphological description of species enable the accurate discrimination of different species, including cryptic ones (Keenan et al., 1998; Lai et al., 2010). There is still so much work needed for the identification. At this juncture, the review of work done by various researchers and insights of DNA barcoding on this Portunid crab species were documented and represented in Figure 7.

This study focuses on the methodology of integrating Truss with DNA barcoding, for which the methodology of latter is depicted in the following Figure 8.

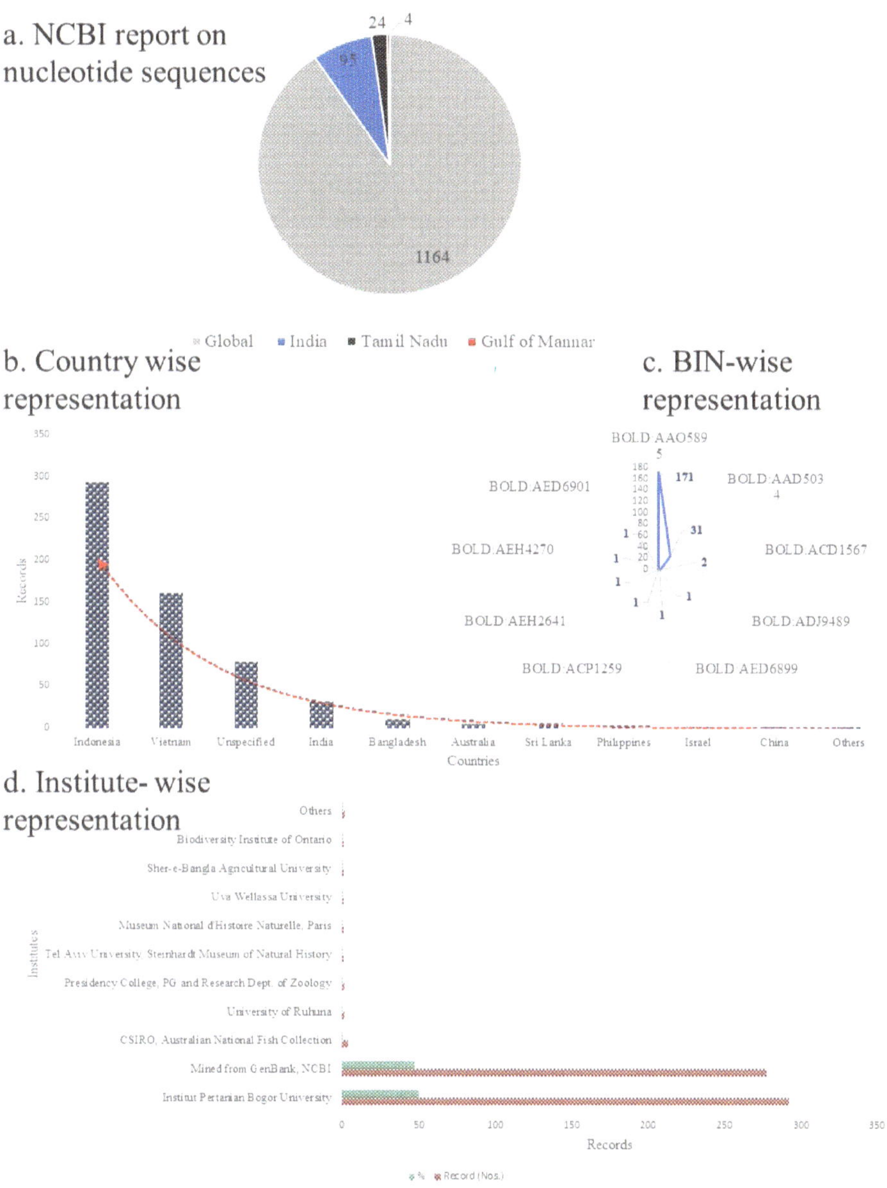

Figure 7: Meta analysis of Portunid crab species on DNA barcoding from NCBI and BOLD systems

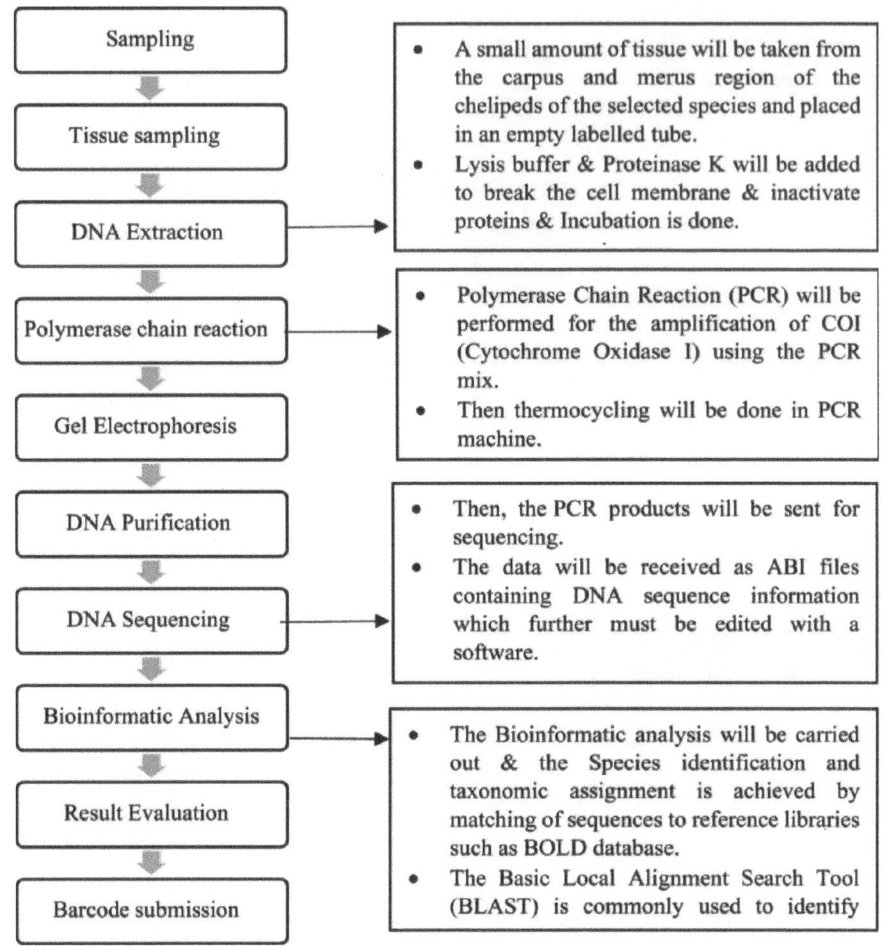

Figure 8: DNA Barcoding

Discussion

After obtaining the outcomes of both of the aforementioned disciplines as two separate sets of data, integration methods might be used. Concordance between both sets of information is sought, according to Integration by Congruence. The evolutionary lineages have fully separated and can be documented as distinct species if both sets agree. According to the Integration through Cumulation method, data is gathered from both the disciplines and evolutionary traits analyzed, and even if one feature is sufficient to explain the species level status, it is taken into account. Thus, the Portunid crab species may be recognized and delimited using these methods, and a useful database can be created.

This Integrative approach using TRUSS (Strauss and Bookstein 1982) and DNA Barcoding (Hebert et al., 2003) will surely decrease the criticisms and skepticism as both the methods have proven reliable, fast and accurate in both identification and delimitation of Crab species. Molecular approach of DNA Barcoding (Hebert et al., 2003) has proven an efficient benefit and aid to the traditional taxonomy. Another advantage of integrative taxonomy is that new species identification and description will be accurately facilitated with better supported evidence acquired from integrating two or more disciplines. The use of DNA barcoding to identify new species has also been heavily criticized (Cicero and Moritz, 2004; Lee, 2004; Ebach and Holdrege, 2005b; Rubinoff et al., 2006). However, a main drawback of this approach is, even today are found a number of views disputing the use of both DNA barcoding and morphology in biodiversity studies, despite of its integrated view.

Conclusion and Future Perspectives

Multiple disciplines must be used to address taxonomic difficulties and in species delimitation to avoid the failures inherent in single disciplines, such as classical taxonomy. A flexible approach based on the usage of information from each discipline independently is offered for procedural consistency in analysis. Each discipline's outcomes are precisely drawn utilising conventional procedural approaches, which are then amalgamated to produce an integrated approach. Traditional taxonomy has been revitalised thanks to DNA barcoding, which recognises the value of sophisticated morphological taxonomy. We conclude that Taxonomy needs to be multidimensional and collaborative to improve species identification and classification and to develop novel protocols to produce the highly essential inventory of life in a reasonable time. Using the advanced tools of identification such as Truss morphometry (Strauss and Bookstein, 1982) and DNA barcoding (Hebert et al., 2003) and integrating the findings enhances the species delimitation and also our knowledge about the evolutionary processes driving their speciation and will eventually integrate the activities of taxonomists and molecular biologists. It's worth repeating that data integration from multiple advanced methodologies will overcome the challenges of precise species identification and delimitation all across the world.

Acknowledgements

The authors are thankful to Tamil Nadu Dr. J. Jayalalithaa Fisheries University, Tamil Nadu, for the necessary support and encouragement.

Conflict of Interest

All the authors declared no conflict of interest.

References

Bhosale, M.M., Pawar, R.A., Bhendarkar, M.P., Sawant, M.S. and Pawase, A.S. (2018). Truss based morphometric approach for the analysis of body shape in portunid crabs (Charybdis feriatus, Portunus pelagicus and P. sanguinolentus) along Ratnagiri coast, India.

Bookstein, F.L. (1982). Foundations of morphometrics. Annual Review of Ecology and Systematics. 13(1): 451-470.

Bookstein, F.L. (1986). Size and shape spaces for landmark data in two dimensions. Statistical Science. 1(2): 181-222.

Boudreau, S.A. and Worm, B. (2012). Ecological role of large benthic decapods in marine ecosystems: a review. Marine Ecology Progress Series. 469: 195-213.

Cadrin, S.X. and Friedland, K.D. (1999). The utility of image processing techniques for morphometric analysis and stock identification. Fisheries Research. 43(1-3): 129-139.

Costa, F.O., DeWaard, J.R., Boutillier, J., Ratnasingham, S., Dooh, R.T., Hajibabaei, M. and Hebert, P.D. (2007). Biological identifications through DNA barcodes: the case of the Crustacea. Canadian Journal of Fisheries and Aquatic Sciences. 64(2): 272-295.

Cywinska, A., Hunter, F.F. and Hebert, P.D. (2006). Identifying Canadian mosquito species through DNA barcodes. Medical and Veterinary Entomology. 20(4): 413-424.

Dayrat, B. (2005). Towards integrative taxonomy. Biological Journal of the Linnean Society. 85(3): 407-417.

Ebach, M.C. and Holdrege, C. (2005). DNA barcoding is no substitute for taxonomy. Nature. 434(7034): 697-697.

Fujita, M.K., Leaché, A.D., Burbrink, F.T., McGuire, J.A. and Moritz, C. (2012). Coalescent-based species delimitation in an integrative taxonomy. Trends in Ecology & Evolution. 27(9): 480-488.

Glaw, F., Koehler, J., De la Riva, I., Vieites, D.R. and Vences, M. (2010). Integrative taxonomy of Malagasy treefrogs: combination of molecular genetics, bioacoustics and comparative morphology reveals twelve additional species of Boophis. Zootaxa. 2383(1): 82.

Godfray, H.C.J. (2002). Challenges for taxonomy. Nature. 417(6884): 17-19.

Goodman, S.M., Maminirina, C.P., Weyeneth, N., Bradman, H.M., Christidis, L., Ruedi, M. and Appleton, B. (2009). The use of molecular and morphological characters to resolve the taxonomic identity of cryptic species: the case of Miniopterus manavi (Chiroptera, Miniopteridae). Zoologica Scripta. 38(4): 339-363.

Hajibabaei, M., Janzen, D.H., Burns, J.M., Hallwachs, W. and Hebert, P.D. (2006). DNA barcodes distinguish species of tropical Lepidoptera. Proceedings of the National Academy of Sciences. 103(4): 968-971.

Hebert, P.D.N., Stoeckle, M.Y., Zemlak, T.S., Francis, C.M. and Godfray, C. (2004). Identification of birds through DNA barcodes. PLoS biology. 2(10): e312.

Hebert, P.D., Cywinska, A., Ball, S.L. and DeWaard, J.R. (2003). Biological identifications through DNA barcodes. Proceedings of the Royal Society of London. Series B: Biological Sciences. 270(1512): 313-321.

Josileen, J., Dineshbabu, A.P., Sarada, P.T., Dash, G., Divipala, I., Kumar, R., Rajkumar, M., Saleela, K.N., Pillai, S.L., Chakraborty, R.D. and Ragesh, N. (2021). Trends in marine crab fishery of India. Marine Fisheries Information Service, Technical and Extension Series. 249: 7-19.

Keenan, C., Davie, P.J. and Mann, D.L. (1998). A revision of the genus Scylla de Haan, 1833 (Crustacea: Decapoda: Brachyura: portunidae). The Raffles Bulletin of Zoology. 46: 217-245.

Kerr, K.C., Lijtmaer, D.A., Barreira, A.S., Hebert, P.D. and Tubaro, P.L. (2009). Probing evolutionary patterns in Neotropical birds through DNA barcodes. PLoS One. 4(2): e4379.

La Salle, J., Wheeler, Q., Jackway, P., Winterton, S., Hobern, D. and Lovell, D. (2009). Accelerating taxonomic discovery through automated character extraction. Zootaxa. 2217(1): 43-55.

Lai, J.C., Ng, P.K. and Davie, P.J. (2010). A revision of the Portunus pelagicus (Linnaeus, 1758) species complex (Crustacea: Brachyura: Portunidae), with the recognition of four species. Raffles Bulletin of Zoology. 58(2).

Leaché, A.D., Koo, M.S., Spencer, C.L., Papenfuss, T.J., Fisher, R.N. and McGuire, J.A. (2009). Quantifying ecological, morphological, and genetic variation to delimit species in the coast horned lizard species complex (Phrynosoma). Proceedings of the National Academy of Sciences. 106(30): 12418-12423.

Lee MS. (2004). The molecularization of taxonomy. Invertebrate Systematics. 18(1): 1-6.

Mallet, J. and Willmott, K. (2003). Taxonomy: renaissance or Tower of Babel. Trends in Ecology & Evolution. 18(2): 57-59.

Marini, M., Suman, A., Farajallah, A. and Wardiatno, Y. (2017). Identifying Penaeus merguiensis de Man, 1888 stocks in Indonesian Fisheries Management Area 573: a truss network analysis approach. Aquaculture, Aquarium, Conservation & Legislation. 10(4): 922-935.

Mayr, E. (2015). Principles of systematic zoology. Scientific Publishers.

Min, X.J. and Hickey, D.A. (2007). DNA barcodes provide a quick preview of mitochondrial genome composition. PLOS one. 2(3): e325.

Mora, C., Tittensor, D.P., Adl, S., Simpson, A.G. and Worm, B. (2011). How many species are there on Earth and in the ocean. PLoS biology. 9(8): e1001127.

Moritz, C., Cicero, C. and Godfray, C. (2004). DNA barcoding: promise and pitfalls. PLoS biology. 2(10): e354.

Nelson, G.J. (1971). " Cladism" as a Philosophy of Classification. Systematic Zoology. 20(3): 373-376.

Ng, P.K., Guinot, D. and Davie, P.J. (2008). Systema Brachyurorum: Part I. An annotated checklist of extant brachyuran crabs of the world. The raffles bulletin of zoology. 17(1): 1-286.

Ovenden, J.R. (1990). Mitochondrial DNA and marine stock assessment: a review. Marine and Freshwater Research. 41(6): 835-853.

Packer, L., Gibbs, J., Sheffield, C. and Hanner, R. (2009). DNA barcoding and the mediocrity of morphology. Molecular Ecology Resources. 9: 42-50.

Padial, J.M. and De la Riva, I. (2007). Integrative taxonomists should use and produce DNA Barcodes.

Padial, J.M., Miralles, A., De la Riva, I. and Vences, M. (2010). The integrative future of taxonomy. Frontiers in Zoology. 7(1): 1-14.

Paramasivam P, Chakraborty RD, Ganesan K, Maheswarudu G. (2017). Stock structure analysis of 'Aristeus alcocki Ramadan, 1938 (Decapoda: Aristeidae)' in the Indian coast with truss network morphometrics. Canadian Journal of Zoology. 96(5): 411-424.

Parsons, K.J., Robinson, B.W. and Hrbek, T. (2003). Getting into shape: an empirical comparison of traditional truss-based morphometric methods with a newer geometric method applied to New World cichlids. Environmental Biology of Fishes. 67(4): 417-431.

Petit, R.J. and Excoffier, L. (2009). Gene flow and species delimitation. Trends in Ecology & Evolution. 24(7): 386-393.

Rebello, V.T., George, M.K., Paulton, M.P. and Sathianandan, T.V. (2013). Morphometric structure of the jumbo tiger prawn, Penaeus monodon Fabricius, 1798 from southeast and southwest coasts of India. Journal of the Marine Biological Association of India. 55(2): 11-15.

Riedel, A., Sagata, K., Suhardjono, Y.R., Tänzler, R. and Balke, M. (2013). Integrative taxonomy on the fast track-towards more sustainability in biodiversity research. Frontiers in Zoology. 10(1): 1-9.

Roe, A.D. and Sperling, F.A. (2007). Population structure and species boundary delimitation of cryptic Dioryctria moths: an integrative approach. Molecular Ecology. 16(17): 3617-3633.

Rohlf, F.J. and Marcus, L.F. (1993). A revolution morphometrics. Trends in Ecology & Evolution. 8(4): 129-132.

Rubinoff, D. (2006). Utility of mitochondrial DNA barcodes in species conservation. Conservation Biology. 20(4): 1026-1033.

Rubinoff, D., Cameron, S. and Will, K. (2006). A genomic perspective on the shortcomings of mitochondrial DNA for "barcoding" identification. Journal of Heredity. 97(6): 581-594.

Števčić, Z. (2005). The reclassification of brachyuran crabs (Crustacea: Decapoda: Brachyura). Natura Croatica: Periodicum Musei Historiae Naturalis Croatici. 14: 1-159.

Schlick-Steiner, B.C., Steiner, F.M., Seifert, B., Stauffer, C., Christian, E. and Crozier, R.H. (2010). Integrative taxonomy: a multisource approach to exploring biodiversity. Annual Review of Entomology. 55: 421-438.

Stankus, A. (2021). State of world aquaculture 2020 and regional reviews: FAO webinar series. FAO Aquaculture Newsletter. (63): 17-18.

Strauss, R.E. and Bookstein, F.L. (1982). The truss: body form reconstructions in morphometrics. Systematic Biology. 31(2): 113-135.

Tautz, D., Arctander, P., Minelli, A., Thomas, R.H. and Vogler, A.P. (2003). A plea for DNA taxonomy. Trends in Ecology & Evolution. 18(2): 70-74.

Thompson, D. W., & Thompson, D. A. W. (1942). On growth and form (Vol. 2). Cambridge: Cambridge University Press.

Turan, C. (1999). A note on the examination of morphometric differentiation among fish populations: the truss system. Turkish Journal of Zoology. 23(3): 259-264.

Walker, J.A. and Bell, M.A. (2000). Net evolutionary trajectories of body shape evolution within a microgeographic radiation of threespine sticklebacks (Gasterosteus aculeatus). Journal of Zoology. 252(3): 293-302.

Ward, R. D., Zemlak, T. S., Innes, B. H., Last, P. R., & Hebert, P. D. (2005). DNA barcoding Australia's fish species. Philosophical Transactions of the Royal Society B: Biological Sciences. 360(1462): 1847-1857.

Wheeler, Q.D. (2007). Invertebrate systematics or spineless taxonomy. Zootaxa. 1668(1): 11-18.

Wheeler, Q.D., Raven, P.H. and Wilson, E.O. (2004). Taxonomy: impediment or expedient. Science. 303(5656): 285-285.

Wilson, E.O. (1985). The biological diversity crisis. BioScience. 35(11): 700-706.

Zelditch, M.L., Fink, W.L. and Swiderski, D.L. (1995). Morphometrics, homology, and phylogenetics: quantified characters as synapomorphies. Systematic Biology. 44(2): 179-189.

Zhang, Z.Q. (2011). Animal biodiversity: an introduction to higher-level classification and taxonomic richness. Zootaxa. 3148(1): 7-12.

10

Need of the Hour To Conserve Portunid Crab Along The Tamilnadu Coast

Adyasha Sahu, V.K. Venkataramani, N. Jayakumar, R. Durairaja, Sudhan C. and Preetysh Nanda Patnaik

Department of Fisheries Biology and Resource Management
Fisheries College and Research Institute, Thoothukudi, Tamil Nadu

Abstract

The Gulf of Mannar is home to a plethora of vegetation and fauna. From which the crustaceans have high commercial importance in local and international markets. Portunid crabs have intensified values regarding nutrition, potential, indirect economic, scientific, and socio-cultural perspectives among crustaceans. These crabs are also in great supply on both domestic and overseas markets. The values of Portunid crab along with its culture practice and importance of fattening is described. The checklist of commercially important portunid crabs along the Tamilnadu coast has been tabulated, including carapace width data and unique characteristics of each portunid crab. The occurrence data of portunid crabs in various ecosystems in Tamilnadu are compiled in table 2. From The checklist data table (Table I, Fig I), it can be deduced that species of the Portunus, Scylla, Charybdis, Podopthalmus, and Thalamita genera are mostly commercially available in the Gulf of Mannar. Based on the occurrence data (table II), it can be concluded that the majority of portunid carbs are found on reefs and muddy shore habitat, while rocky shore habitat has a lower number of crabs. The table III displays the occurrence data of Portunid crab along Tamilnadu coast from different stations are given and it is concluded that Portunid crabs are abundant in the state. Table IV and V shows the abundance data of Portunid crab along Gulf of Mannar and Nagapattinum coast year wise respectively and it is concluded that Portunid crabs are under increasing fishing pressure along Tamilnadu coast. To meet these rising demands, we look at ways to manage portunid crabs in a sustainable way in this article, including sea ranching, banning certain types of gear, regularising mesh size, establishing a minimum size at capture, and conducting stock assessment studies. Aside from these, Eco labelling is a market-based mechanism that promotes the sustainable use of natural resources. The blue MSC designation is only given to wild fish or seafood fisheries that have been approved by the MSC Fisheries Standard, which is a set of conditions for sustainable fishing. Choosing seafood with the blue MSC designation ensures a sustainable fishery that has been independently certified. The blue MSC's high-quality care, direction, and management help ensure that fish stocks and habitats are finely balanced, and that the fishing community's livelihood is secure. Blue crab conservation has benefited from ecolabelling and the Blue MSC label.

Keywords: Tamilnadu coast, Portunid Crab, occurrence and abundance data, Management strategies, Ecolabeling, Blue MSC level

Background

The Gulf of Mannar is one of the world's richest marine biodiversity areas, as well as the first biosphere reserve in the region. The EEZ (Exclusive Economic Zone) of this region is around 15,000 square kilometres, with commercial fishing taking place in about 5,500 square kilometres (Arnold John M. 2017). Throughout the year, both mechanised trawlers and non-mechanized vessels fish in this area (Arnold John M 2001).

Crustaceans are the most key members of the benthic organism strata and have a higher nutritional value for humans, and a large number of tiny species contribute to the complexity and sustainability of tropical ecosystems. (Hendrickx 1995). The Brachyuran crab is one of them, and it dominates many estuary regions where salinity and temperatures can vary substantially on a daily basis. (Ng et al. 2008). There are around seven hundred genera and five thousand to ten thousand species of brachyuran crabs in the globe. (Kaestner 1970; Melo 1996; Ng 1998; Martin and Davis 2001; Sternberg and Cumberlidge 2001; Ng et al. 2008; Yeo et al. 2008). There are seven hundred five numbers of brachyuran crab species in India, divided into twenty-eight families and two hundred seventy genera (Venkataraman and Wafar 2005). Four hundred four crabs from twenty-six families and one hundred fifty-two genera were found along the Tamil Nadu coast (Kathirvel 2008). On India's south coast, the Gulf of Mannar, Palk Bay, Nagapattinam, and Puducherry landings produce the most crab. (Rao et al. 1973).

Only 15 species of edible marine crabs have been identified in Indian waters, and they live in coastal waters and adjacent brackishwater ecosystems, sustaining commercial fisheries. (Radhakrishnan 1979; Varadharajan et al. 2009; Varadharajan 2012). A total of 210 crab species have been documented in the Gulf of Mannar. (CMFRI 1969a and 1998).

Introduction

In Indian waters, individuals of the Portunidae family contributed the majority of the crab to the crab fisheries (Prasad and Thampi 1952; Pillai and Nair 1973; CMFRI 1998, 2000). Five different genera of Portunid crabs have been reported (*Scylla, Portunus, Charybdis, Lupocyclus,* and *Thalamita*). Though most portunid species are edible, a few have been found to be commercially important in large or smaller fisheries throughout the Tamilnadu coast. They are *Charybdis affinis, C. anisodon, C. annulatata, C. bimaculata, C. feriatus, C. helleri, C. hongkongensis, C. japonica, C. lucifera, C. natator, C. smithi, C. vadorum, C. variegate, C. truncate, Podopthalmus vigil, Portunus gladiator, P. glacilimanus, P. haani, P. hastatoides, P. pelagicus, P. sanguinolentus, P. trituberculatus, Scylla serrata, S. olivacea.*

Due to the rising demand for food, India's bioresources and ecosystem services are under increasing strain. (Kannupandi et al. 2003; Varadharajan et al. 2009). It is one of the major suppliers of marine crustaceans to the global market. (Anon 1982). In terms of the value of the fisheries it supports, the crab is the third most lucrative crustacean, after shrimp and lobster. (Mohammed Saved and Rajeev Rahavan 2001). Crustacean fishing accounts for almost 60% of the foreign cash generated in India's seafood industry, as it is a rapidly growing industry with a big domestic and international demand for crabmeat. (Manisseri Mary and Radhakrishnan 2003; Dinakaran and Soundarapandian 2009). Locally, crab species are consumed, and pasteurised crabmeat products are widely exported to the US and European markets. (Camacho and Apya 2001; Ramano and Zeng, 2008). The objective of this paper is understanding the occurrence and abundance data of Portunid crab and accordingly manage sustainably through practices and recently developed MSC technique.

Values of Portunid crab

Crabmeat is high in protein, vitamin A and D, minerals, glycogen, and essential amino acids. Anti-inflammatory components in crabmeat can aid with asthma and chronic fevers. (Raja 1981). The portunid crab has also nutritional values as it contains a notable quantity of protein, vitamins (Vit-B12), and various nutrients. It's high in omega-3 fatty acids, which can help prevent heart disease and cancer. Besides, the soup of *Portunus sanguinolentus and P. pelagicus* can be used to recovery from typhoid and malaria. For the cure of diarrhoea and dysentery, *Scylla serrata* can be used. The crab is an important part of the food chain in terms of energy transfer. They also aid in the recycling of nutrients, which increases the soil's richness through ploughing. The antibacterial, antileukemic, anticoagulant, and cardioactive characteristics of the Portunid crab have been obtained. In India, the portunid crab has some socio-cultural significance. Christians are said to avoid eating *Charybdis feriatus* because it has religious significance, and *Portunus sanguinolentus*, with three reddish circular spots ringed by a whitish ring, is said to resemble Lord Shiva, who has three eyes.

Gear-craft Particulars of Crab Catch State-wise

In different maritime states different types of gears are used for different species. In Gujarat, stake net is used for *Scylla serrata* and cast net, gill net and drag net are used for *Portunus pelagicus*. In the state of Maharashtra, seine net is used for *S. serrata*, hoop net is used for *P. pelagicus*, hooked iron or steel rods are used for *P. sanguinolentus*. In Karnataka state, shore seine is used for *P. pelagicus*, cast net is used for *P. sanguinolentus*, gill net is used for *S.*

serrata. In Kerala, shore seine is used for *P. sanguinolentus*, boat seine is used for *P. pelagicus*, gill net is used for *S. serrata*, cast net is used for *Charybdis annulata*, drag net and stake net is used for *C. natator*. In Tamilnadu, boat seine is used for *P. pelagicus*, shore seine is used for *P. sanguinolentus*, gill net and crab net are used for *S. serrata*. In Andhrapradesh, gill net and drag net are used for *P. pelagicus* and *S. serrata* respectively. In Odisha and West Bengal line with baits and seine net are used for *S. serrata* and *P. pelagicus* respectively.

Crabbers are fishing boats that are primarily and extensively used to catch crabs. These boats are also known as 'fish trap' boats because they are equipped with special entrapping aids for catching crustaceans.

Crab Culture and Importance of Fattening

Mud crabs may survive in a broad range of salinities during their growing phase, so crab farming takes place in brackishwater areas of all salinity gradients. Salinity 10-34 ppt, temperature 23-30°C, dissolved oxygen above 3 ppm, and pH 8.0-8.5 are the most desirable ranges of water quality parameters. There are two ways to farm mud crabs commercially. In one method, juvenile crabs are grown to marketable size in earthen ponds for 3-6 months, while in the other method, medium or large crabs, preferably post-moulted crabs (soft-shelled crabs or water crabs), are reared in cages, pens, or ponds for 20 to 30 days (or even shorter) until the shells harden with additional weight gain. The former is known as "grow-out operation," while the latter is known as "crab fattening."

Crab fattening primarily involves holding young female crabs until their gonads form and fill the mantle cavity, or holding post-moult or water crabs for brief intervals until they “flesh out”. Due to an increase in demand for gravid females and large hard-shelled crabs in seafood restaurants from Hong Kong to Indonesia, this type of crab farming has become very popular throughout Asia. Newly moulted crabs larger than 550 g and about 15 cm Carapace Width are obtained alive from commercial catches in countries like India and fattened in ponds, cages, or pens. The ponds used for this purpose are smaller (0.1-0.2 ha) and have a 1 to 1.5 m depth of water holding capacity. Fencing, a water exchange facility, and other grow-out conditions are all described in the fattening system. Normally, the stocking density is 1 crab/1 to 3 m^2. The fattening period lasts 20-30 days, after which they fully 'flesh out.' The crabs are harvested once the shell has hardened sufficiently and before the next moult. The ponds are prepared for the next fattening cycle after the harvest is completed. Crab fattening/hardening can be profitable if stocking and harvesting are done repeatedly. Cages made of arecanut palm splits in the size of 3 x 3 x 1.5 m, with lid, are increasingly being used in the open backwaters of States like Kerala, in addition to fattening in pond.

Mud crab fattening is the best method for small-scale aquaculture because:

- Turnover is quick, resulting in a short investment-to-return period.
- Because fattened crabs do not moult, they can be stocked in greater numbers (15 crabs/ m^2) than grow out crabs (1 crab/ m^2), resulting in lower losses due to cannibalism.
- When compared to grow-out systems, a short production time reduces the risk of losing crabs to disease, resulting in a high fattening survival rate (>90%)

Table 1: Checklist of Commercially important Portunid crab along Tamilnadu coast

Taxonomic Position	Genus	Species	Common names	Vernacular names (Tamil)	Carapace Width (CW) in mm	Special characters
Kingdom- Animalia Phylum- Arthropoda Subphylum- Crustacea Class- Malacostraca Order- Decapoda Suborder- Reptantia Infraorder- Brachyura Family-Portunidae	*Portunus*	*P. pelagicus*	Blue swimmer crab	Pulli nandu	56-170	Males-Bluish colour marking Females- Dull green or greenish brown
		P. sanguinolentus	Three spot swimming crab	Mukkannu nandu	51-165	Chelae elongated in male
	Scylla	*S. serrata*	Giant Mud crab	Pachal nandu	-	H- shaped gastric groove deep; Relatively broad frontal lobes
	Charybdis	*C. natator*	Ridged swimming crab	Vari nandu	46-95	Carapace densely covered with very short pubescence which is absent on granulated ridges in anterior half
		C. lucifera	Yellowish brown crab	Manjal nandu	41-105	Posterior border of cephalothorax curved, forming curved postero-lateral junction;
		C. feriatus	Crucifix crab	Siluval nandu	81-120	Carapace ovate; 5 distinct teeth on each anterolateral margin
		C. anisodon	Two spined arm swimming crab	_	-	Antennal flagellum excluded from orbit.
	Podopthalmus	*P. vigil*	Long eyed swimming crab	Neendakon nandu	51-205	Anterior margin much broader than posterior margin; Orbits very broad; Eyes very long, reaching to or extending beyond edge of carapace
	Thalamita	*T. integra*	Not designated	Sirumal nandu	-	Cardiac ridges not present on carapace; protogastric region smooth or granular but without ridge; male first pleopod without recurved tip,
		T. crenata	Mangrove swimming crab	_	66-70	Rounded lobes; Front with 6 equal-sized; Ridges low but distinct

[CW: Source; CMFRI, 2019 & FAO identification manual]

Figure 1 : Checklist of Portunid crab in Tamilnadu coast

Table 2: Occurrence data of Portunid crab based on different ecosystems along Tamilnadu ✓ - Present; ×- Absent

Species	Coral reef	Mangrove ecosystem	Seaweed/ Seagrass	Open water	Muddy shore	Sandy shore	Rocky shore	Tidal mudflats/ Estuary/ Creek	Reference
Portunus pelagicus	✓	✓	✓	✓	✓	✓	×	✓	Ravichandran S, Soundarapandian P, Kannupandi T. Zonation and distribution of crabs in Pitchavaram mangrove swamp, southeast coast of India. Indian J. Fish. 2001 Apr;48(2):221-6.
P. sanguinolentus	✓	✓	×	×	✓	✓	×	✓	
Scylla serrata	×	✓	✓	✓	✓	×	×	✓	Ramesh, R., Nammalwar, P. and Gowri, V.S., 2008. Database on coastal information of Tamilnadu. Institute for Ocean Management, Anna University, Chennai. Report submitted to Environmental Information System (ENVIS), Department of Environment, Government of Tamilnadu, Chennai, 133pp.
Charybdis natator	✓	×	✓	✓	×	✓	✓	✓	
C. lucifera	✓	×	×	×	×	×	×	×	Vidhya V, Jawahar P, Karuppasamy K. Growth and mortality parameters of the three-spot crab, Portunus sanguinolentus (Herbst, 1783) from Gulf of Mannar, Southeast Coast of India.
C. feriatus	✓	×	×	×	✓	✓	×	×	
C. anisodon	✓	×	✓	×	✓	×	×	×	Asphama AI, Amir F, Malina AC, Fujaya Y. Habitat preferences of blue swimming crab (Portunus pelagicus). Aquacultura Indonesiana. 2015;16(1):10-5.
Podopthalmus vigil	×	×	×	×	✓	✓	×	×	Paramasivam K, Venkataraman K, Venkatraman C, Rajkumar R, Shrinivaasu S. Diversity and Distribution of Sea Grass Associated Macrofauna in Gulf of Mannar Biosphere Reserve, Southern India. In Marine Faunal Diversity in India 2015 Jan 1 (pp. 137-160). Academic Press.
Thalamita integra	✓	×	×	×	×	×	×	×	
T. crenata	×	✓	✓	×	✓	×	✓	×	

Table 3: Occurrence data of Portunid crab along Tamilnadu coast at different stations S: Station, +: Present, _: Not recorded

Species	SI	SII	SIII	SIV	SV	SVI	SVII	SVIII	SIX	SX	SXI
	Gulf of Mannar (2015-16)				Puducherry coast (2011-12)					Mudasal Odai and Nagapattinam (2011-12)	
Portunus pelagicus	+	+	+	+	+	+	+	+	+		
P. sanguinolentus	+	+	+	+	+	+	+	+	+	+	+
P. gladiator	+	_	_	_	+	+	+	_	_	_	+
Scylla serrata	+	+	+	+	+	+	+	+	+	_	_
S. tranquebarica	+	+	+	+	+	+	+	+	+	_	_
Charybdis natator	+	+	+	+	+	+	+	+	_	+	+
C. lucifera	+	+	_	_	+	+	+	_	_	+	+
C. feriatus	+	+	+	+	+	+	+	_	_	+	+
C. annulata	+	+	_	_	_	_	_	_	_	_	_
C. japonica	+	+	_	+	_	_	_	_	_	_	_
C. variegate	_	_	+	_	+	+	_	_	_	_	_
Podopthalmus vigil	+	+	+	+	+	+	+	+	+	+	+
Thalamita integra	_	+	_	+	_	_	_	_	_	_	_
T. crenata	+	+	_	+	+	+	_	+	+	+	+
Total	12	12	8	10	11	11	9	7	6	6	7

[*Source*: From "Annotated check list of the brachyuran crabs (Crustacea: Decapoda) from Gulf of Mannar region, south east coast of India" by V Vidhya, P Jawahar and K Karuppasamy and from "The global science of crab biodiversity from Puducherry coast, south east coast of India" by D. Varadharajan, P. Soundarapandian, N. Pushparajan and from "Brachyuran crabs' diversity in Mudasal Odai and Nagapattinam coast of south east India" by Kollimalai Sakthivel, Antony Fernando]

Table 4: Abundance data of Portunid crab along Gulf of Mannar

Species	**Catch data (Kg) at Gulf of Mannar (2015 June -16 June)**	**Catch data (Kg) at Gulf of Mannar (2007-08)**
Portunus pelagicus	22254.21	27,599.875
P. sanguinolentus	7359.921	581.05
Charybdis natator	6832.508	581.05
Scylla tranquebarica	-	290.525
Total	36,446.639	29,052.5

Source: From "Population Biology and Stock Assessment of Selected Portunus Species of Gulf of Mannar" by V. Vidya & "Current Status of Crab Fishery in the Artisanal Sector along Gulf of Mannar and Palkbay Coasts" by M. Rajamani and A. Palanichamy

Table 5: Abundance data of Portunid crab along Nagapattinum coast

Species	**Catch data (Kg) at Nagapattinum coast (2017 Jan-17 Dec)**	**Catch data (Kg) at Nagapattinum coast (2009 April -10 March)**
Portunus Pelagicus	1006	17000
P. sanguinolentus	1021	16650
P. gladiator	487	1800
Scylla serrata	961	2900
S. tranquebarica	1069	8100
Charybdis natator	653	3885
C. feriatus	723	5600
C. lucifera	358	3200
C. variegate	265	3100
C. granulate	498	2800
C. truncate	391	3000
Podopthalmus vigil	696	5350
Total	8128	73385

Source: From "Portunid Crab Fishery Resources from Nagapattinam Coast, South East Coast of India" by D Varadharajan and P Soundarapandian & From "Diversity of Commercially Important Marine Crabs in Nagapattinam Coastal Area, Tamilnadu, India" by U. Sathiya and V. Valarmathi

The checklist of commercially important crab species from the coast of Tamilnadu are listed in (Table I, Figure1), along with their taxonomic locations, common names, and vernacular names. These crabs' carapace width data comes from the CMFRI 2019 manual, and the identifiable characters of these commercially important crabs come from the FAO identification sheet.

Table II shows the distribution of these commercially important crab species in the Gulf of Mannar, based on ecosystems such as coral reefs, mangrove ecosystems, sea weed or seagrass ecosystems, open water, muddy shore, sandy shore, rocky shore, and tidal mudflats/ estuary/ creek. The data on the occurrence of these crabs in these ecosystems comes from the FAO identification sheet and the reference in the column.

Table III displays that, Vidya et al., conducted a comparative study of occurrence data along the Gulf of Mannar in 2015-16 by taking four stations (Station I, II, III, IV) i.e., Therespuram, Vellapatti, Vedalai, Periyapattinam respectively. Varadharajan, Soundarapandian and Pushparajan conducted a comparative study of occurrence data along the Puducherry coast in 2011-12 by taking five different station covering Puducherry coast. Sakthivel and Fernando conducted a comparative study of occurrence data along the Nagapattinum coast in 2011-12. Station X and station XI represent the Mudasal Odai and Nagapattinam coast respectively.

Table IV comprises the abundance data of Portunid crab along Gulf of Mannar in the year of 2015-16 conducted by V. Vidya and in the year 2007-08 conducted by Rajamani and Palanichamy.

Table V comprises that the abundance data of Portunid crab along Nagapattinum coast in the year of 2017-18 conducted by U. Sathiya and V. Valarmathi. In the year of 2009-10 the abundance data of Portunid crab along Nagapattinum coast is conducted by Varadharajan and Soundarapandian.

Why to Conserve?

Portunid crabs have heavily exploited, with fishing mortality rates (F) over double that of natural levels (M) because of high fishing pressure. Fishing pressure adds additional stresses to the population and can cause growth to slow, resulting in reductions in the size at which maturity is reached, leading to lowered reproductive potential (Jorgensen 1990; Trippel 1995, de Lestang et al. 2003a). The fisheries experts have developed management plans aiming to protect this reproductive potential, and minimum size limits fixed to well above the size at which maturity is reached (Roberston and Kruger 1994, Johnston et al. 2014a, Zairion et al. 2015b), restrictions on the capture of ovigerous female crabs throughout the reproductive season or entire year

(Johnston et al. 2014b, Anon 2016) and male-only harvest strategies (Grubert et al. 2014).

Climate change and excess nutrients in marine habitats are expected to have a negative impact on blue crabs and their catches. The natural ecosystems of Portunid in the Gulf of Mannar have been harmed by adverse environmental causes and human involvement, and it is vital to discover feasible crab fishing fleets.

Strategies for Sustainable Management

The management of portunid fisheries varies significantly between and within countries, depending on jurisdictional boundaries, with estuarine and coastal fisheries often controlled by state or provincial governments (Criddle 2008; Australian 2016). Most management strategies implement both input and output controls, limiting the fishing effort and the methods used, and the total landings and using size restrictions (Helser and Kahn 2001; Lipcius and Stockhausen 2002; Grubert et al. 2014; Johnston et al. 2014b).

Fishing and boat licenses, seasonal closures, and protected areas are commonly used input controls (Lipcius et al. 2003; Johnston et al. 2011b), while minimum size limits, total allowable catches (commercial), daily bag limits (recreational), and protection of females in general or ovigerous females are commonly used output controls (Helser and Kahn 2001; Svane and Hooper 2004; Johnston et al. 2014a).

Eco-label and its Scheme

Eco-labelling is a market-based mechanism that encourages the use of natural resources in a sustainable manner. The eco-label is a tag or label attached to a product that verifies that it was made in an environmentally responsible manner. Environmental labelling has been practised for many years around the world and is characterised as "making important environmental information available to appropriate consumers" (U.S. EPA. 1993, 1998). In the fisheries sector, there are a variety of eco-labelling and certification systems, each with its own set of criteria, assessment procedures, transparency level, and agency sponsoring. In1997, the Marine Stewardship Council (MSC) was established by WWF (World Wildlife Fund) and Unilever as one of the first scientifically designed eco-labelling systems. The MSC Fisheries Standard is based on the Code of Conduct for Responsible Fisheries of the United Nations Food and Agriculture Organization (FAO) (Josileen Jose 2018). The short-neck clam fishery (*Paphia malabarica*) in Ashtamudi is India's first MSC-certified fishery, and Asia's third. (CMFRI 2018)

MSC Certification

The Marine Stewardship Council (MSC) is a non-profit organisation that establishes a sustainable fishing standard. Fisheries that would like to show they are well-managed and sustainable in comparison to the science-based MSC standard are assessed by a panel of experts who are not affiliated with either the fishery or the MSC. The MSC's mission is to contribute to the health of the world's oceans by recognising and rewarding sustainable fishing practises, influencing people's seafood purchasing decisions, and working with partners to transform the seafood market to a sustainable basis through its ecolabel and fishery certification programme, for which the MSC receives royalties for licencing it to products. MSC accreditation may offer increased branding, greater visibility, improved communication with stakeholders, a road for advancements, protected livelihoods, access to new markets, secure markets, and promotional opportunities.

The MSC Fisheries Standard is used to assess the management and sustainability of a fishery. The Standard reflects the most current knowledge of internationally recognised fisheries science and management. A fishery's certified catch can be sold with the blue MSC label once it has been successfully certified to the Fisheries Standard. Conformity Assessment Bodies (CABs) – also known as certification bodies, are accredited independent certifiers who assess fisheries.

The Ashtamudi short-neck clam fishery (*Paphia malabarica*) is the first MSC-certified fishery in India, and only the third in Asia. WWF-India, the Central Marine Fisheries Research Institute (CMFRI), and the Kerala State Fisheries Department collaborated to achieve this milestone, working closely with the local fishing community.

A product bearing a blue fish label, become a part of virtous cycle that helps to protect the productivity and health of oceans. The working of cycle is given Fig. (2).

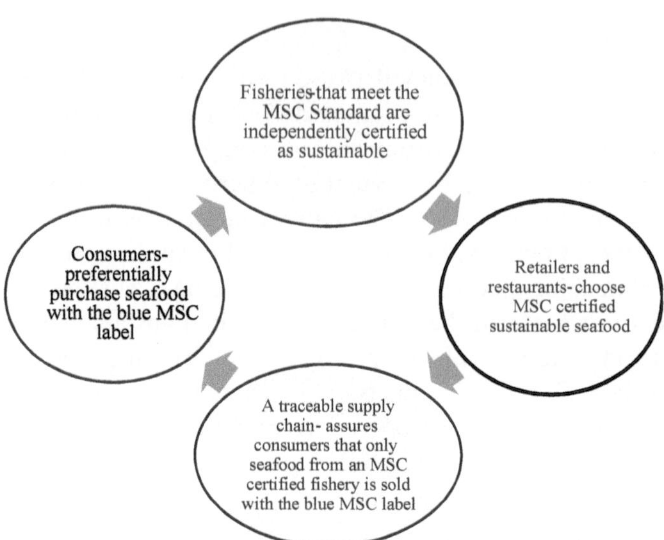

Figure 2 : Virtous cycle

Blue Swimming Crab (BSC)

Several more Indian species are currently being certified, and the MSC identified a few Indian species in 2017, including the Blue Swimming Crab, Portunus pelagicus, which is on the Priority-I list. The Indian Crab Meat Processors Association (CMPA) has continued work in this area, which began in late 2013 with the following aims.

To accomplish an MSC Fishery Improvement Project (FIP) for the Gill net Blue Swimmer Crab.

To conduct a fishery assessment, gather stakeholders, and formalise a multi-faceted Work Plan aimed at ensuring the industry's long-term viability.

To implement fishery management systems that ensure productive blue swimmer crab populations and economic viability for those who rely on the resource now and in the future.

The demand for Portunus pelagicus, also known as Blue Swimming Crab (BSC), is steadily rising, and Tamil Nadu leads the country in marine crab landings, especially for blue swimming crab production (FigureIII). The Palk Bay and Gulf of Mannar areas of the Ramanathapuram, Pudukkottai, and Thanjavur districts of the state are main landing areas for BSC. Crab merchants and crab fishermen in the area rely heavily on BSC products, which are the most important internationally traded commodity from the region (Josileen et.al., 2018).

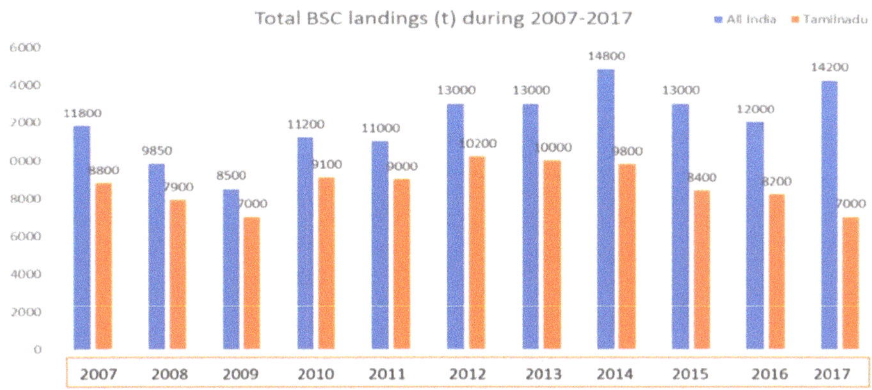

Figure 3 : Total BSC landings in ton during 2007-2017 [*Source*: CMFRI]

The overall trend of the BSC is clearly declining, and now is the time to take the necessary steps to control and manage fishing in order to ensure a sustainable fishery for future use.

On the 'Blue swimming crab fishery improvement programme and launch of an action plan,' hosted by Crab Meat Processors Association, the MSC identified the Flower Crab, Mandapam Flower Shrimp (species captured in the Mandapam area of Rameswaram island), and Kanyakumari Lobster. Various MSC departments, the Fisheries Department, the Central Marine Fisheries Research Institute (CMFRI), the Marine Products Export Development Authority (MPEDA), the Coastal and Marine Protected Areas (CMPA), Fisheries Universities, crabmeat exporters, the Worldwide Fund for Nature, country boat fishermen, and other stakeholders took part in this workshop programme. (Crab meat Association 2018) The 'Blue MSC designation' would guarantee science-based sustainable fishing, as well as a robust export market and a good pricing value for the fishermen.

MSC Certification for Sustainable Seafood

The certification body will use a detailed, public, and rigorous process to determine whether or not your fishery meets the MSC Fisheries Standard. A full assessment takes an average of 12 months and a minimum of 8 months. Overview of the full assessment process is given in Figure (3)

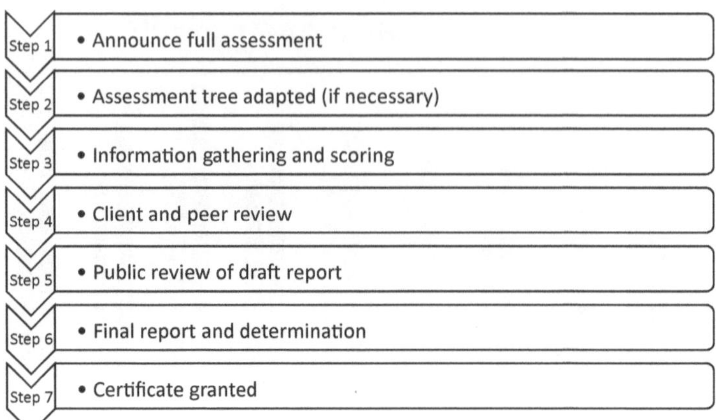

Figure 4 : Overview of full assess

Fisheries receive certification from an impartial assessor, and NGOs and others can participate in the process in a variety of ways. Fisheries must improve continuously under the Standard procedures until they reach the best sustainable practice. If fisheries do not improve within a certain time frame, their certificates are stopped until they meet the MSC Standard's requirements. MSC-certified fisheries are frequently at the forefront of global developments and best practices.

Various Sustainable Management Practices are Given Below

Sea Ranching

In recent years, stock improvement and ocean ranching have been acknowledged for their ability to increase and sustain coastal fisheries (Oshima 1984; Bartley 1999; Liao 1999, 2002). To avoid overfishing and achieve sustainable harvesting, (Muthiga 1986) advocated releasing juveniles, ovigerous females, and fresh moulted crabs. The crab fishery has been in operation since the 1950s, and its size of the stocks has decreased since 2001, owing primarily to the harvest of immature and ovigerous female crabs (Radhakrishnan et al. 2005). Concerns have been raised about the harvesting of brooders from the ocean. This issue must be solved immediately.

Minimum Size at Capture

The major approach for avoiding recruitment and growth overfishing is to fix the size of the first capture. It also aids the animal in reproducing naturally at least once during its lifetime. According to the MSC label, fishermen should only collect crabs with a carapace length greater than 90 mm and should avoid catching berried female crabs carrying egg clusters. (CMFRI 2018).

Ban on the use of Certain Gears

The fishermen should be enlightened on the fishing method and type of nets used to catch the crabs at the MSC label. Some fishing methods such as 'Trawling because detrimental effects on the environment, as it can modify the benthic habitat (Engel and Kvitek 1998; Watson et al. 2006). However, other methods cause less noticeable effects. When baited traps are lost or left in the water, they continue to catch and kill crabs and other creatures, a phenomenon known as "ghost fishing" (Guillory 1993). To achieve a sustainable level, trawl nets should be avoided whenever possible.

MSC-accredited trawling fisheries have improved their procedures significantly. Monitoring and charting fishing areas, avoiding fishing in recognised sensitive or protected areas, and prohibiting bottom trawling during spawning are just a few of the measures in place. Rockhoppers (balls or discs, usually made of rubber) can help limit the amount of contact between the net and the seabed. Crustaceans including lobsters and crabs are caught in stationary traps or pots, which are commonly composed of wood, wire netting, or plastic.

Stock Assessment Studies

Many specialists expressed concern about the scarcity of stock assessment studies for major crustaceans in different parts of the world. To design a better management strategy for preserve the crab resource in the Gulf of Mannar, biological data must be collected. Based on length, spawning potential ratio (SPR) assessment established by Hordyk et al. (2015a) is another promising tool for informing managers of portunid fisheries (2015b). Better collaboration among stakeholders is needed to conserve crab resources.

Regularization of the Mesh Size

Fishers involved in multi-day fishing have expanded their fishing activities to deeper seas, which has resulted in an increase in crab landings. However, in order to maintain the crab fishery, the cod-end mesh size of trawl nets must be increased to 40 rows or higher. (CMFRI 2018)

Results and Discussion

Based on the occurrence data (Table 2), it can be concluded that the majority of portunid carbs are found on reefs and muddy shore habitat, while rocky shore habitat has a lower number of crabs. The comparison of the three-occurrence data in (Table 3) revealed that the Portunid crab is abundant in the Gulf of Mannar. As a result, long-term management of Portunid crab in the Gulf of Mannar is critical. The (Table 4) shows that the abundance of Portunid crab increases slightly from the year 2007-08 to the year 2015-16. However, in both the case, Portunus pelagicus is abundant. The (Table 5) shows that there is high reduction from the year 2009-10 to 2017 along Nagapattinum coast. Therefore, the sustainable management of Portunid crab along Tamilnadu coast id highly needed.

The great waves need our protection. These great oceans are the abode and diversity of unique species and bear one tenth of the world's population. However, this marine ecosystem is under tremendous stress. Intolerable fishing threatens fish populations, ocean habitats, coastal fishing communities, and economies. Choosing seafood with the blue MSC designation, on the other hand, can endure the independently certified sustainable fishery. The blue MSC's qualitative care and direction and management implementation aid the fish stock and habitats are fine enough and that the fishing community's livelihood is safe. To be certified by MSC, fisheries are separately evaluated by specialized and authorized scientists and marine experts to ensure they meet environmentally defendable fishing standards. The thorough analysis makes sure that they withstand their expected heights. MSC certified fisheries reduce their effect on the whole marine environment to secure health and prosperity, flourishing oceans for the future. Processors, retailers, and restaurants must ensure that MSC-certified seafood does not come into contact with non-certified seafood, products with the blue MSC label range from pickled herring to luxury caviar. Purchasing MSC labelled seafood helps to raise an inducement for more fisheries, retailers, and restaurants to produce and sell certified sustainable seafood that leads to a healthy marine ecosystem.

In addition to the aforementioned points relating to the conservation and management of the crab fishery, there are a few strategies that can be useful in managing crab resources, such as regulation of fishing areas, prohibition of destructive gears, prohibition of the fishing season, raising awareness among fishermen through audio-visual advertisements, promotion of artificial reefs affecting a code of conduct for responsible fishing, and enforcement of laws such as fishing regulations. Anyone who disobeys the law will be penalised or punished.

Conclusion

30 to 35 tonnes of blue crabs were caught in Tamilnadu's district, and the meat was mostly sold to western countries on daily basis. (Kathavarayan, Deputy Director of Fisheries 2018). The export of crab meat was carried out by seven seafood businesses, and they mostly profited from the Palk Bay crab catch. Four million tonnes of crabs worth $40 million were shipped from the country every year, with 90 percent of the catch coming from Vellapatti in the

Gulf of Mannar to Adhiramapattinam in Palk Bay (T Kathavarayan, Deputy Director of Fisheries 2018). The MSC Fisheries Standard has been certified by fisheries responsible for 12% of all marine catch since its inception in 1997. Certification aids in the expansion and maintenance of viable fish populations. To keep their certification, fisheries have made over 1,200 improvements to their management and performance. More than 38,000 locations, including supermarkets, restaurants, fishmongers, and hotels, have been certified to sell MSC-certified seafood. The blue MSC label is now found on over 25000 products on the market. MSC-certified blue crabs would go a huge way toward ensuring the survival of country boat fishermen. The greatest way to assure a year-round sustainable fishery and increase yield quality is to prohibit fishing and commercialization of undersized and berried crabs. (Josileen 2007). CMFRI has proposed a minimum legal size (MLS) for fishing for Kerala's fishery resources. (Mohammed et al. 2014).

Conflict of Interest

There is no conflict of interest.

Acknowledgement

The authors are sincerely grateful to Dr. J. Jayalalithaa Fisheries University, Tamil Nadu, for providing adequate help and inspiration.

References

Arnold, J.M., Summers, W.C., Gilbert, D.L., Manalis, R.S., Daw, N.W. and Lasek, R.J., 1974. A guide to laboratory use of the squid Loligo pealei.

Asphama, A.I., Amir, F., Malina, A.C. and Fujaya, Y., 2015. Habitat preferences of blue swimming crab (Portunus pelagicus). Aquacultura Indonesiana, 16(1), pp.10-15.

Australia, G., 2008. Canberra. Geoscience Australia in association with Air services Australia.

Bartley, D.M., 1999. Marine ranching: a global perspective.

Brown, C.W., Hood, R.R., Long, W., Jacobs, J., Ramers, D.L., Wazniak, C., Wiggert, J.D., Wood, R. and Xu, J., 2013. Ecological forecasting in Chesapeake Bay: Using a mechanistic–empirical modeling approach. Journal of Marine Systems, 125, pp.113-125.

Camacho, A.S. and Aypa, S.M., 2001. Research needs and data on production of portunid crabs in the Philippines. Asian Fisheries Science, 14(2), pp.243-245.

CMFRI, F., 1982. Trends in marine fish production in India-1981. Marine Fisheries Information Service, Technical and Extension Series, 41, pp.1-33.

CMFRI, K., 1998. CMFRI Annual Report 1997-98.

CMFRI, K., 2000. CMFRI Annual Report 1999-2000.

CMFRI, K., 2018. CMFRI Annual Report 2017-2018.

Criddle, K.R., 2008. the evolution of rights-based management in Alaska. Case studies in fisheries self-governance, (504), p.369.

de Lestang, S., Hall, N. and Potter, I.C., 2003. Changes in density, age composition, and growth rate of Portunus pelagicus in a large embayment in which fishing pressures and environmental conditions have been altered. Journal of Crustacean Biology, 23(4), pp.908-919.

Dinakaran, G.K. and Soundarapandian, P., 2009. Mating behaviour and broodstock development of commercially important blue swimming crab, Portunus sanguinolentus (Herbst). Indian Journal of Science and Technology, 2(4), pp.71-75.

Engel, J. and Kvitek, R., 1998. Effects of otter trawling on a benthic community in Monterey Bay National Marine Sanctuary. Conservation Biology, 12(6), pp.1204-1214.

Grubert, M., M. Leslie, and D. Bucher., 2014. Mud Crab Scylla seratta, S. olivacea. Fisheries Research and Development Corporation, Canberra.

Guillory, V., 1993. Ghost fishing by blue crab traps. North American Journal of Fisheries Management, 13(3), pp.459-466.

Helser, T.E. and Khan, D.M., 2001. Stock Assessment of Delaware Bay blue crab (Callinectes sapidus) for 2001. Department of Natural Resources and Environmental Control Delaware, 41 pp. Division of Fish and Wildlife. Delaware Division of Fish and Wildlife, pp.1-41.

Hendrickx, M.E., 1995. Checklist of brachyuran crabs (Crustacea: Decapoda) from the eastern tropical Pacific. Bulletin de l'Institut royal des Sciences naturelles de Belgique, Biologie, 65, pp.125-150.

Hordyk, A., Ono, K., Sainsbury, K., Loneragan, N. and Prince, J., 2015. Some explorations of the life history ratios to describe length composition, spawning-per-recruit, and the spawning potential ratio. ICES Journal of Marine Science, 72(1), pp.204-216.

Istiak, S.M., 2018. Study for assessing mud crab (Scylla serrata, Forskal, 1755) market chain and value-added products development in Bangladesh. Bangladesh Journal of Zoology, 46(2), pp.263-273.

James, P.S.B.R., Thomas, P.A., Pillai, C.S.G. and Achari, G.P., 1969. Catalogue of Types and of Sponges, Corals, Polychaetes, Crabs and Echinoderms in the Reference Collections of the Central Marine Fisheries Research Institute. CMFRI Bulletin, 7, pp.1-66.

Johnston, D., Chandrapavan, A., Wise, B. and Caputi, N., 2014. Assessment of blue swimmer crab recruitment and breeding stock levels in the Peel-Harvey Estuary and status of the Mandurah to Bunbury Developing Crab Fishery.

Johnston, D., Harris, D., Caputi, N. and Thomson, A., 2011. Decline of a blue swimmer crab (Portunus pelagicus) fishery in Western Australia—History, contributing factors and future management strategy. Fisheries Research, 109(1), pp.119-130.

Johnston, D., Wesche, S., Noell, C. and Johnson, D., 2014. Blue Swimmer Crab Portunus armatus (formerly Portunus pelagicus). Fisheries Research and Development Corporation, Canberra.

Joseph, M.M. and Jayaprakash, A.A., 2003. Status of exploited marine fishery resources of India. Kochi: Central Marine Fisheries Research Institute, 157.

Josileen, J., 2018. Marine Crab Resources of India with Facts on Life Cycle and Biology In: ICAR Sponsored Winter School on Recent Advances in Fishery Biology Techniques for Biodiversity Evaluation and Conservation, 1-21 December 2018, Kochi.

Josileen, J., Maheswarudu, G., Jinesh, P.T., Sreesanth, L., Ragesh, N. and Divya, P.R., 2018. New record of Charybdis (Goniohellenus) omanensis septentrionalis Turkey & Spiridonov, 2006 (Brachyura, Decapoda, Portunidae) from the Arabian Sea, Kerala, India. Crustaceana, 91(4), pp.389-402.

Jørgensen, T., 1990. Long-term changes in age at sexual maturity of Northeast Arctic cod (Gadus morhua L.). ICES Journal of Marine Science, 46(3), pp.235-248.

Kannupandi, T., Vijayakumar, G. and Soundarapandian, P., 2003. Yolk utilization in a mangrove crab Sesarma brockii (de man). Indian J. Fish, 50(2), pp.199-202.

Kumaraguru, A.K., Joseph, V.E., Marimuthu, N. and Wilson, J.J., 2006. Scientific information on Gulf of Mannar-A bibliography.

Liao, I.C., 1999. How can stock enhancement and sea ranching help sustain and increase coastal fisheries? In: BR Howell, E. Moksness and T. Svasand (eds.).

Liao, I.C., 2002. Roles and contributions of fisheries science in Asia in the 21st century. Fisheries science, 68(sup1), pp.3-13.

Lipcius, R.N., Stockhausen, W.T., Seitz, R.D. and Geer, P.J., 2003. Spatial dynamics and value of a marine protected area and corridor for the blue crab spawning stock in Chesapeake Bay. Bulletin of Marine Science, 72(2), pp.453-469.

Lipcius, R.N. and Stockhausen, W.T., 2002. Concurrent decline of the spawning stock, recruitment, larval abundance, and size of the blue crab Callinectes sapidus in Chesapeake Bay. Marine Ecology Progress Series, 226, pp.45-61.

Manisseri, M.K. and Radhakrishnan, E.V., 2003. Marine crabs.

Martin, J.W. and Davis, G.E., 2001. An updated classification of the recent Crustacea (Vol. 39, p. 129). Los Angeles: Natural History Museum of Los Angeles County.

Melo-Filho, G.A.S.D., 1996. Manual de identificacao dos brachyura (caranguejos e siris) do litoral brasileiro (No. 595.3 (81) MEL).

Mohamed, K.S., Zacharia, P.U., Maheswarudu, G., Sathianandan, T.V., Abdussamad, E.M., Ganga, U., Pillai, S.L., Sobhana, K.S., Nair, R.J., Josileen, J. and Chakraborty, R.D., 2014. Minimum Legal Size (MLS) of capture to avoid growth overfishing of commercially exploited fish and shellfish species of Kerala. Marine Fisheries Information Service; Technical and Extension Series, (220), pp.3-7.

Muthiga, N., 1986. Edible crabs of Kenya.

Ng, P.K., Guinot, D. and Davie, P.J., 2008. Systema Brachyurorum: Part I. An annotated checklist of extant brachyuran crabs of the world. The raffles bulletin of zoology, 17(1), pp.1-286.

Oshima, Y., 1984. Status of fish farming and related technological development in the cultivation of aquatic resources in Japan. In Proceedings of ROC–Japan Symposium on Mariculture. TML Conf. Proc (Vol. 1, pp. 1-11).

Paramasivam, K., Venkataraman, K., Venkatraman, C., Rajkumar, R. and Shrinivaasu, S., 2015. Diversity and distribution of sea grass associated macrofauna in Gulf of Mannar biosphere reserve, Southern India. In Marine Faunal Diversity in India (pp. 137-160). Academic Press.

Pillai, K.K. and Nair, N.B., 1973. Observations on the breeding biology of some crabs from the southwest coast of India. J. mar. biol. Ass. India, 15(2), pp.745-770.

Radhakrishnan, C.K., 1979. Studies on portunid crabs of Porto Novo (Crustacea: Decapoda: Brachyura). PhD Thesis, Annamalai University, India.

Radhakrishnan, E.V., Deshmukh, V.D., Manisseri, M.K., Rajamani, M., Kizhakudan, J.K. and Thangaraja, R., 2005. Status of the major lobster fisheries in India. New Zealand Journal of Marine and Freshwater Research, 39(3), pp.723-732.

Raghu Prasad, R. and Tampi, P.S., 1952. Account of the Fishery and fishing methods for Neptunus pelagicus (Linnaeus) near Mandapam. Journal of the Zoological Society of India, 3(2), pp.335-339.

Raja, S., 1981. The edible crab Scylla serrata and fishery in indo-pacific region. A review (Doctoral dissertation, M. Sc. Dissertation, Annamalai University, India).

Rajamani, M. and Palanichamy, A., 2010. Current Status of Crab Fishery in the Artisanal Sector along Gulf of Mannar and Palk Bay Coasts.

Ramesh, R., Nammalwar, P. and Gowri, V.S., 2008. Database on coastal information of Tamilnadu. Institute for Ocean Management, Anna University, Chennai. Report submitted to Environmental Information System (ENVIS), Department of Environment, Government of Tamilnadu, Chennai, 133pp.

Rao, K.V., 1973. Distribution pattern of the major exploited marine fishery resources of India.

Rao, P.V., Thomas, M.M. and Rao, G.S., 1973. The crab fishery resources of India.

Ravichandran, S., Soundarapandian, P. and Kannupandi, T., 2001. Zonation and distribution of crabs in Pichavaram mangrove swamp, southeast coast of India. Indian Journal of Fisheries, 48(2), pp.221-226.

Robertson, W.D. and Kruger, A., 1994. Size at maturity, mating and spawning in the portunid crab Scylla serrata (Forskål) in Natal, South Africa. Estuarine, Coastal and Shelf Science, 39(2), pp.185-200.

Romano, N. and Zeng, C., 2008. Blue swimmer crabs: emerging species in Asia. Global Aquaculture Advocate, 11, pp.34-36.

Sakthivel, K. and Fernando, A., 2012. Brachyuran crabs' diversity in Mudasal Odai and Nagapattinam coast of south east India. Arthropods, 1(4), p.136.

Savad, A.M. and Raghavan, P.R., 2001. Mud crab-culture and fattening techniques, status and prospects. Seafood Exp. J, 32(11), pp.25-29.

Suseelan, C., 1996. Crab culture and crab fattening. CMFRI Bulletin-Artificial reefs and Sea farming technologies, 48, pp.99-102.

Svane, I. and Hooper, G., 2004. Blue swimmer crab (Portunus pelagicus) fishery. Fishery Assessment Report to PIRSA, for the Blue Crab Fishery Management Committee. SARDI Aquatic Sciences Publication No. RD03/ 0274.

Swanson, C.J., 1970. Bibliography: Invertebrate Zoology. BioScience, 20(6), pp.380-382.

Trippel, E.A., 1995. Age at maturity as a stress indicator in fisheries. Bioscience, 45(11), pp.759-771.

United States Environmental Protection Agency., 1993. Status report on the use of Environmental worldwide, office of pollution prevention toxics, EPA 742-R-9-93-001, September.

United States Environmental Protection Agency., 1998. Environmental labelling: Issues, policies, and practices worldwide, Office of prevention, pesticides and toxic substances, EPA 742-R-98-009, December.

V. Vidya., 2016. Population Biology and Stock Assessment of Selected Portunus Species of Gulf of Mannar, B.F.Sc. Fisheries College and Research Institute Tamilnadu Fisheries University, Thoothuludi-628008

Varadharajan, D., Soundarapandian, P., Dinakaran, G.K. and Vijakumar, G., 2009. Crab Fishery resources from Arukkattuthurai to Aiyammpattinam, South east coast of India. Current Research Journal of Biological Sciences, 1(3), pp.118-122.

Varadharajan, D., Soundarapandian, P. and Pushparajan, N., 2013. The global science of crab biodiversity from Puducherry coast, south east coast of India. Arthropods, 2(1), p.26.

Varadharajan, D. and Soundarapandian, P., 2012. Commercially important crab fishery resources from Arukkattuthurai to pasipattinam, South east coast of India. J Marine Sci Res Dev, 2(110), p.2.

Varadharajan, D. and Soundarapandian, P., 2013. Portunid crab fishery resources from Nagapattinam coast, south east coast of India. Journal of Marine Science Research and Development, 3(3), pp.1-3.

Vidhya, V., Jawahar, P. and Karuppasamy, K., 2017. Annotated check list of the brachyuran crabs (Crustacea: Decapoda) from Gulf of Mannar region, south east coast of India. Journal of Entomology and Zoology Studies, 5(6), pp.2331-2336.

Von Sternberg, R. and Cumberlidge, N., 2001. On the heterotreme-thoracotreme distinction in the Eubrachyura de Saint Laurent, 1980 (Decapoda Brachyura). Crustaceana, pp.321-338.

Wafar, M., Venkataraman, K., Ingole, B., Ajmal Khan, S. and LokaBharathi, P., 2011. State of knowledge of coastal and marine biodiversity of Indian Ocean countries. PLoS one, 6(1), p.e14613.

Walton, M.E., Le Vay, L., Lebata, J.H., Binas, J. and Primavera, J.H., 2006. Seasonal abundance, distribution and recruitment of mud crabs (Scylla spp.) in replanted mangroves. Estuarine, Coastal and Shelf Science, 66(3-4), pp.493-500.

Wardiatno, Y. and Fahrudin, A., 2015. Sexual maturity, reproductive pattern and spawning female population of the blue swimming crab, Portunus pelagicus (Brachyura: Portunidae) in East Lampung Coastal Waters, Indonesia. Indian Journal of Science and Technology, 8(7), pp.596-607.

Yeo, D.C., Ng, P.K., Cumberlidge, N., Magalhaes, C., Daniels, S.R. and Campos, M.R., 2007. Global diversity of crabs (Crustacea: Decapoda: Brachyura) in freshwater. In Freshwater animal diversity assessment (pp. 275-286). Springer, Dordrecht.

Section 3
Fish Processing Fisheries Engineering and Fishing Technology

11

Pathogenicity of Salmonella and Their Impact in Seafood Rejection

[1]Thamizhselvan Surya, [2]Balasubramanian Sivaraman [3]Geevaretnam Jeyasekaran, [4]Robinson Jeyashakila and [5]Shanmugam Sundhar

[1&5]Department of Fish Quality Assurance and Management, Fisheries College and Research Institute, TNJFU, Tuticorin, Tamil Nadu

[2]Department of Fish Processing Technology, Fisheries College and Research Institute, TNJFU, Tuticorin, Tamil Nadu

[3]Tamil Nadu Dr. J. Jayalalithaa Fisheries University, Nagapattinam, Tamil Nadu

[4]Dean, Dr. MGR Fisheries College and Research Institute, TNJFU Ponneri, Tamil Nadu

---

Abstract

Salmonella is a food borne pathogen causing negative impacts in food safety and also in export. It causes typhoid and non-typhoid salmonellosis (NTS) through food contamination. It is a ubiquitous and hardy bacterium that can survive several weeks in a dry environment. Being a pathogen it survives in a wider range of environment and shown resistance towards a range of environmental stresses (acid stress, cold stress, temperature, starvation and oxidative stress) and anti-microbial peptides. These resistance makes them more strong and virulent towards host and outside environment. The worldwide scenario of seafood rejection reports that the Salmonella is the second major reason for the rejection in international markets. Pathogenicity, virulence, resistance of Salmonella might be the reason for most of the outbreaks and food rejection. Hence, there is a need to know nook and corners of Salmonella in these aspects. This chapter covers the detailed note on pathogenicity, virulence mechanism, virulence factors, stress of Salmonella and the impact of this food borne pathogen in seafood export and risk in human health.

Keywords: *Salmonella*, Seafood borne pathogen. Virulence, Pathogenicity, Outbreak

Introduction

Salmonella is a gram-negative, rod shaped bacteria, facultative anaerobe and non-spore forming. Motile by peritrichous flagella, except *S.* Gallinarum & *S.* Pullorum. *Salmonella* is a genus consists two species called *S.* enterica and *S.* bongori and six subspecies viz., enterica I, salamae II, arizonae IIIa, diarizonae IIIb, houtenae IV, indica VI and bongori V. the sub species were categorized into serotypes based on serological characters like somatic (O) - lipopolysaccharides (LPS) on the external surface of the bacterial outer membrane, flagellin (H) – globular protein peritrichous flagella and capsular (Vi) – polysaccharide - O acetylated **α** 1-4, linked N- acetyl galactosaminuronic acids occurs only in *S.* Typhi, *S.* Paratyphi and *S.* Dublin. *Salmonella* plays an important role in affecting the safety of foods as it survives in a wider range of environmental conditions (Finn et al., 2013). Seafood are often contaminated with *Salmonella* (Duran and Marshall, 2005) and their presence causes negative impact on seafood export. During 2020-21, India exported 1.15 MMT of seafood worth US$ 5.96 billion mainly to the USA, China and EU, which are the major importers of Indian seafood (MPEDA, 2022). Shrimp is the major exported seafood commodity. In recent years, seafood rejection happens frequently due to the presence of *Salmonella* followed by allegedly used antibiotics (Business Standard, 2018; Undercurrent News, 2020 and Southern Shrimp Alliance, 2019). In September 2021, the Food and Drug Administration and Centre for Disease Control in USA have reported multistate outbreak of *Salmonella* Weltevreden associated with frozen shrimp of Avanti seafood, India (FDA, 2021). From 2018 to 2020, totally 219 consignments of India were rejected and the majority of the rejections was due to the presence of *Salmonella* (>35%) (MPEDA, 2020).

Salmonellosis

Salmonellosis is the term collectively given to typhoidal and non-typhoidal illnesses of *Salmonella* and are of greater concern in public health in developing countries. It is a major cause of diarrheal disease globally and is estimated to cause 93 million enteric infections and 155,000 diarrheal deaths each year (Majowicz et al., 2010). Being an intracellular pathogen, Non Typhoidal *Salmonella* (NTS) infections affects infants, elderly people and immunocompromised people to (Scallan et al., 2011). Most people who get ill from *Salmonella* infection have diarrhea, fever, and stomach cramps. Food is the main source for most of these illnesses. Non-typhoidal *Salmonella* (NTS) are major seafood borne pathogens likely to have been associated with seafood consumption especially shrimps (Tusevljak et al., 2012). *Salmonella* is a foodborne bacteria causes around 1.35 million infections and 420 deaths in the United States every year (CDC, 2021).

Distribution

Globally totally >2500 serotypes were found based on their serological characters and the Figure1 shows the graph of serotypes distribution around the world.

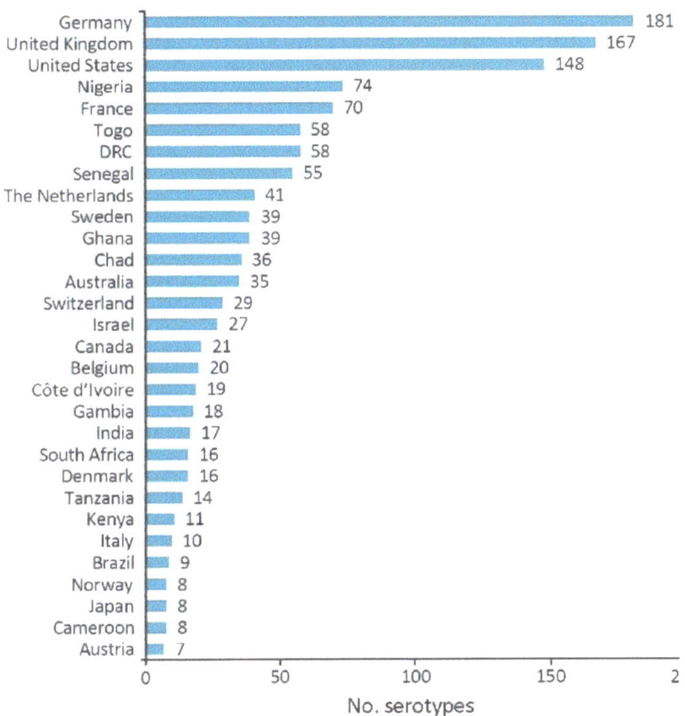

Figure 1 : World-wide distribution of *Salmonella* serotypes

Incidence of *Salmonella* in Seafood

The quality of seafoods can be compromised by contamination with biological agents such as pathogens. Earlier in India, the presence of *Salmonella* in seafoods have been reported by few workers from 1980s (Varma et al., 1985; Gopalakrishna Iyer and Shrivastava, 1989). Kumar et al. (2003) reported the presence of *Salmonella* in tropical seafoods of Mangalore and Cochin, India, respectively. High number of *Salmonella* incidences have been reported in seafood worldwide, in association with the symptoms of fever, nausea, vomiting and diarrhoea (Ling et al., 2002; Asai et al., 2008; Kumar et al., 2009). Koonse et al. (2005) reported the presence of *Salmonella* in the intestinal tract of warm-blooded animals. The presence of *Salmonella* in the aquaculture was mainly due to the introduction of faecal bacteria in the culture ponds (Shabarinath et al., 2007). Ponce et al. (2008) isolated the *Salmonella* enterica serovars Saintpaul and Newport from Indian seafood samples imported into US. Kumar et al. (2009) studied the distribution and phenotypical characterization of *Salmonella* serovars, which were isolated from the samples of water, fish, crustaceans and molluscs from India and found out that Weltevreden and Rissen were the two prominent serovars in seafoods. In 2021 a multistate outbreak of *Salmonella* Thompson infections was reported in relation with processed seafood at Colorado, United States. The environmental samples collected from the concerned seafood industry were also reported to carry *Salmonella* Thompson (FDA, 2021).

Impact of *Salmonella* on Seafoods

Most of Indian seafoods are processed as block frozen and mainly exported to Europe, Japan, and the USA. Out of the total export of frozen seafood, shrimps contributed approximately 68% by value. The frozen shrimp product exports have faced many problems in the past due to the presence of *Salmonella* (Abdallah et al., 2014). During 2020-21, India exported 1.15 MMT of seafood worth US$ 5.96 billion mainly to the USA, China and EU with shrimp as the major exported seafood (MPEDA, 2022). In which frozen shrimp contributes about 68% followed by frozen fish with 10% in total export value. Despite the export value of the Indian frozen shrimp in importing countries faces rejection due to *Salmonella* contamination. USFDA has refused 60 entry lines of shrimp for *Salmonella* in 2018, of which the majority originated from India," said the Southern Shrimp Alliance (SSA), a US-based body of shrimp fishers and processors (Business Standard, 2018). The USFDA also refused six entry lines of shrimp from India due to the presence of *Salmonella* in January 2019 (Southern Shrimp Alliance, 2019). In late 2019, the USFDA refused 11 entry lines of shrimps due to *Salmonella*, including two from India, seven from Indonesia, and one each from the Philippines and Vietnam (Undercurrent News, 2019).

Salmonella Outbreaks Associated with Foods

In 2021 a multistate outbreak of *Salmonella* Thompson infections was reported in relation with processed seafood at Colorado, United States. The environmental samples collected from the concerned seafood industry were also reported to carry *Salmonella* Thompson (FDA, 2021). In Europe, out of 4,362 food-borne outbreaks, *Salmonella* spp. accounts for 21.8% of all outbreaks in 2015 (Eng et al., 2015).

Salmonella Infection in Host

Pathogenesis of Infection

Pathogenesis of *Salmonella* is governed by various pathogenicity island i.e. SPIs, the secretion system of the bacterium and virulence plasmids. These SPIs are acquired through horizontal gene transfer (HGT) during the bacterial evolution and majorly contribute to survival, virulence, and dissemination of pathogens. So far 23 SPIs have been described but SPI-I and SPI-II are majorly important virulence determinant of *Salmonella* (Hurley et al., 2014). In *Salmonella* majority of virulent genes are present in horizontally acquired pathogenicity islands called *Salmonella* pathogenicity islands (SPIs) and plays an important role in infection, their intracellular survival and also there are various regulatory systems present in the *Salmonella*.

Fimbrae H antigen – adhesion

↓

Salmonella invades epithelial cells

↓

Colonizes the Payer's patches associated lymphoid tissue.

↓

Phagocytised by immune cells such as macrophages and neutrophils.
(Key surface markers)

↓

Sub epithelial stromal cell produce cytokines RANKL, Adhesion to specific M cells

↓

Salmonella-containing macrophages enter the mesenteric lymph node, then they are shuttled to the liver and spleen.

Membrane ruffling

Salmonella has distinct fimbriae – high adherence on intestinal epithelial cells which invades via epithelial cells and causes membrane ruffling. It usually invades Via M cells (microfold cells).

*Ruffling: Disorder or Disarrange

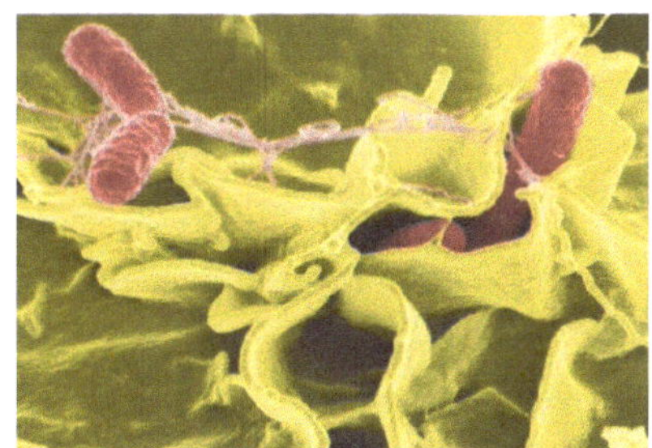

Figure 2 : Membrane ruffling

Host – Pathogen interaction

Innate immunity- recognizes LPS – TLR4, Bacteria lipoprotein –TLR2, Flagellin – TLR5, *TLR: Toll-like receptors (TLRs) are a class of pattern recognition receptors (PRRs) that initiate the innate immune response

Adaptive immunity- Final stage of infection, Th1-type CD40T cells response, production of specific antibodies by B cells. * The T helper cells (Th cells) a type of T cell that play an important role in the immune system, particularly in the adaptive immune system.

Pathogenicity Islands

Pathogenicity islands (PAIs) are gene clusters incorporated in the genome, chromosomally or extra chromosomally, large clusters (30-200 kb) with a distinct GC content.

SPI-1 enables Bacteria Invasion into Host Epithelial Cells

Phenotype

Encodes proteins to assemble a complex type III secretion apparatus. Delivers effectors: to mediate invasion that directly engage host cell signalling pathways. Associated effector genes: sopA,sopB, sopD, invF.

Salmonella Pathogenicity islands – SPI 1

- HilC- RtsA –Hil D: core part of regulation network to control HilA (central regulator of SPI 1) (Hyperinvasion locus)
- HilA activates 2 genes: InvF, sicA
- InvF- Transcriptional activator of SPI 1 T3SS. (It need sicA)
- BarA/sirA- 2 component regulator activates invF.
- glnZ – (glutamine synthase) vital for growth and virulence of *Salmonella.*

Down Regulation

- CsrA- global regulatory RNA binding protein. Post transcriptionally down regulates hilD expression.
- H-NS – DNA binding protein inhibitis HilA- RtsA –Hil .
- Hil E-down regulates expression of SPI-1 by inactivation Hil F.
- Fliz – fimbrial gene occurs under growth condition where H-Ns doen't perfrom.
- Global regulatory system: ArcAB- promotes bac intracellular survival (Arabinose operon regulatory protein).

SPI 1: Secreted Proteins

- Pro-inflammatory proteins: Sip C- form channel in host membrane.
- Sip A- polymerization actin filaments.

- Sop E- Activates Rho family proteins Cdc 4, rac & rho induces membrane ruffles & macropinocytosis.
- Induce inflammation via NF-KB.
- SopB –Increases accumulation of myo inositol phosphate & efflux of chloride & water. (Diarrhea)

Anti-inflammatory Proteins

- Spt P- inactivation Rho GTPase signalling, aposes sopE.
- SSPH 1 & AvrA proteins inhibits NF-KB action.(Nuclear factor-keppa b).

Salmonella has distinct fimbriae – high adherence on intestinal epithelial cells

- Invades epithelial cells - causes membrane ruffling
- Invades Via M cells (microfold cells).
- Bacteria endocytosis mediated – SPI1 T3SS (causes ruffling mediated).

Activates T3SS complex macro molecular machine

Chaperon – helps to direct the effector proteins to host cells.

Macrophages invasion occurs through bacterial mediated macropinocytosis

Adaptation inside depends on PhoP/PhoQ regulatory system

SPI-2 Encodes a Second TTSS that is Essential for Intracellular Growth, and Necessary in the Systemic Phase of Infection

- The smaller 14.5-kb portion containing a group of five ttr genes: Tetrathionate reduction and seven ORFs of unknown function.
- The larger 25.3-kb portion harbours genes important for virulence function. Four types of genes:
  ssa - type III secretion system apparatus),
  ssr - secretion system regulators),
  ssc - secretion system chaperones,
  sse - secretion system

SPI-3 Enterica (*serovar* Typhimurium)

Contains ten ORFs organized into six transcriptional units, two of these genes, mgtC & B encoding the high-affinity magnesium transport system, are required for intramacrophage survival and virulence.

SPI -4 *Enterica (serovar* Typhimurium)

Has 18 putative ORFs, encodes a type I secretion system that mediates toxin secretion, much like *E. coli* hemolysin secretion. SPI 4-dependent mechanism of inducing apoptosis may act in concert with the binding of SipB to caspase-1.

SPI-5 *Enterica (serovar* Typhimurium)

Encode effectors that are induced by distinct regulatory cues and targeted to different T3SS:

- SopB is secreted by T3SS of SPI1
- PipB is translocated by the SPI-2
- TTSS to the *Salmonella*-containing vacuole and *Salmonella*-induced filaments.
- SPI 9- Encodes T3SS & mediate adhesion
- SPI 7- Host specific serovar *S.* Typhi & absent in Typhimurium
- SPI 5- Encodes SopB

Virulence of *Salmonella*

The degree of pathogenicity of a pathogen (bacteria, fungi, or viruses) and is determined by its ability to invade and multiply within the host. Virulence of *Salmonella* encoded by horizontally acquired pathogenicity islands (PAIs) (Hacker and Kaper, 2000)

- Adhesins
- Invasins
- T3SS
- FUR Ferric Uptake Regulator
- Toxin

Virulence Factors, Genes, Functions

The properties (i.e., gene products) that enable a microorganism to establish itself within a host and enhance its potential to cause disease.

Table 1: Virulence factors and responsible genes of Salmonella

Virulence factors	Responsible genes	Functions
Agf (Thin aggregative fimbriae (or curli)) Adherence	csgA, csgB, csgC, csgE, csgF, csgG	Adhesion
Lpf (Long polar fimbriae) Fimbrial, Adherence.	lpfA, lpfB, lpfC, lpfD, lpfE	Attachment to the Payer's patches.
MisL- Adherence; Nonfimbrial structure	misL	Extracellular matrix adhesin involved in intestinal colonization
Pef (Plasmid-encoded fimbriae) – Adherence, Fimbrial	pefA, pefB, pefC, pefD	Mediate binding of the bacteria to the microvilli of enterocytes
RatB- Adherence, Non fimbrial structure	ratB	Involved in intestinal colonization and persistence.
ShdA- Adherence; Nonfimbrial structure;	ShdA	Binding to extracellular matrix proteins fibronectin and collagen I, possibly by mimicking the host ligand heparin
SinH - Adherence; Nonfimbrial structure	sinH	Involved in intestinal colonization and persistence
Type 1 fimbriae	fimA, fimC, fimD, fimF, fimH, fimI, fimW, fimY, fimZ.	Adherence, Fimbrial
MgtBC- Magnesium uptake	mgtB, mgtC	MgtC is essential for intracellular survival/ virulence
Fur (ferric uptake regulator) -Regulation	fur	Iron-regulated genes. Required for acid-induced activation of atr genes.

Adhesion

- Attachment of bacteria to target cell
- Encodes by fimbriae - (pef) plasmid encoded fimbriae
- Long polar fimbriae (lpf)
- Thin aggregative fimbriae (Agf)
- Auto transpoters Mis L & Shd A
- Each mediate adhesion to particular kind of cells due to specificity for receptors.

Adhesion Mechanism

Attachment of bacteria to target cell are encoded by plasmid encoded fimbriae (pef), long polar fimbriae (lpf), thin aggregative fimbriae (Agf).

Stress Undergone by *Salmonella* in Host Environment

1 Acidic pH of stomach: acid tolerance response ATR
2. Intracellular passage of gut epithelia: (SPI2 mediated apoptosis helps *Salmonella* to cross the epithelial lining)
3. Immune responses generated by host: Deviating host defence by retaining one bacterium per SCV to reduce the count of lysosome per SCV.
4. Starvation inside host: Tetrathionate production induced inflammation during *Salmonella* infection, tetrathionate aids in competing with gut micro-flora for electron source and hence better survival.
5. Anti-microbial peptides: Change in lipid A to prevent cationic peptides
6. Others: Bile salts, secretory IgA, mucus

Stress Undergone

1. Acidic pH of stomach: ATR
2. Intracellular passage of gut epithelia: (SPI2 mediated apoptosis helps *Salmonella* to cross the epithelial lining)
3. Immune responses generated by host: Deviating host defence by retaining one bacterium per SCV to reduce the count of lysosome per SCV.

   SPI-2 prevents maturation of SCV into Phagolysosome.
4. Starvation inside host: Tetrathionate production induced inflammation during *Salmonella* infection, tetrathionate aids in competing with gut micro-flora for electron source and hence better survival.
5. Anti microbial peptides: Change in lipid A to prevent cationic peptides
6. Others: Bile salts, secretory IgA, mucus

Stress Management Mechanism

1. Sigma factor
2. Two-component regulatory system
3. Quorum sensing
4. Signalling

Sigma Factor

- Proteins that regulate transcription of gene in bacteria.
- Environmental stress activates σ
- σ factors specifically binds with RNA polymerase

Table 2 : Examples for sigma factor and its respective stress

Sigma factor	Regulator	Expression in respective stress
σ^{38}	rpoS	Starvation
σ^{24}	rpoE	Extreme heat stress extracellular protein
σ^{32}	rpoH	Heat shock proteins

Two Component Regulatory System

Pho P/ Pho Q: regulates group of genes to the survival of the bacterial cells, within the macrophages. Responds to the levels of Mg^{2+} and Ca^{2+} Signals from host environment. Trigger the autophosphorylation of the kinase sensor Pho Q, followed by the phosphorylation of PhoP (PhoQ →PhoP).

*PhoP-activated genes (pag)

- *pag* C - promotes survival within macrophages
- *pag* B is a part of an operon that include the polymyxin resistance locus *pmrA* and *pmrB* genes involved in regulation of multiple factors involving virulence.

Table 3 : Two component regulatory system

Regulatory system	Function
PhoP/PhoQ	Survival of bacteria in macrophages
EnvZ/OmpR	Outer membrane porins in stress
SsrA/ Ssr B	Control SPI-2 gene expression

Siderophores

- *Salmonella* siderophores to compete with host transferrin, lactoferrin and ferritin ligands for utilizing available iron.
- Interferes the functions of electron transport chain and enzymes with iron as co-factor.
- *Salmonella* sequesters Fe^{3+} ions by high affinity phenolate enterochelin (enterobactin).
  *Called as Salmochelin.
- The *fur* gene regulates the synthesis of bacterial sideropores.

Fur – Ferric Uptake Regulator

Transcriptional repressor that regulates gene expression (Troxell et al., 2011, Hantke., 1981, 1984). *Salmonella* contains a number of iron- responsive genes that allow for the uptake and storage of iron, with regulation mediated primarily through the FUR.

Table 4 : FUR regulating genes and their functions

Genes	Functions	Reference
Bfr	Bacterioferritin (Bfr), an iron storage protein and the second gene in an operon with bfd.	McClelland et al., 2001, Ochsner et al., 2002
Bfd	Regulatory or redox component complexing with Bfr in iron storage and mobility	McClelland et al., 2001, Ochsner et al., 2002
sitA	*Salmonella* iron transporter	Zhou et al., 1999
fhuA	Outer membrane protein receptor or transporter for ferrichrome.	Tsolis et al., 1995, Hantke., 1981
fepA-	Outer membrane porin, receptor for ferric enterobactin	Tsolis et al., 1995, Hantke., 1981

Type 3 Secretion System (T3SS): (Coburn et al., 2007)

- Needle like bacterial machinery
- Virulence of several Gram negative pathogens
- Needle 9 Kda in size.
- 60-80 nm in length
- 8 nm external width

T 3SS Proteins Grouped into Three

1. Structural protein: Builds the base, inner rod and needle.
2. Effector protein: secreted into host cells and promote infection.
3. Chaperons: Bind effectors in bacteria cytoplasm. Suppress host cell defense protect them from aggregation & degradation direct them towards needle complex

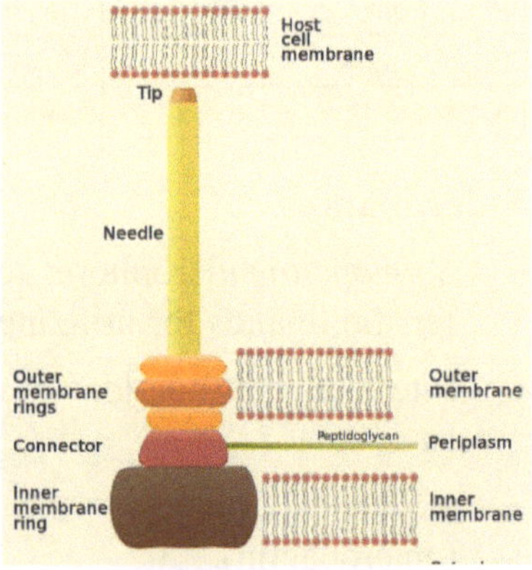

Figure 3 : T3SS

T 3SS Functions – Injectisome

- Delivers protein to host cells
- Needle like probe to detect eukaryotic organism & secrete proteins to infect
- Starts at cytoplasm of bacteria crosses two membrane of bacteria
- Basal body: Part anchored in membrane
- Needle complex similar with bacterial flagella

- Base composed of rings
- The translocated effectors - modify host cell function by disrupting the normal cell-signaling processes

Genes responsible for the function of T3SS are as follows

i) Prg I- Needle monitor
ii) Prg J- Inner rod
iii) Sip D- Needle tip protein
iv) Sip B- Translocator
v) Sic A- Chaperons For 2translocators
vi) InvC- ATPase
vii) InvJ- Ruler protein
viii) SpaS- Switch
ix) InvE- Gate keeper

Vi antigen- Immune Evasion

Related genes: tviA, tviB, tviC, tviD, tviE, vexA, vexB, vexC, vexD, vexE.

- Increase resistance to host peroxide.
- Inhibits the opsonization of the C3b factor to bacterial surface LPS that induce the macrophage phagocytosis of *Salmonella.*
- Vi capsule protects from phagocytosis – intracellular survival

Toxin stn

Enterotoxin – *Salmonella* -structurally and immunologically - cholera toxin and heat-labile enterotoxin in *E. coli.* The release of toxin into the cytoplasm of infected host cells precipitates an activation of adenyl cyclase localized in the epithelial cell membrane and increase the concentration of cyclic AMP in host cells.

Resistance to Antimicrobial Peptides by Mig-14 (Macrophage-inducible gene-14)

An inner membrane-associated protein, necessary for bacterial proliferation in the liver and spleen. Mig-14 -induced within macrophages and is under the control of the global regulator PhoP. Prevent the penetration of the inner membrane by cathelin-related anti-microbial peptide (CRAMP). This resistance is important to the survival of *Salmonella* in systemic sites during both acute and persistent infection (Brodsky et al., 2002).

Conclusion

- Virulence-associated SPI-1 is widely involved in interactions between *Salmonella* and its hosts.
- SPI-1 affects the whole process of pathogenesis, including pathogen invasion, proliferation, and host responses.
- Deeper study on SPI-1 and its complex regulatory network may contribute to drug investigation and *Salmonella* infection control.
- Type 3SS can be used to deliver antigen and elicit immune response against cancer.
- Attenuated *Salmonella* induces cell mediated immune responses.

References

Abdallah, M., Benoliel, C., Drider, D., Dhulster, P., Chihib, N.E., 2014. Biofilm formation and persistence on abiotic surfaces in the context of food and medical environments. Arch. Microbiol. 196, 453–472.

Asai, Y., Kaneko, M., Ohtsuka, K., Morita, Y., Kaneko, S., Noda, H., Furukawa, I., Takatori K., Harakudo Y., 2008. Salmonella prevalence in seafood imported into Japan. J. Food Prot. 71, 460-1464.

Brodsky, I.E., Ernst, R.K., Miller, S.I. and Falkow, S., 2002. mig-14 is a Salmonella gene that plays a role in bacterial resistance to antimicrobial peptides. Journal of Bacteriology, 184(12), 3203-3213.

Bussiness standard, Seafood source 2018, retrived from https://www.business standard. com/article/companies/more-indian-shrimp-shipments-refused-entry-into-us-due-to-Salmonella-118080901057_1.html.

CDC,2021:https://www.cdc.gov/Salmonella/index.html#:~:text=CDC%20estimates%20 Salmonella%20bacteria%20cause,%2C%20fever%2C%20and%20stomach%20cramps.

Coburn, B., Sekirov, I. and Finlay, B.B., 2007. Type III secretion systems and disease. Clinical microbiology reviews, 20(4), pp.535-549.

Duran, G.M., Marshall, D.L., 2005. Ready to eat shrimp as an international vehicle of antibiotic resistant bacteria. J. Food Prot. 68, 2395-2401.

Eng, S.K., Pusparajah, P., Ab Mutalib, N.S., Ser, H.L., Chan, K.G. and Lee, L.H., 2015. Salmonella: a review on pathogenesis, epidemiology and antibiotic resistance. Frontiers in Life Science, 8(3), 284-293.

FDA, 2021. https://www.fda.gov/food/outbreaks-foodborne-illness/outbreak-investigation-Salmonella-thompson-seafood-october-2021.

Finn, S., Hinton, J.C., McCLURE, P.E.T.E.R., Amezquita, A., Martins, M. and Fanning, S., 2013. Phenotypic characterization of Salmonella isolated from food production environments associated with low–water activity foods. J. Food Prot. 76, 1488-1499.

Gopalakrishna Iyer, T.S., Shrivastava, K.P., 1989. Incidence and low temperature survival of Salmonella in fishery products, Fishery Tech. 26, 39-42.

Hacker, J. and Kaper, J.B., 2000. Pathogenicity islands and the evolution of microbes. Annual Reviews in Microbiology, 54(1), 641-679.

Hantke, K., 1981. Regulation of ferric iron transport in Escherichia coli K12: isolation of a constitutive mutant. Molecular and General Genetics MGG, 182(2), 288-292.

Hurley, D., McCusker, M.P., Fanning, S., Martins, M., 2014. Salmonella–host interactions–modulation of the host innate immune system. Front. Immunol. 5, 481.

Koonse, B., Burkhardt III, W.I.L.L.I.A.M., Chirtel, S., Hoskin, G.P., 2005. Salmonella and the sanitary quality of aquacultured shrimp. J. Food Prot. 68, 2527-2532.

Kumar, H.S., Sunil, R., Venugopal, M.N., Karunasagar, I., Karunasagar, I., 2003. Detection of Salmonella spp in tropical seafood by polymerase chain reaction. Int. J. Food Microbiol. 2787, 1–5.

Kumar, R., Surendran, P., K., Thampuran, N., 2009. Analysis of antimicrobial resistance and plasmid profiles in Salmonella serovars associated with tropical seafood of India. Foodborne Pathog. Dis. 6, 621-625.

Ling, M.L., Goh, K.T., Wang, G.C.Y., Neo, K.S., Chua, T., 2002. An outbreak of multidrug-resistant Salmonella enterica subsp. enterica serotype Typhimurium, DT104L linked to dried anchovy in Singapore. Epidemiol. Infect. 128, 1-5.

Majowicz, S.E., Musto, J., Scallan, E., Angulo, F.J., Kirk, M., O'brien, S.J., Jones, T.F., Fazil, A., Hoekstra, R.M., 2010. International Collaboration on Enteric Disease "Burden of Illness" Studies. 2010. The global burden of nontyphoidal Salmonella gastroenteritis. Clin. Infect. Dis. 50, 882-889.

McClelland, M., Sanderson, K. E., Spieth, J., Clifton, S. W., Latreille, P., Courtney,L., et al. (2001). Complete genome sequence of Salmonella enterica serovar Typhimurium LT2. Nature 413, 852–856.

MPEDA, 2022 retrieved from https://mpeda.gov.in/page_id=5581

Ochsner, U.A., Wilderman, P.J., Vasil, A.I. and Vasil, M.L., 2002. GeneChip® expression analysis of the iron starvation response in Pseudomonas aeruginosa: identification of novel pyoverdine biosynthesis genes. Molecular microbiology, 45(5), 1277-1287.

Ponce, E., Khan, A.A., Cheng, C.M., Summage-West, C., Cerniglia, C. E., 2008. Prevalence and characterization of Salmonella enterica serovar Weltevreden from imported seafood. Food Microbiol. 25, 29-35.

Scallan, E., Hoekstra, R.M., Angulo, F.J., Tauxe, R.V., Widdowson, M.A., Roy, S.L., Jones, J.L., Griffin, P.M., 2011. Foodborne illness acquired in the United States—major pathogens. Emerg. Infect. Dis. 17, 7.

Shabarinath, S., Kumar, H.S., Khushiramani, R., Karunasagar, I., Karunasagar I., 2007. Detection and Characterization of Salmonella Associated with Tropical Seafood. Int. J. Food Microbiol. 114, 2227-233.

Southern Shrimp Alliance, 2019. https://www.shrimpalliance.com/fda-cracks-down-on-indian-shrimp-to-begin-2019/

Troxell, B., Fink, R.C., Porwollik, S., McClelland, M. and Hassan, H.M., 2011. The Fur regulon in anaerobically grown Salmonella enterica sv. Typhimurium: identification of new Fur targets. BMC microbiology, 11(1), 1-19.

Tsolis, R.M., Adams, L.G., Ficht, T.A. and Bäumler, A.J., 1999. Contribution of Salmonella typhimurium virulence factors to diarrheal disease in calves. Infection and immunity, 67(9), 4879-4885.

Tusevljak, N., Rajic, A., Waddell, L., Dutil, L., Cernicchiaro, N., Greig, J., Wilhelm, B.J., Wilkins, W., Totton, S., Uhland, F.C., Avery, B., 2012. Prevalence of zoonotic bacteria in wild and farmed aquatic species and seafood: a scoping study, systematic review, and meta-analysis of published research. Foodborne. Pathog. Dis. 9, 487-497.

Undercurrentnews, 2019. https://www.undercurrentnews.com/2019/11/06/us-fda-rejects-13-lines-of-shrimp-in-october-over-Salmonella-antibiotics/

Varma, P.R.G., Mathen, C., Mathew, A., 1985. Bacteriological quality of frozen seafood for export with special reference to Salmonella. Proceedings of Harvest and Post-harvest technology of fish, Soc. Fish. Technol. Cochin, India, 665-667.

Zhou, D., Hardt, W.D. and Galán, J.E., 1999. Salmonella typhimurium encodes a putative iron transport system within the centisome 63 pathogenicity island. Infection and immunity, 67(4), 1974-1981.

12

Overview of Surimi Technology and Its Challenges and Opportunities

Mohammed Akram Javith S.

Department of Post-Harvest Technology, Fisheries Resources, Harvest and Post-Harvest Management Division, ICAR-CIFE, Mumbai, Maharashtra

Abstract

Surimi is a crude myofibrillar protein concentrate prepared by washing minced, mechanically deboned fish muscle to remove sarcoplasmic constituents and fat, followed by mixing with cryoprotectants (usually polyols) to prevent protein denaturation during frozen storage. The species 'Alaskan Pollock' has been the preferred source of surimi production all around the world from time immemorial. Most of the seafood products, especially surimi and related products are among the world's most popular foods today and have a distinct place in stores. Surimi can be used to prepare imitation products like, shrimp analogue, crab analogue, crab leg and also to prepare various food products such as pasta, sausages , bakery products, restructured products. The quality of surimi produced has a direct relationship with the quality of fish used for production; hence necessitating proper maintenance of raw material quality is crucial from the point of catch until further processing. The major problems for the seafood industry concern utilization of natural resources. They include large requirements of freshwater, the negative impact on the environment as a consequence of discharging processing water that has not been adequately treated, and the poor utilization of fish resources. However, recently several cost effective and environmental concern researches have been done on surimi to overcome the challenges in surimi production.

Keywords: Surimi, Raw material, Surimi preparation, Quality assessment, Challenges and Opportunities.

Introduction

Surimi and surimi based seafoods are traditional products of Japan and occupy an important position in the dietary culture of the country. Today, the largest producers of surimi are the United States, Japan and Thailand. Surimi is also manufactured in China, Vietnam and Malaysia. The process of making surimi originated in Southeast Asia and was further developed in Japan in the 16th century. Surimi is a product of Japanese origin, derived from a traditional Japanese way of using and preserving fresh fish. The word 'surimi' comes from the Japanese words 'suru' meaning 'to process' and 'mash/mi' meaning 'meat'. Technically, surimi is the stabilized myofibrillar protein which is obtained by mechanically deboned fish flesh, which is washed, mixed with cryoprotectants, and stored frozen. Washing not only removes fat and undesirable matters such as blood, pigments and odoriferous substances but also increases the concentration of myofibrillar protein. A fish-based product serving the raw material for preparation of analog of seafoods like crab, lobster, scallop & other shellfish. Presently, surimi production uses 2–3 million metric tonnes of fish from around the world, amounting to 2–3 percent of the world fisheries supply.

Raw Material for Surimi Production

According to gel-strength Alaska Pollock has been the predominant fish species used for surimi production. In India, for the preparation of surimi, several authors were utilized the mince of barracuda (*Sphyraena spp.*), threadfin bream(*Nemipterus japanicus*), croaker, Lizard fish, prawn (Metapenaeus dobsonii) and tilapia (*Oreochromis mossabicus*).

It is important to distinguish between surimi and surimi product:

- **Surimi**: frozen block of fish protein made from different fish species,
- **Surimi-based product** or **surimi product**: fresh or frozen final product, such as the popular imitation crabmeat sticks made of surimi mixed with other raw material.

The Principal Steps in the Surimi Production are

- Sorting the fish by species and by size as necessary.
- De-heading, gutting , de-boning , skinning and filleting the fish to separate the flesh .
- Mincing the fish flesh.
- Washing the fish flesh (many times) to remove undesirable water –soluble materials, such as fats , inorganic salts and some proteins.
- Refining the fish flesh to remove any residual materials such as skin, bones and scales.
- De-watering the fish flesh in a screw press.

- Mixing the fish flesh with cryoprotective compounds, such as sugar.
- After that it involves shaping, packing, freezing, master packing and storage in cold store condition

Surmi Production Method

Raw Material Selection

The important criteria in selection of fish for surimi are:

- Low cost
- White meat
- Non oily
- Abundantly available year round
- Good gelling ability.

Raw Material Receiving

Fresh fish whole, received in insulated vehicle, is unloaded and taken into raw-material receiving room through chute doors. Organoleptic inspection and temperature check of the raw material is carried out to ensure the quality of fish .Fish is drained and weighed. Raw material is iced properly and if necessary stored in the chill room (2 -5 °C).

Sorting & Cleaning

Before processing can begin, the fish must be sorted according to size. This is done so that when the fish are processed by machine the processing speed can be increased and the yield of fillets is raised. Sorting can be performed automatically using either roller or caterpillar type apparatus.

Heading, Gutting and Filleting

It involves beheading, evisceration and excision of the backbone, yielding a boneless fillet. The presence of viscera, gills, heart etc. will affect the quality of surimi and therefore these should be removed. These will influence the quality and quantity of mince.

Mincing

All deboning machines depend on the principle of forcing soft portions through a perforated plate or screen into the interior of a drum while leaving bones, hard cartilage and skin on the exterior to be scraped away. To prevent the skin from passing through, the diameter of the drum perforations should not be larger than 3-4 mm. Deskinned and deboned fillets give cleaner minced meat because blood, membrane and other contaminants have been removed.

Washing

Effiicient washing is the most important step in surimi processing as it will ensure maximum gelling and a colourless and odourless product. Many of the problems with colour, taste and odour, which develop in minced meat are minimized or eliminated when washed. Approximately two thirds of minced fish meat consists of myofibrillar proteins, which are the primary components in the formation of a three-dimensional gel structure. The remaining one-third consists of blood, myoglobin, fat and sarcoplasmic proteins, which impede the final quality of the surimi gels, and is removed by washing.

Dewatering

The dewatering is carried out by screening, by using dehydrators (screw press) or by centrifuging to about 5-10% solids. This process is repeated two or three times. Before the finai dewatering under a screw press, undesirable particles, such as fine bones, scales and connective tissues are removed by the refiner. The screw press which commonly has 0.5-1.2 mm perforation, draws water out with compression to a level of 82-85% moisture, which makes the product to look similar to a fish fillet. It is also common to use a 0.1-0.3% salt mixture of NaCl and CaCl, to facilitate the removal of water from the screw press.

Stabilizing Surimi with Cryoprotectants

The addition of cryoprotectants is important to ensure maximum functionality of frozen surimi because freezing induces Protein denaturation and aggregation. Sucrose and sorbitol, alone or mixed at 9% w/w to dewatered fish meat, serve as the primary cryoprotectants in the manufacture of surimi. In addition a mixture (1:1) of sodium tripolyphosphate and tetrasodium pyrolyphosphate at 0.2-0.3% is commonly used as a synergist to the cryoprotective effects of carbohydrate additives. Cryoprotectants were orginally incorporated into the dewatered meat by a kneader. At present, silent cutters are used because they uniformly distribute cryoprotectants faster and temperature increases during chopping are less.

Rapid Freezing and Packaging

After mixing with the cryoprotectants, the fish meat are formed into a block and put each blocks in a plastic bag and then placed on to a stainless steel tray for freezing. Surimi blocks are placed in a contact plate freezer and held for approximately 2.5 hours or until the core temperature reaches -25°C. After inspection of frozen surimi blocks with a metal detector they are packed in cardboard boxes.

Cold Storage

The packed cardboard boxes are shifted to cold storage and kept at the temperature below -18°C.Generally 24% recovery of raw materials will be achieved after cold storage.

Surimi

Frozen surimi block

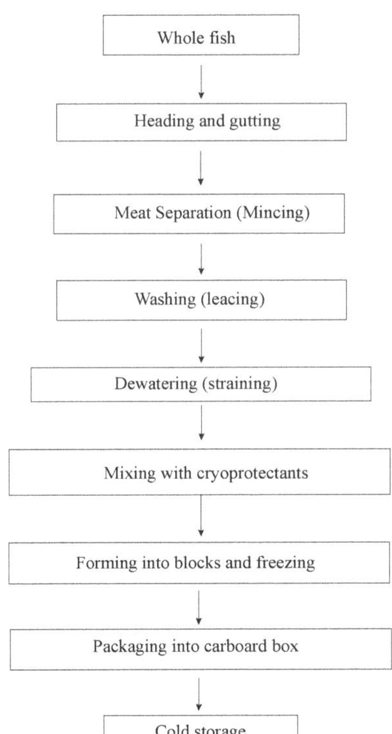

Importance of Cryoproctectants in Surimi

Fish proteins are highly susceptible to freeze denaturation causing it to lose its gel-forming ability . The mechanism of stabilization of fish muscle protein during frozen storage by cryoprotectant is by interaction and bonding with the protein molecules via functional groups on the surfaces. Thus each protein molecule gets covered by hydrated cryoprotectant molecules resulting in increased hydration and decreased aggregation of the proteins. The cryostabilization technique used in surimi manufacture effectively prevents such freeze denaturation. The cryostabilization of fish muscle proteins of surimi is affected by two elements 1) physicochemical factors(of which the effect of leaching the fish mince is most important) and 2) chemical factors, which include the effects of cryoprotectant compounds such as sucrose, sorbitol and phosphates. Physicochemical factors important to the cryoprotection of surimi would include effects of leaching, temperature tolerance of the myofibrillar proteins, pH effects and the freezing/frozen storage/thawing procedures.

Gels prepared with surimi having a low level of cryoprotectant had a sponge-like microstructure with relatively large ice crystal voids, whereas gels made with a high level of cryoprotectant had a more compact and uniform network with smaller and more numerous ice crystal voids. In regards to variation in crystal size within a single sample, a wide difference was noted at the low level of cryoprotectant. The denser microstructure can be attributed to a strong protein-water interaction leading to less randomness and more uniformity of matrix; this causes a more elastic nature in the gel. Formation of a more compact and uniform network at a high level of cryoprotectant resulted from setting of high amounts of solubilized myofibrillar proteins that had been well cryoprotected. That is, the higher level of cryoprotectant produced firmer and more cohesive gels with greater water binding and freeze-thaw stability.

Hexoses(Glucose and fructose) and disaccharides (Sucrose and lactose) were the most effective croprotectants because the pentoses (xylose and ribose) has free aldehyde group and these group bring about the amino carbonyl reaction and so chemically reactive substances are not suitable as cryoprotectants. **Sucrose** has hydroxyl as functional group which can form hydrogen bond around the protein network at low temperature and hydrogen bond is strongest at low temperature so it will prevent the protein from freeze denaturation. **Glycitols such as sorbitol** is widely used in surimi processing because of its excellent cryoprotective properties,relatively low cost and low sweetness. **Phosphate** is added to surimi as a cryoprectant at 0.25-0.3%. The most likely explanation is its function as a metal chelator and/or antioxidant. In addition,because of the strength of the phosphate in raising pH, the holding/ binding of the gel improves and better salt solubilization of myofibrillar protein results. It enhances the cryoprotective effect of sugar. Disodium phosphate, sodium tripolyphosphate or tetrasodium pyrophosphate etc. are used in @ 0.1 to 0.3% by weight.

Quality Assessment Methods of Surimi

Codex criteria for frozen surimi (FAO/WHO 2005) were used for quality assessment of fish surimi

Raw Surimi Tests

Moisture

A sample for moisture content was taken from the interior part of an surimi block to ensure that there was no freezer burn (surface dehydration) of the sample. The test sample was put in a polyethylene bottle, sealed and thawed so that the temperature of the sealed sample increased to room temperature. The AOAC method was used to measure the moisture of the samples. Calculation of the moisture was done according to the following formula to the first decimal place.

$$\text{Moisture (\%)} = \frac{\text{Pre dry weight (g) - after dry weight (g)}}{\text{Pre dry weight}} \times 100$$

pH

To measure pH, 90 ml of distilled water was added to 10 g of the test sample and homogenised. The pH of the suspension was measured with a glass electrode pH meter. Sodium bicarbonate was used for adjusting the pH of the samples.

Objectionable Matter

The term "objectionable matter" as used here means skin, small bones and any objectionable matter other than fish muscle. In this method 10 g of the test sample is spread to the thickness of 1 mm or less, and the number of visible objectionable matters more than 2 mm in diameter is noted. Objectionable matter smaller than 2 mm shall be counted as one half but any objectionable matter smaller than 1 mm shall be disregarded.

Cooked Surimi Tests

Determination of Gel Strength by Puncture Test

The gel strength or gel forming ability is one of the most important factors for quality evaluation of surimi . The puncture test, which is a convenient and easy method to use, was applied to determine this attribute (Park 2004).

The test should be performed between 24 and 48 hours after cooking and after equilibrium with room temperature had been reached. The casing of the inspection sample of cooked gel was removed and samples were cut into test specimens, 15-25 mm in length. They were measured using a Texture Analyzer. A spherical plunger, 5 mm in diameter was dropped at 60 mm/minute. The test specimen was placed on the sample deck of the tester so that the centre of the test specimen would come just under the plunger. The penetration force in g and the deformation at breakage in mm was measured and recorded. Six test specimens were prepared from the same inspection sample of surimi gel and each of them was tested. The average values for all the samples were calculated.

Whiteness

`The colour and whiteness of surimi gel is another important factor which affects the quality of this product especially for East Asian markets. The inspection sample of surimi gel was cut into flat and smooth slices 15 mm in thickness or more. The samples were evaluated immediately with a colour-difference meter instrument by measuring the values of L*(lightness), a* (red-green colours) and b* (yellow-blue colours) to the first decimal place. Three or more sliced pieces were tested. Whiteness, as an index for the general appearance of surimi gel, can be calculated as: **Whiteness = L* - 3b*.**

Determination of water holding capacity by measuring expressible moisture

Water holding capacity of surimi is easily determined by measuring expressible moisture of cooked surimi gel. A small amount of test sample (around 2 g) was placed between 6 filter papers and pressed by pressure equipment (Texture Analyzer) under a fixed pressure (10 kg/cm^2). The expressible water is calculated according to the following formula to the first decimal place:

$$\textbf{Expressible water (\%)} = \frac{\text{Pre pressed weight - After pressed weight}}{\text{Pre pressed weight}} \times 100$$

Water holding capacity is calculated as follows:

$$\textbf{Water holding capacity (\%)} = \frac{\text{Expressible water content}}{\text{Total moisture content of pre pressed sample}} \times 100$$

Sensorial tests for measuring gel strength and elasticity of surimi

Sensory evaluation is convenient and easy to use for the determination of gel strength and elasticity of surimi if it is performed by trained panellists. Folding the test samples by hand and biting samples by the front teeth are commonly used in Japan and South East Asia for sensory valuation of surimi. The folding test should be conducted by folding a 5 mm thick slice of gel slowly in half and in half again while examining it for signs of structural failure (cracks). Three or more slice pieces of the same inspection sample were folded completely in half for 5 seconds and changes in the shape were evaluated using five stage merit marks according to Table 1. The average values for three trials were calculated.

Table 1: Scoring of folding test.

Merit	Mark property
5	No crack occurs even if folded in four.
4	No crack occurs if folded in two but a crack(s) occur(s) if folded in four.
3	No crack occurs if folded in two but splits if folded in four.
2	Cracks if folded in two.
1	Splits into two if folded in two.

The biting test was done by biting 5 mm thick slices of the gel sample with the front teeth to evaluate the resilience and elasticity of cooked surimi. More than three sliced pieces of the same inspection sample were tested by a panellist. Scoring of the biting test is given in Table 2.

Table 2: Scoring of biting test.

Merit mark	Gel quality
10	Extremely strong
9	Very strong
8	Strong
7	Slightly strong
6	Fair
5	Slightly weak
4	Weak
3	Very weak
2	Extremely weak
1	Incapable to form gel

Use of Surimi

Surimi is a useful ingredient for producing various kinds of processed foods. It allows a manufacturer to imitate the texture and taste of a more expensive product, such as lobster tail, using a relatively low-cost material. Surimi is an inexpensive source of protein. In Asian cultures, surimi is eaten as a food in its own right and seldom used to imitate other foods. In Japan, fish cakes (kamaboko) and fish sausages, as well as other extruded fish products, are commonly sold as cured surimi. In Chinese cuisine, fish surimi, often called "fish paste", is used directly as stuffing or made into balls. Fried, steamed, and boiled surimi products also are found commonly in Southeast Asian cuisine. In the West, surimi products usually are imitation seafood products, such as crab, abalone, shrimp, calamari, and scallop. Several companies do produce surimi sausages, luncheon meats, hams, and burgers. Some examples include Salmolux salmon burgers and SeaPak surimi ham, salami, and rolls. A patent was issued for the process of making even higher-quality proteins from fish such as in the making of imitation steak from surimi. Surimi is also used to make kosher imitation shrimp and crabmeat, using only kosher fish such as Pollock.

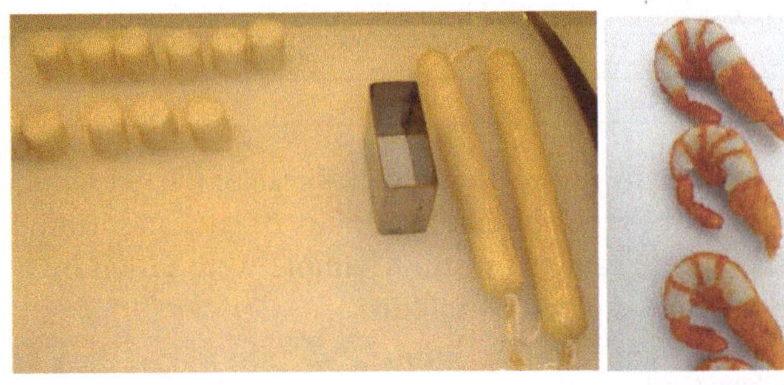

Surimi sausage Shrimp analogue

Fish ball from surimi

Fish burger from surimi

Challenges and Opportunities

In general, the major problems for the seafood industry concern utilization of natural resources. They include large requirements of freshwater, the negative impact on the environment as a consequence of discharging processing water that has not been adequately treated, and the poor utilization of fish resources (Morrissey et al., 2005). Surimi processing can be divided into 2 main stages. The 1st phase (heading, gutting, deboning, and mincing) prepares the fish mince for the 2nd one (washing and refining of mince), and all stages consume water. Surimi processing from white-flesh fish requires a large amount of chilled water because of the extensive washing. The mean consumption for washing is about 27m^3 per ton of surimi (Afonso and B′orquez 2002). The minimum might be less than 10 to 15 L of water for shore-side operations and less than 5 to 7 L for at-sea operations to produce 1 kg of surimi (Park and Lin 2005); and water is often discharged carrying proteins, oils, and other organic materials. Considering that 2 to 3 million metric tons of fish per year (2% to 3% of the world fisheries supply) are used for surimi production the worldwide consumption of water, and the contamination and loss of valuable components are notable. In addition, the use of water is becoming more expensive; therefore, the industry is interested in the reduction of water usage and the improvement of washing efficiency. In fact, during the last 20 y the better ratios of water/meat and the washing cycles have been linked to a more effective washing process. Excessive washing increases the water requirements and also the wastewater, and it results in the loss of myofibrillar proteins. Therefore, increasing the washing time and the number of washing cycles with a lower water/meat ratio would be attainable with the same washing effect with less water (Park and Lin 2005). However, wastewater is the biggest problem for the surimi industry and, therefore, developing the efficient methods for wastewater treatment and protein recovery provide additional income opportunities to the suimi industry

The commercial demand for the white-fleshed fish is higher than for others, and therefore the industry mainly depends on them. The surimi industry also demands white fish mainly because of the importance of the whiteness and textural properties of the resulting products. At the same time, numerous species are underutilized because they are linked to some technological problems. Even conventional processing methods such as canning, salting, drying, and smoking incur limitations with some of the so-called "less valuated fish," which include pelagic fish species, for example, when they are too small. However, according to FAO (2007) the catch of these small species is increasing, whereas some of the most valuable species such as Alaska pollock (*Theragra chalcogramma*) are declining. Hence, the manufacture of surimi can be an alternative to revalue and make use of these fish that are unwanted or unsuitable for other processes.

Recent Research Related to Surimi Technology

Javith et al. (2022) studied the effect of histidine on gelation properties of low salt surimi from tilapia The addition of histidine to low salt surimi gels increased the breaking force and deformation and resulted in the highest gel strength (458 g.cm). Expressible moisture was less in the histidine sample and proteolytic degradation was higher in the control sample indicating that autolysis did not occur in histidine added sample during gel setting. Protein patterns revealed cross-linking of myosin heavy chain in gels with added histidine. Fourier transformed infrared spectra implied that histidine-induced unfolding of proteins occurred after heating. Histidine added samples exhibited a dense and compact microstructure, whereas the control gel was loose. It was reported that the addition of histidine could yield high-quality gels from tilapia surimi in low salt conditions.

Priyadarshini (2018) investigated the effect of different washing methods and natural additives on the quality and stability of tilapia surimi. In this study surimi was prepared by single washing cycle with cold water (T-1), alkaline saline solution(T-2), and with calcium chloride & salt (T-3), respectively, and compared with conventionally washed (CW) surimi. From the results, it was observed that compared to conventional washed surimi, alkaline saline washed surimi with single washing cycle exhibited significantly ($p < 0.05$) highest gel strength of 60.72 N.mm. Significant ($p < 0.05$) decrease in expressible moisture content was also observed in T-2, which suggested higher water holding capacity compared to other treatments. Heat-induced surimi gels exhibited highest L* value ($p < 0.05$), followed by raw surimi in all the treatments. Washing with single cycle by T-2 solution had not only improved the quality of surimi but also can reduced the wastage of water that otherwise released into the environment without further treatment. Application of natural additives such as coconut husk and cluster bean extract increased the gel strength and overall acceptability score for single washed tilapia surimi gel. The optimized concentrations of coconut husk and cluster bean extracts were incorporated into shrimp analogue to evaluate their keeping quality at refrigerated and frozen storage. The shrimp analogues were stable and safe for human consumption for 12 days, 16 days and 20 days for control, T-1 and T-2 samples respectively at refrigerated temperature and 120 days in frozen storage. From this investigation it can be concluded that, single washing step in surimi production with incorporation of natural additives as gel enhancer not only reduced the over exploitation of water resources but also valorized the coconut husk reducing the environmental pollution.

Hassan (2018) studied the utilization of *Pangasius* surimi processing waste as by-products and nutraceuticals. He reported that *Pangasius* wastes generated during its surimi production i.e., skin, bone, head, viscera, surimi wash water can be converted into valuable by-products such as collagen, gelatin, protein hydrolyasate, fish bone powder, oil and sarcoplasmic protein with better quality, functionality and stability by using suitable methods whish was optimized in his study.

Devi (2017) developed the food packaging film using gelatin extracted from surimi refiner discharge and reported that surimi refiner discharge can be exploited as a new source of collagen and gelatin production. Packaging films can be successfully prepared from gelatin recovered from refiner discharge and pulse light can be used as one of the quick methods to modify the properties of gelatin films.

Conclusion

Surimi can be used as raw material for a variety of food products and imitation seafood products. Surimi processing technology has been developed to benefit the less developed countries, which could use the fish resources in their countries. Use of underutilized fish resources for surimi production resulting industry more sustainable and profitable. But on the other hand, surimi is associated with the use of declining fish stocks, large volumes of freshwater, high levels of contaminated wastewater, and poor use of whole fish for human foods. The introduction of novel species in the industry could enable processors to maintain a profitable level of business and minimize some environmental impacts, particularly if they are able to process a range of raw materials with variable qualities. In addition to legislation and enforcement pressure, the effort from the fish industry to increase efficiency and to improve its operations footprints is likely to result in more sustainable practices. With the possibility of the use of more advanced technologies, including HHP, ohmic and microwave heating, advanced decanting and filtration, among others, improvement of current practices and the development of value-added products, mainly using mince from low-cost fishery resources, the surimi industry will have a good future.

Conflict of Interest

There is no conflict of interest

References

Afonso MD, B´orquez R. 2002. Review of the treatment of seafood processing wastewaters and recovery of proteins therein by membrane separation processes: prospects of the ultrafiltration of wastewaters from the fish meal industry. Desalination 142:29–45

Anwar, C., Tsao, C.Y. and Hsiao, H.I., 2013. Effect of cryoprotectants on the quality of surimi during storage at-20°C. Annals Food Science and Technology, 14: 199-205.

Bhargavi priyadarshini M., 2018. Effect of different washing methods and natural additives on the quality and stability of Tilapia surimi. Ph.D. Thesis, ICAR – Central Institute of Fisheries Education (University Under sec. 3 of UGC Act 1956), Panch Marg, off Yari Road, Versova, Mumbai-400 061.

Campo-Deaño, L., Tovar, C.A. and Borderías, J., 2010. Effect of several cryoprotectants on the physicochemical and rheological properties of suwari gels from frozen squid surimi made by two methods. Journal of Food Engineering, 97:457-464.

FAO/WHO. 2005. Codex code for frozen surimi. In: Park, J.W. editor, Surimi and Surimi Sea Food, Boca Raton: Taylor and Francis Group.P.869-885.

FAO, Food and Agriculture Organisation. 2007. The state of world fisheries and aquaculture 2006. Rome, Italy: Food and Agriculture Organization of the United Nations, Electronic Publishing Policy and Support Branch. 162 p.

Hanjambam mandakini devi., 2017Development of food packaging film using gelatin extracted from surimi refiner discharge. Ph.D. Thesis, ICAR – Central Institute of Fisheries Education (University Under sec. 3 of UGC Act 1956), Panch Marg, off Yari Road, Versova, Mumbai-400 061.

Hassan, M.A., 2018. Utilization of Pangasius surimi processing waste as by-products and nutraceuticals. Ph.D. Thesis, ICAR – Central Institute of Fisheries Education (University Under sec. 3 of UGC Act 1956), Panch Marg, off Yari Road, Versova, Mumbai-400 061.

Javith S, M.A., Gunasekaran, J., Xavier, K.M., Nayak, B.B., Krishna, G. and Balange, A.K., 2022. Influence of histidine on gelation properties of low sodium surimi from tilapia (Oreochromis niloticus). International Journal of Food Science & Technology. https://doi.org/10.1111/ijfs.15802

Jeyakumari, A., 2014. Surimi and surimi based products. Central Institute of Fisheries Technology, Cochin.

Lanier, T.C., 1992. Measurement of surimi composition and functional properties. Surimi technology, pp.123-163.

Martín-Sánchez, A.M., Navarro, C., Pérez-Álvarez, J.A. and Kuri, V., 2009. Alternatives for efficient and sustainable production of surimi: A review. Comprehensive Reviews in Food Science and Food Safety, 8:359-374.

Morrissey MT, Lin J, Ismond A. 2005.Waste management and by-product utilization. In: Park JW, editor. Surimi and surimi seafood. 2nd ed. Boca Raton, Fla.: Taylor & Francis Group. p 279–323.

Park, J. W. and Lanier, T. C. 2000. Processing of Surimi and Surimi Seafood. In Marine Freshwater Products Handbook; R. E. Martin, Ed.; Technomic Publishing Company: Lancaster, NH.

Park JW, Lin TM. 2005. Surimi: manufacturing and evaluation. In: Park JW, editor. Surimi and surimi seafood. 2nd ed. Boca Raton, Fla.: Taylor & Francis Group. p 33–106.

Parvathy, U. and George, S., 2014. Influence of cryoprotectant levels on storage stability of surimi from Nemipterus japonicus and quality of surimi-based products. Journal of Food Science and Technology, 51: 982-987.

Shaviklo, G.R., 2006. Quality assessment of fish protein isolates using surimi standard methods. The United Nations University, fisheries training programme.

Yoon, K.S. and Lee, C.M., 1990. Cryoprotectant effects in surimi and surimi/mince-based extruded products. Journal of Food Science, 55:1210-1216.

13

Nucleic Acid Amplification: Alternative Methods of Polymerase Chain Reaction

Nahida Quyoom, Rajendiran Rajeshkannan, A Pradeep and Darshan Pawaskar

Department of Fish Biotechnology, Fish Genetics and Biotechnology Division ICAR-Central Institute of Fisheries Education (ICAR-CIFE), Mumbai Maharashtra

---

Abstract

Nucleic acid amplification is an essential molecular tool in basic research as well as in various application-oriented fields such as gene cloning, clinical medicine development, disease diagnosis etc. Polymerase chain reaction is one of the in vitro methods for amplification of gene of interest. For its simplicity, easy methodology and validated standard operating procedure, it is the standard method of nucleic acid amplification. However, PCR has few limitations, including high cost, sensitivity to certain classes of contaminants & inhibitors etc. These limitations has opened a new portal to alternative methods such as Loop mediated isothermal amplification (LAMP), Nucleic acid sequence-based amplification (NASBA), Rolling circle amplification (RCA), Recombinase Polymerase amplification (RPA) and Selfsustained sequence replication (3SR) etc. The majority of alternative techniques are isothermal precluding require for thermal cyclers. Recently, RNA-guided CRISPR/Cas nuclease-based nucleic acid detection has been widely used for the development of next-generation molecular diagnostics technology due to its high sensitivity, specificity and reliability. However, most of these alternative methods have relatively complex principles than that of PCR but they provide better results.

Keywords: Nucleic acid amplification; PCR; Loop-mediated isothermal amplification (LAMP); Nucleic acid sequence-based amplification (NASBA), Selfsustained sequence replication (3SR), Rolling circle amplification (RCA), Recombinase Polymerase amplification (RPA)

Introduction

PCR is an important molecular tool used widely in different applied research fields such as Infectious disease diagnosis, clinical medicine development and gene cloning etc., (Fang et al., 2008). Among different amplification methods, PCR was the first to be developed and has been the method of choice for its simplicity, easier methodology, extensively validated standard operating procedure and availability of reagents and equipment's. However, it also has a good number of limitations such as advanced equipment, high cost, sensitivity to certain classes of contaminants and inhibitors requirement of thermal cycling etc. (Fakruddin, 2011). These limitations of PCR allowed to develop the alternative methods such as loop mediated isothermal amplification (LAMP) (Notomi et al., 2000), nucleic acid sequence-based amplification (NASBA) (Compton, 1991), self-sustained sequence replication (3SR) (Guatelli et al., 1990), rolling circle amplification (RCA) (Lizardi et al., 1998) etc., most of these techniques are isothermal that don't require thermal cycler. These methods have many advantages over PCR in terms of speed, cost, scale or portability. Recently, RNA-guided CRISPR/Cas nuclease-based nucleic acid detection method has been widely used to develop the next-generation molecular diagnostics technology due to its high sensitivity, specificity and reliability (Yin et al., 2019).

Loop Mediated Isothermal Amplification (LAMP)

LAMP is a simple, rapid, cost-effective and specific isothermal nucleic acid amplification method developed by Notomi et al., 2000. This technique depends on the auto-cycling strand displacement deoxyribonucleic acid (DNA) synthesis, which is performed at 60-65^{0}C for 45-60 min in the presence of *Bacillus stearothermophylus* (Bst) DNA polymerase, certain primers deoxyribonucleotide triphosphate (dNTPs), and the target DNA template. It also has a straightforward visual amplicon detection system. The LAMP method makes use of a DNA polymerase having high strand displacement activity and a group of four precisely generated primers (two inner and two outer primers) that recognize six different DNA sequences on the target. The Production of precursor material, cycling amplification and elongation, and recycling are the three fundamental phases that make up the LAMP amplification reaction's mechanism (Figure 1). Without using expensive equipment's, significant level of precision can really be attained. Compared to conventional PCR and real-time PCR, Sample preparation is straightforward and includes fever steps (Notomi et al., 2000). Change in the fluorescence of the reaction tube can be visualized without the use of expensive specialized equipment as the signal detection system is highly sensitive. There is no need for substantial DNA purification since LAMP has a one-step amplification that takes only 30-60 min and is more resistant than PCR to different inhibitory compounds that are found in clinical samples. Ribonucleic acid (RNA) sequences can be amplified with high efficiency using LAMP in conjunction with reverse transcription (RT). It is highly sensitive, able to locate DNA in the reaction mixture with as little as six copies. LAMP has the potential to be useful in basic research on medicine and pharmacy, point-of-care testing, environmental hygiene and cost-effective diagnosis of infectious diseases (Fakhruddin, 2011). Like PCR, LAMP is also suitable for DNA sequencing, in terms of both Sanger sequencing and Pyro sequencing (Fakhruddin, 2011).

The four primers must hybridize with the target DNA in the first step in order for the LAMP reaction to be effective. This is one of many factors that affect the LAMP reaction's effectiveness. The formation of stem-loop DNA structure from a dumb-bell is crucial for LAMP cycling. Since this technique has a limiting step for amplification is strand displacement DNA synthesis, there is a chance that size of the target can also affect the effectiveness of LAMP, which is should be less than 300bp. Appropriate DNA polymerase selection is also critical for LAMP efficiency. Chemicals that destabilize the DNA helix have been found to markedly elevate amplification efficiencies in LAMP (Notomi et al., 2000).

Advantages of LAMP

LAMP holds great promise to be a method of choices for nucleic acid amplification for its unique advantages. Some of these advantages are,

1. It uses simple cost-effective reaction equipment's (Parida et al. 2008).
2. Both amplification and identification of nucleic acid sequences can be completed in single step.
3. The amplification efficiency is very high
4. The reaction proceeds rapidly as there is no need for initial heat denaturation of the template DNA
5. Isothermal amplification techniques have advantage of tolerance to some inhibitory materials such as a culture media and few biological compounds that also affects the ability of PCR (Kaneko et al. 2005). In LAMP, DNA purification is unnecessary because it is rarely affected by the different components of clinical samples than PCR (Nagamine et al. 2001).
6. Compared to PCR, LAMP is unique, rapid and straightforward to perform.
7. As LAMP technique produces a huge amount of DNA, the products can be identified by simple turbidity.
8. Expensive equipment is not required to provide a significant level of precision when compared to PCR technique.

9. It has extremely high level of amplification efficiency because time loss is not there for thermal change, according to its isothermal reaction.
10 LAMP has ability to amplify specific sequences of DNA under isothermal conditions, thereby precluding the need for a thermal cycler.
11. The approach is less costly
12. The amplification products can be visualized directly.
13. It has only one group of primers for the target DNA amplification
14. LAMP is less prone to the formation of irrelevant DNA than PCR.
15. LAMP in combination with reverse transcription, can amplify RNA sequences with high efficiency.
16. It is extremely sensitive, able to identify DNA with as little as 6 copies in the reaction mixture (Notomi et al., 2000).

Disadvantages

In LAMP primer designing is more difficult than PCR. Also, most detection methods are not sequence specific and it is difficult to run multiplex LAMP reaction in a single tube.

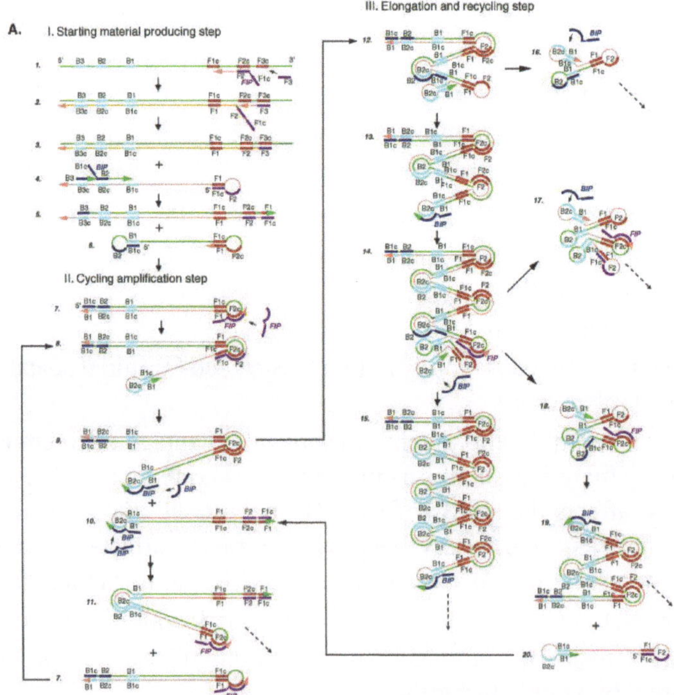

Figure 1: Schematic description of Loop Mediated Isothermal Amplification (LAMP) assay (Notomi et al., 2000).

Nucleic Acid Sequence Based Amplification (NASBA)

NASBA, also known as 3SR (Guatelli, 1990) and transcription arbitrated amplification (Gill, 2008), is an isothermal transcription-based amplification system that is particularly designed for the identification of RNA targets. The complete amplification reaction is performed at the temperature of 41°C. Constant temperature is sustained during the amplification reaction, allowing each and every steps of the reaction to progress immeadiately amplification intermediates are formed. Formation of RNA copies from a provided DNA product by multiple transcription is responsible for the exponential kinetic of the NASBA process. It is more effective than DNA-amplification techniques that are restricted to binary raises per cycle (Sooknanan, 1995). In this method a set of three enzymes (avian myeloblastosis virus reverse transcriptase, RNase H and T7 DNA dependent RNA polymerase) is used, leading to main amplification product of single-stranded RNA (Figure 2) (Deiman et al., 2002). In this technique, a labelled oligonucleotide primer and dideoxy technique using RT are used to sequence the RNA product directly. The target sequence length is limited to 100-250 nucleotides for efficient amplification. NASBA amplicon detection includes the use of enzymatic bead-based detection and electro chemiluminescent (ECL) detection, enzyme-linked gel assay, molecular beacon technology and fluorescent correlation spectroscopy (Sergentet et al., 2008). NASBA shows significant analytical sensitivity than reverse transcription-polymerase chain reaction (RT-PCR) theoretically in identification of pathogen and clinical applications, offering it a well-developed diagnostic tool (Manojkumar et al., 2006).

Advantages of NASBA

NASBA has various advantages compare to other mRNA amplification methods. It can amplify more than 10^9 copies of nucleic acid sequence within 90 minutes with a help of three-enzymes.

1. NASBA is an isothermal reaction carried out at 41°C, precludes the requirement for the thermal cycler (Sergentet et al., 2008).
2. The prime benefit of NASBA is the production of single stranded RNA amplicons that could be employed directly in additional rounds of amplification or can be probed for detection without strand separation or denaturation (Deiman et al., 2002)
3. NASBA is specifically designed to detect RNA and can selectively amplified RNA in the presence of DNA background (Fakhruddin, 2012).
4. DNA amplification by NASBA has also been shown using primers directed against easily obtainable DNA regions including plasmid DNA, low-melting point sequences or single stranded regions (Voisset et al., 2000).

5. The constant temperature sustained during the amplification reaction permits each and every step of the reaction to begin as soon as amplification intermediate becomes available. It has more systematic exponential kinetics process inherently than other DNA amplification techniques which are restricted to binary improves per cycle (Sooknanan and Malek, 1995).
6. Since an RNA virus has RNA as the genomic material, it can be diminishing the chances of contamination and decreasing the time by removing an unnecessary step is reverse transcription (RT) by using RNA-based amplification technique (Loens et al., 2005).
7. It can be used to quantify the DNA virus replication mechanism by identifying late mRNA expression.
8. It helps to diagnose the sequences of human mRNA with absence of the DNA contamination risk.
9. Gene expression studies could be accomplished without DNAses or intron flanking primers.

Disadvantages of NASBA

1. The main concern for NASBA, is RNA integrity (Loens et al., 2005)
2. The amplification reaction is isothermal at 41°C, but a single melting step preceding to the amplification process is an essential to perform annealing of the primers to the target. (Deiman et al., 2002)
3. The specificity of the reactions depends on thermo labile enzymes, so the reaction temperature cannot exceed 42°C without compromising it.
4. It is required the amplified RNA target sequence length should not be more than 120–250 nucleotides because it has chances to amplify the shorter or longer sequences with less efficiently (Loens et al., 2005).

Strand Displacement Amplification (SDA)

SDA (Walker, 1993) is an isothermal amplification method, using four different primers of which a primer having a restriction site (a recognition sequence for HincII exonuclease) is annealed to the DNA template. Primer is elongated using exonuclease-deficient fragment of Escherichia coli DNA polymerase 1 (exo-Klenow). Every SDA cycle comprises of (1) binding of primer to a displaced target fragment, (2) primer/target complex extension by exo-Klenow, (3) nicking the resultant hemiphosphothioate HincII site, (4) dissociation of HincII, and (5) nick extension and repositioned of the downstream strand by exo-Klenow (Figure 3) (Walker, 1993). With this technique, 10^9 copies of the target DNA can be synthesized in a single reaction in less than an hour at high temperatures. Only semi-quantitation is possible by this method. A major limitation of SDA is its inability to amplify long target sequences efficiently (Walker, 1993). SDA forms the base for few commercial identification tests such as BDProbeTec (Becton Dickinson, Franklin Lakes, NJ, and USA) and has been recently evaluated to detect the *Mycobacterium tuberculosis* directly from clinical specimens (McHugh et al., 2004). The robustness and effectiveness of this technology still remains to be proven in the large clinical studies. A recent development for real-time sequence specific DNA target screening utilizing the fluorogenic reporter probes has been reported (Nadeau et al., 1999).

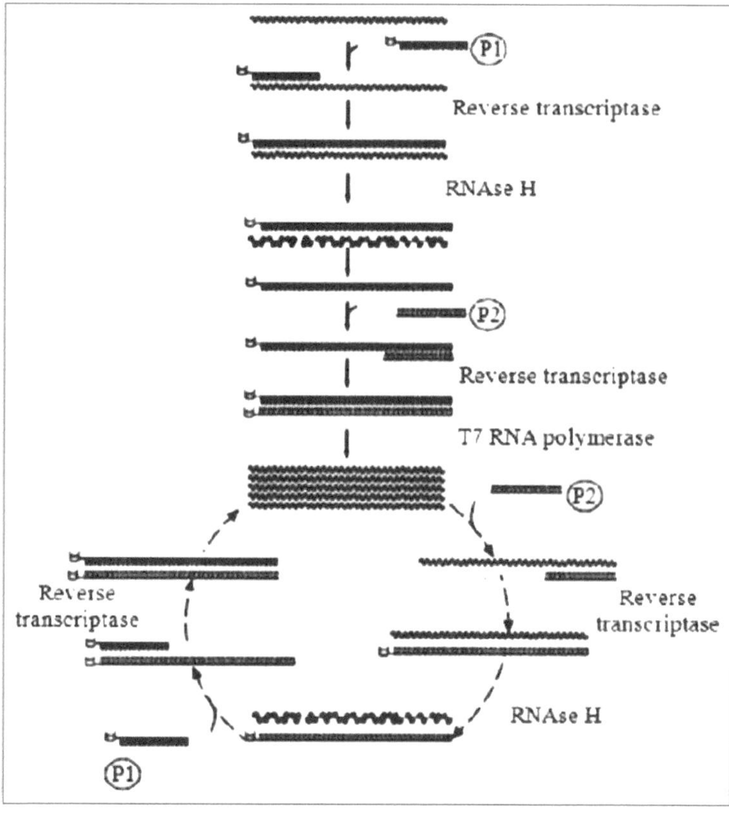

Figure 2: Principles of NASBA (Fakruddin et al., 2013)

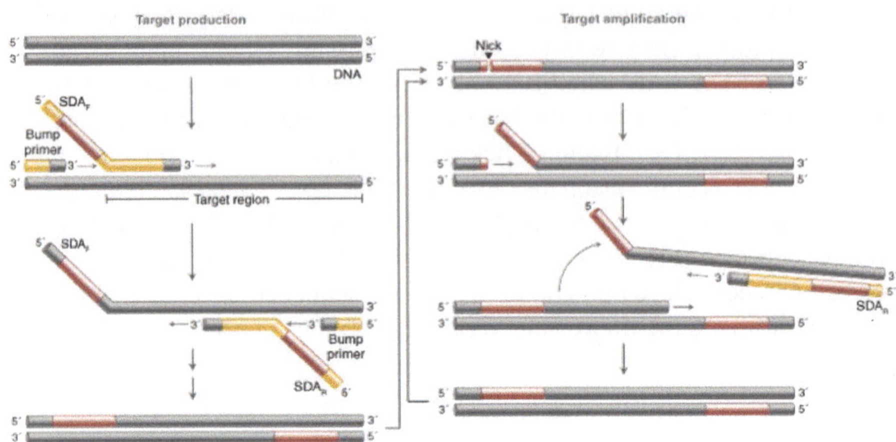

Figure 3: Principle of Strand Displacement Amplification (Walker, 1993)

Multiple Displacement Amplification (MDA)

The MDA is an isothermal, strand-displacing method which is based on the use of the highly processive and strand-displacing DNA polymerase from bacteriophage Ø29, and modified random primers to amplify the entire genome with high-fidelity (Hawkins, 2002 and Hughes, 2005). It can amplify all DNA in a sample from a very small amount of starting material (Dean et al., 2001). MDA involves incubating Ø29 DNA polymerase, dNTPs, random hexamers and denatured template DNA at 30°C for 16-18 h. At 65°C for10 min the enzyme is inactivated and the product DNA can be directly employed in downstream applications (Figure 4) (Morisset et al., 2008). Compared to PCR no repeated cycling is required, but a short initial denaturation followed by the amplification step of 6-18 h and a final inactivation of the enzyme is needed. This method can also be used to produce highly pure DNA, to generate capture probes for microarrays, or even to amplify stored DNA (Lasken et al., 2003) and could be the method of choice when limited amounts of sample are available. Yields and sensitivity are high about 20-30 µg of DNA can be obtained from 1-10 copies of human genomic DNA barely (Dean et al., 2002). MDA can also be performed directly from biological samples such as tissue culture cells and crude whole blood (Lasken et al., 2003). The utility of MDA has not been fully assessed for use in applications such as forensics, sample archiving, and single cell clinical diagnostics (Morisset et al., 2008).

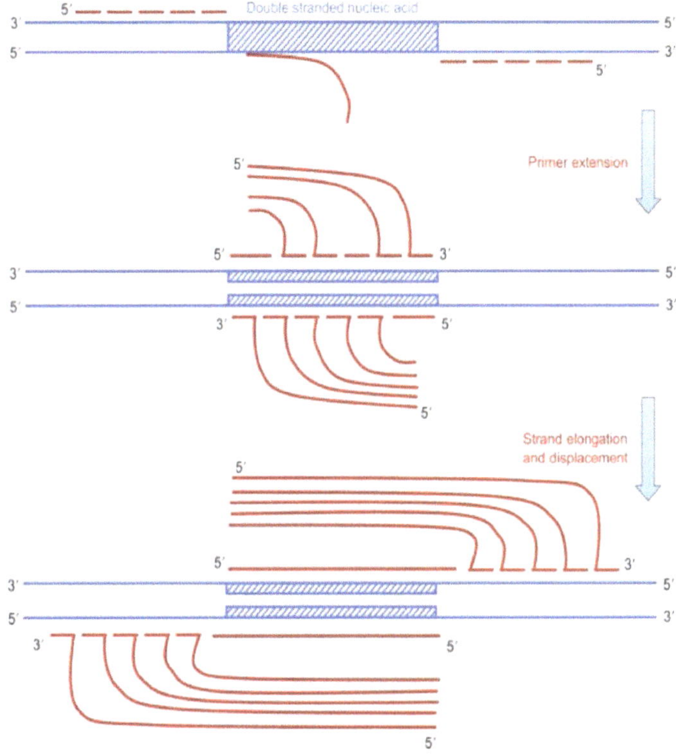

Figure 4: Principle of multiple displacement amplification (Chang-Hui Shen, 2019)

Rolling Circle Amplification (RCA)

RCA is an isothermal nucleic acid amplification technique (Lizardi et al., 1998, Schweitzer et al., 2000) that amplifies the probe DNA sequences more than 10^9-fold both in solution and on the solid phase at a single temperature. It has the ability to detect down to a few target-specific circularized probes in a test sample. In RCA reaction, many rounds of isothermal enzymatic synthesis are involved; Ø29 DNA polymerase extends a circle-hybridized primer by continuously progressing around the circular DNA probe of several dozen nucleotides to replicate its sequence over and over again (Figure 5) (Demidov, 2002). A main advantage of RCA is that, this is resistant to contamination and unlike some other isothermal technologies; it requires little or no assay optimization. The surface-bound amplification products obtained in RCA offer important advantages to in situ or microarray hybridization assays and is suitable in cases where it is essential to maintain morphological information. In linear RCA, the product of amplification remains coupled to the target molecule. RCA amplification allows the localization of signals, thus representing single molecules with specific genetic traits (Dean et al., 2001) or biochemical features (Schweitzer et al., 2000). RCA reactions can express an excellent sequence precision which is used for mutation detection or genotyping and allows identifying DNA markers on the excessive unrelated background (Mothershed et al., 2006). Compared with PCR, the RCA have capability of extreme multiplexity and there are less amplification errors, thus permitting contamination-resistant identification of target molecules in a various kind of testing formats. The efficiency and simplicity of RCA, along with ease and accuracy of quantitation, makes it suitable for miniaturization and automation in the high-throughput analysis (Gusev et al., 2001).

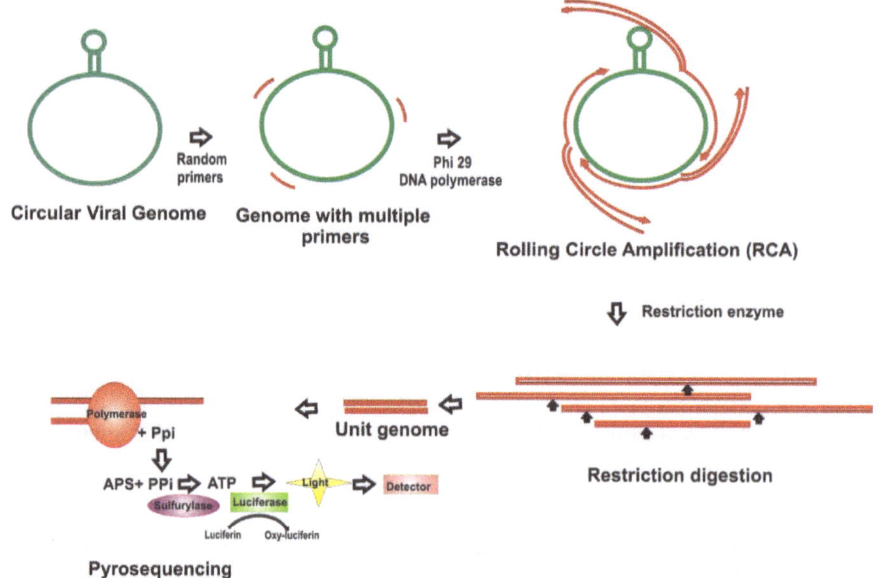

Figure 5: Multiply-primed rolling circle amplification (Ramesh et al., 2013)

Ramification Amplification Method (RAM)

RAM is a new isothermal nucleic acid amplification technique that uses a unique probe (C-probe) that is circular in which the 3' and 5' ends are aggregated in proximity by incorporate to a target. Those two ends are linked covalently with a help of T4 DNA ligase in a target-dependent approach, then resulting a closed DNA circle. Bacteriophage Ø29 DNA polymerase elongates the coupled forward primer along with the C-probe and relocates the downstream strand when the excess of primers exists (forward and reverse primers) and by continuously rolling over the closed circular DNA, it generates a multimeric ssDNA, analogous to the "rolling circle" replication of bacteriophages in vivo.

The single stranded multimeric DNA so formed then provides as a template where multiple reverse primers hybridize, extend and displace downstream DNA and generate a massive ramified (branching) DNA complex (Figure 6) (Zhang et al., 1998). This ramification process proceeds until all ssDNAs are double-stranded and producing an exponential amplification that differs from the non-exponential RCA previously discussed. The name of this technique refers to that fact that various ramification (branching) points, strand displacement and primer extension all contribute to the amplification power. By using a bacteriophage DNA polymerase, Ø29 DNA polymerase, having intrinsic high processivity, it is able to achieve significant amplification within 1 h at 35°C (Beals et al., 2010). RAM is just an isothermal amplification technique that produces vast multimeric products, which helps amplification products efficiently localize in cells while preserving cell morphological properties, making it the best choice for in situ amplification (Hsuih et al., 1996).

Advantages of RAM

The RAM technique presents many advantages than other amplification methods:

1. The reaction can be carried out in isothermal conditions without the need of a thermo cycler because the primers attach easily to ssDNAs that the DNA polymerase displaces.
2. Generic primers replicate all probes with equal efficiency, making a better multiplex ability than conventional PCR, both ends of the probe can be ligated regardless of the nature of target (DNA or RNA), eliminating the need for RT for detecting RNA and creating a uniform assay format for both RNA and DNA detection and
3. Both probe termini are required to hybridize with perfect matching during ligation, permitting the detection of a single-nucleotide polymorphism.

It can be used in clinical laboratories for the identification of genes and infectious agents in various areas, such as Hematology, Oncology, infectious disease, Pathology, forensics, blood banks and genetic disease. In addition, it also has great potential for use in field tests and doctors' offices due to its simple and isothermal amplification format.

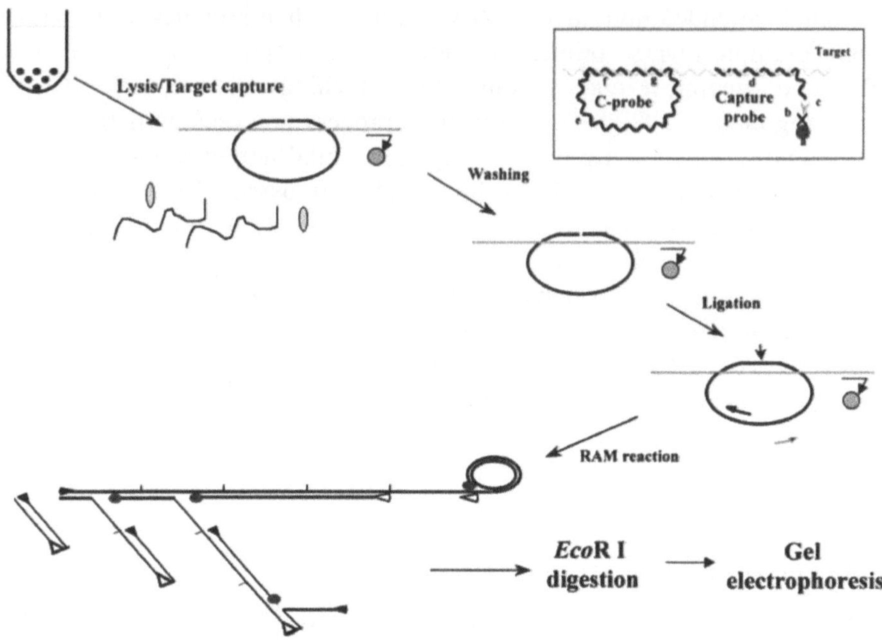

Figure 6: Schematic representation of ramification amplification of ligated circular probe (Fakruddin et al., 2013)

Recombinase Polymerase Amplification (RCA)

RCA represents an adaptable alternative to polymerase chain reaction (PCR) for fast and portable nucleic acid detection (Figure 7). The reaction works optimally at a temperature of around 37°C, capacity of replicating as low as 1 -10 DNA target copies within 20 min. Enzymes known as Recombinase, (T4 UvsX protein) form complexes with oligonucleotide primers assisted by the loaded factor (T4 UvsY) to form a nucleoprotein filament. The resulted complex pairs with their homologous sequences in duplex DNA, forming a D-loop. Polymerase initiates DNA amplification from the primer, when the target sequence is available. Single-stranded DNA binding (SSB) protein binds to the displaced DNA strand, stabilizing the resultant D loop. Once initiated, the amplification process proceeds rapidly, in order that began with just a pair of target copies of DNA, the significantly specific DNA amplification attains accessible levels within minutes. Regardless of template type, the RPA amplicon should be below 500 nucleotides in length for efficient amplification and unlike PCR the length of primer is long (23-35 n).

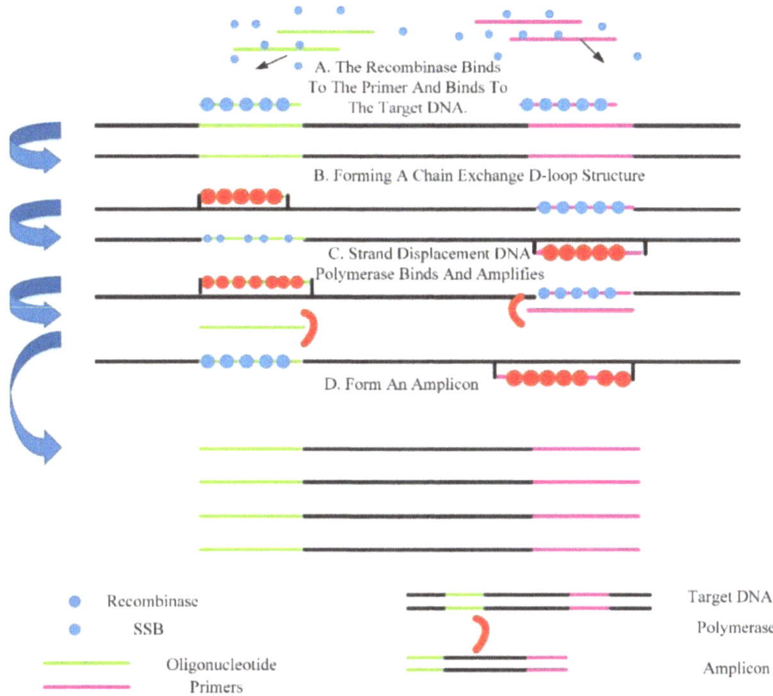

Figure 7: Recombinase Polymerase Amplification (*Lili and Hongli*, 2019)

Cas9 nickase-based Amplification Reaction

Cas9 nickase based amplification is also an isothermal amplification method, combining an RNA-guided Cas9n with precise target site recognition and single-strand nicking abilities and exo- Klenow polymerase with strand-displacement ability. It occurs at a constant temperature of 37°C in single step. Cas9 nuclease can be converted into a DNA nickase when inactivating HNH or RuvC with D10A (HNH) or H840A (RuvC) mutation, offering a versatile tool to introduce a nick at a single strand of duplex DNA. An exo-Klenow polymerase then complements the cut strand, beginning at the first nick and moving toward the second nick, setting the old DNA strand free as it moves. The new DNA is nicked and complemented by the Cas9 nickase complex. The short single DNA strands that are released by this process become the beginning point for further amplification in a second cycle.

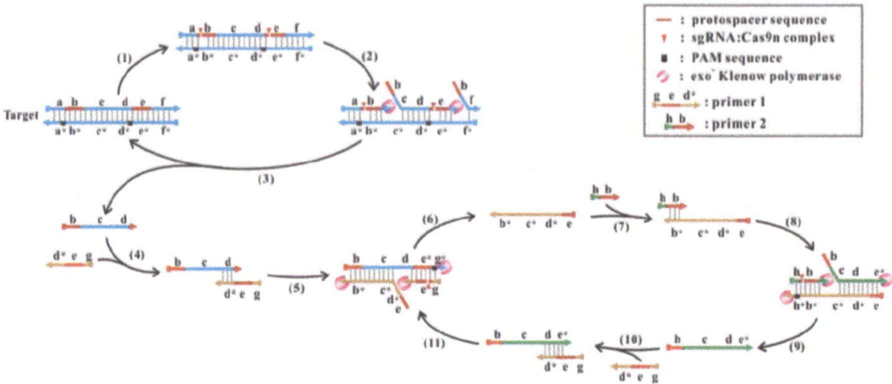

Figure 8: Schematic representation of the Cas9nAR for amplification of a DNA fragment of interest from genomic DNA (Yin et al., 2019).

The working principle of Cas9nAR is illustrated in figure 8. It is composed of two circuits operating in series. The first circuit (steps 1 to 3) is designed to obtain a desired ssDNA sequence from a genomic DNA target. In step 1, a pair of sgRNA: Cas9n complexes targeting domains **b** and **e** (protospacer, denoted in red colour), respectively, introduces two nicks at a desired strand of target genomic DNA. Note that a 5'-NGG-3' PAM (denoted as black rectangle) next to the domains **b** and **e** in the noncomplementary strand should be required for the activation of sgRNA: Cas9n complex. In step 2, exo-Klenow polymerase extends the nick at domain **b** and displaces the nicked ssDNA from domains **b** and **e**. Note that the nick extension at domain **e** is also subjected to the sgRNA: Cas9n complex, theoretically resulting in constant cleavage. The second circuit (steps 4 to 11) is designed to amplify the revealed nicked ssDNA (domains **b** to **d**) from circuit 1 via primer 1 and primer 2. Primer 1 is designed with domain **d*** complementary to domain **d** of the nicked ssDNA at 3' end, domain **e** complementary to domain **e*** of the guide sgRNA, and a 5'-CCN-3' sequence in domain **g**. In step 4, primer 1 attaches to the 3' end of nicked ssDNA to form a duplex with 5' overhangs. Exo- Klenow polymerase extends both 3' ends of the duplex, leading to the generation of the complementary strand of nicked ssDNA and a PAM sequence for Cas9n activation (step 5). Subsequently, sgRNA:Cas9n complex nicks at domain **e** and extends at the nick to displace the synthesized fragment (step 6), which is the template for primer 2. Primer 2, designed with an intact sequence of domain **b** at the 3' end and a 5'-CCN-3' sequence in domain **h**, couples to the displaced ssDNA (step 7) to start a new cycle of extension, nicking, and displacement (steps 8 to 11). Thus, priming, extension, nicking, and displacement steps cycle repeatedly on primer 1 and primer 2-assisted nicked ssDNA amplification from genomic

DNA target. The progression of Cas9nAR can be monitored in real-time by the intercalation dye SYBR Green I, and the fluorescence intensity is directly proportional to the concentration of dsDNA products.

Compared with classical isothermal DNA amplification methods, Cas9nAR exhibits enormous features as follows: It provides a simple DNA amplification scheme performed at one temperature (37°C) for the entire process without the need for DNA denaturation by a heating step. It is a method to amplify a target sequence without length limit, due to the sgRNA: Cas9n complexes. It is highly specific to discriminate a single-nucleotide polymorphism within a gene, due to the site-specific cleavage of sgRNA: Cas9n complexes and sequence-dependent exo- Klenow polymerase extension. This method offers simplicity in primer design and universality in application. Cas9nAR has a great deal of potential to establish itself as a standard assay for the identification and quantification of nucleic acids in basic and applied research due to its exceptional sensitivity and specificity as well as its easy-to-implement, quick and isothermal properties.

Conclusion

We covered the working principle and efficiency of other methods of nucleic acid amplification in this chapter in order to get around PCR's drawbacks. To identify microorganisms, aquatic organisms, and other animals and plants at the species level as well as their physiological responses to external stimuli at the molecular level, it is crucial to locate the different nucleic acid sequences in the target. In basic and applied studies where the PCR is not appropriate, we can amplify the nucleic acid sequences using these alternate approaches.

References

Beals TP, Smith JH, Nietupski RM, Lane DJ. A mechanism for ramified rolling circle amplification. BMC Mol Biol 2010; 11:94.

Chang-HuiShen, 2019. Chapter 9 - Amplification of Nucleic Acids Diagnostic Molecular Biology, Pages 215-247. https://doi.org/10.1016/B978-0-12-802823-0.00009-2.

Dean FB, Nelson JR, Giesler TL, Lasken RS. Rapid amplification of plasmid and phage DNA using Phi 29 DNA polymerase and multiplyprimed rolling circle amplification. Genome Res 2001; 11:10959.

Deiman B, van Aarle P, Sillekens P. Characteristics and applications of nucleic acid sequencebased amplification (NASBA). Mol Biotechnol 2002;20:16379.

Demidov VV. Rollingcircle amplification in DNA diagnostics: The power of simplicity. Expert Rev Mol Diagn 2002; 2:5428.

E. Deltcheva, K. Chylinski, C. M. Sharma, K. Gonzales, Y. Chao, Z. A. Pirzada, M. R. Eckert, J. Vogel, E. Charpentier, Nature 2011, 471, 602- 607.

Fakruddin, M.D., 2011. Loop mediated isothermal amplification (LAMP)–an alternative to polymerase chain reaction (PCR). Bangladesh Res. Publ. J, 5(4).

Fakruddin M, Mannan KS, Chowdhury A, Mazumdar RM, Hossain MN, Islam S, et al. Nucleic acid amplification: Alternative methods of polymerase chain reaction. J Pharm Bioall Sci., 2013;5:245-52.

Gill P, AbdulTehrani H, Ghaemi A, Hashempour T, Amiri VP. Molecular detection of mycobacterium tuberculosis by tHDAELISA DIG detection system. Int J Antimicrob Agents 2007;29:5701.

Guatelli JC, Whitfield KM, Kwoh DY, Barringer KJ, Richman DD, Gingeras TR. Isothermal, in vitro amplification of nucleic acids by a multienzyme reaction modeled after retroviral replication. Proc Natl Acad Sci U S A 1990;87:7797.

Gusev Y, Sparkowski J, Raghunathan A, Ferguson H Jr, Montano J, Bogdan N, et al. Rolling circle amplification: A new approach to increase sensitivity for immunohistochemistry and flow cytometry. Am J Pathol 2001; 159:639.

Hawkins TL, Detter JC, Richardson PM. Whole genome amplification – Applications and advances. Curr Opin Biotechnol 2002; 13:657.

Hsuih TC, Park YN, Zaretsky C, Wu F, Tyagi S, Kramer FR, et al. Novel, ligationdependent PCR assay for detection of hepatitis C in serum. J Clin Microbiol 1996; 34:5017.

Kaneko H., T. Iida, K. Aoki. (2005). Sensitive and rapid detection of herpes simplex virus and varicella-zoster virus DNA by loop-mediated isothermal amplification. J. Clin. Microbiol. 43(7): 3290-3296.

Lasken RS, Egholm M. Whole genome amplification: Abundant supplies of DNA from precious samples or clinical specimens. Trends Biotechnol 2003; 21:5315.

M. Jinek, K. Chylinski, I. Fonfara, M. Hauer, J. A. Doudna, E. Charpentier, Science 2012, 337, 816-821; L. Cong, F. A. Ran, D. Cox, S. Lin, R. Barretto, N. Habib, P. D. Hsu, X. Wu, W. Jiang, L. A. Marraffini, F. Zhang, Science 2013, 339, 819-823; F. A. Ran, P. D. Hsu, C. Y. Lin, J. S. Gootenberg, S. Konermann, A. E. Trevino, D. A. Scott, A. Inoue, S. Matoba, Y. Zhang, F. Zhang, Cell 2013, 154, 1380-1389.

Manojkumar R, Mrudula V. Applications of realtime reverse transcription polymerase chain reaction in clinical virology laboratories for the diagnosis of human diseases. Am J Infect Dis 2006; 2:2049.

McHugh TD, Pope CF, Ling CL, Patel S, Billington OJ, Gosling RD, et al. Prospective evaluation of BDProbeTec strand displacement amplification (SDA) system for diagnosis of tuberculosis in nonrespiratory and respiratory samples. J Med Microbiol 2004; 53:12159.

Morisset D, Stebih D, Cankar K, Zel J, Gruden K. Alternative DNA amplification methods to PCR and their application in GMO detection: Review. Eur Food Res Technol 2008; 227:128797.

Mothershed EA, Whitney AM. Nucleic acidbased methods for the detection of bacterial pathogens: Present and future considerations for the clinical laboratory. Clin Chim Acta 2006; 363:20620.

Nadeau JG, Pitner JB, Linn CP, Schram JL, Dean CH, Nycz CM. Realtime, sequencespecific detection of nucleic acids during strand displacement amplification. Anal Biochem 1999; 276:17787.

Nagamine K., K. Watanabe, K. Ohtsuka, T. Hase, T. Notomi. (2001). Loop mediated isothermal amplification reaction using a non-denaturated template. Clin. Chem. 47: 1742-1743.

Niu Lili and Liu Hongli, 2019. Amplification Techniques of Recombinase & Polymerase and their Application in Parasite Detection. ChinaBiofilms, E3S Web of Conferences 131, 01023. https://doi.org/10.1051/e3sconf/201913101023.

Notomi T., H. Okayama , H. Masubuchi, T. Yonekawa, K. Watanabe, N. Amino, T. Hase. (2000). Loop-mediated isothermal amplification of DNA. Nucl. Acid Res. 28: e63.

Parida M.M. (2008). Rapid and real-time detection technologies for emerging viruses of biomedical importance. J. Biosci. 33: 617–628.

Ramesh S V, Ramesh B, Admane N, Gupta G K & Husain S M. 2013. Multiply Primed Rolling Circle Amplification (MPRCA) of Yellow Mosaic Disease From Infected Soybean In Central Detected The Presence Of Mungbean Yellow Mosaic Indian Virus-[Sb]. Durban South Africa World Soyabean Research Conference (WSRC IX).

Schweitzer B, Kingsmore S. Combining nucleic acid amplification and detection. Curr Opin Biotechnol 2001; 12:217.

Sergentet TD, Montet MP, VernozyRozand C. Challenges to developing nucleic acid sequence based amplification technology for the detection of microbial pathogens in food. Rev Med Vet 2008; 159:51427.

Sooknanan R, Malek LT. NASBA: A detection and amplification system uniquely suited for RNA. BioTechnol 1995;13:5634.

Walker G T, Fraiser M S, Schram J L, Little M C, Nadeau J G, Malinowski D P. Strand displacement amplification – An isothermal, in vitro DNA amplification technique. Nucleic Acids Res 1992; 20:16916.

Xue-en Fang, Jian Li and Qin Chen. (2008). One new method of Nucleic Acid Amplification-Loop-mediated Isothermal Amplification of DNA. Virologica Sinica. 23(3): 167-172. doi: 10.1007/s12250-008-2929-8

Yin, B.-C., Wang, T., Liu, Y., Sun, H.-H., & Ye, B.-C. (2019). A Universal, Isothermal DNA Amplification Method based on RNA-Guided Cas9 Nickase. Angewandte Chemie. doi:10.1002/ange.201901292

Zhang DY, Brandwein M, Hsuih TC, Li H. Amplification of targetspecific, ligationdependent circular probe. Gene. 1998; 211:27785.

14

Recent Trends in Packaging

Angela Brighty. R.J

Post-Harvest Technology, FRHPHM, ICAR-Central Institute of Fisheries Education Mumbai, Maharashtra

Introduction

The increase in population has led to the demand for good quality food. Consumer food demand is a key factor in the development of many agricultural and food policy and high-quality food products with longer shelf lives is increasing all the time. Over recent years, Packaging has revolutionised the marketing and delivery of food goods, especially fishery products. The introduction of modern packaging technology and innovative packaging material processes paved the way for convenience products. In 2018, the value of fish and fish products exported was USD 164 billion because of improved chilling, packaging, and transportation technology has resulted in increased commerce in live, fresh, and chilled fish, which accounted for around 10% of global fish trade and 78% of the total amount shipped was for human consumption (Sofia 2020). When both fish and terrestrial meat exports for human consumption are considered, fish exports have been larger in value terms since 2016. (51% versus 49%). Furthermore, increased utilisation of fisheries and aquaculture products minimises loss and waste, which can assist reduce strain on fisheries resources and support sector sustainability.

Fishery products are highly perishable due to their higher water content, neutral pH, presence of non-protein nitrogenous substances and high content of unsaturated fatty acids. Post-harvest accelerates deterioration through autolysis, melanosis, oxidation of fat and spoilage due to bacterial activity resulting in loss of appearance, taste, texture, colour, odour and flavour. Appropriate packaging technology can be used to prevent these losses and maintain the quality of the product. Only the packaging material or method will not determine the quality so the fish used in the process must be of good quality before processing and packaging. The main function of packaging is to protect the product from undesirable environmental conditions such as heat, oxygen, light, or contamination. Packaging of fish products helps in extending the shelf life, improves quality attributes, and enhances consumer palatability, and product seal ability. Food packaging has traditionally been used for preserving food, communication with consumers, convenience to use and as containment for the product. Distinct items have different packaging requirements; therefore, it is critical to select appropriate packaging material accordingly.

Modified Atmosphere Packaging

MAP was established in 1927 as a method of increasing the shelf-life of apples by storing them in atmospheres with lower O_2 and higher CO_2 concentrations. In the 18th century, Kolbe was the first to research and identify the preserving effect of CO_2 on meat, and Coyne was the first to use MAP to fisheries goods as early as the 1930s. MAP is versatile and also called gas flushing because air is removed but different combinations of gases are used to preserve the product. This is called active MAP. In the case of passive MAP air is not filled in fixed proportion but depending upon the respiration rate of the product the atmosphere in the package will change and so the film is chosen accordingly. MAP technology reduces biochemical processes like lipid oxidation and endogenous enzyme activity thereby preventing bacterial growth causing spoilage. Carbon dioxide (CO_2), oxygen (O_2), and nitrogen (N2) are the most widely employed gases in MAP technology (N2) and are utilised in various combinations, and their functions vary in MAP. The volume of gas to volume of food product (G/P ratio) should be 2: 1 or 3: 1 in most cases (gas: food product). MAP uses natural components of air, no synthetic chemicals as a preservative, no toxic residue is left in the product and is environmentally friendly. Marks and Spencer launched MAP for beef in the United Kingdom in 1979; the popularity of this product led to the development of MAP for ham, fish, cooked shellfish, and sliced cooked meats years later. The effect of these gases depends on concentration, food product, and storage temperature. The solubility of CO_2 has an inverse relationship with temperature and the bacteriostatic effect is dependent on its concentration which acts by altering the permeability and pH of the bacterial cell. If the product absorbs too much CO_2 because CO_2 is soluble in both fat and water in the product, the volume inside the packaging will be minimized, leading to a vacuum package appearance known as 'pack collapse'. Carbon dioxide (CO_2) levels have been suggested to be higher (60%) in fatty fish such as salmon, trout, and tuna and lower in lean fish such as cod, hake, and haddock (Speranza et al. 2009). O_2 is an odourless gas used in fishery products to inhibit the anaerobic organisms, and prevent the reduction of TMAO to TMA and their application is limited because of their ability to oxidize unsaturated lipids present in the fish products and growth of psychrophilic bacteria causing spoilage in presence of O_2. The combination of and is used to retain the colour pigments of oxymyoglobin and eliminate botulism (Babic Milijasevic, Milijasevic, and Djordjevic 2019). Nitrogen is an inert gas used to prevent the growth of aerobic bacteria in presence of O_2 and it has been reported that N2 in fish products packaged in MAP retards oxidative rancidity (Soccol and Oetterer 2003). In general, it is utilised to substitute O_2 and CO_2 in MAP because of low solubility, which helps to prevent pack collapse. For fatty fish, the gas ratio is 60% CO_2+40% N_2 and for lean fish is 40% CO_2+30% O_2+30% N_2.

Reducing Atmosphere Packaging (RAP)

Reducing atmosphere packaging (RAP) is an advancement of MAP. In this technique in addition to CO_2, $O_2$and N_2 gases reducing gas like hydrogen is used in the preservation of the product. Hydrogen is a colourless and flavourless gas permitted to be used as a food additive with a code of E949. Hydrogen gas will reduce the oxidoreduction potential (ORP, Eh) in the package so the oxidation reactions will be reduced by the interaction between free electrons and hydroxyl radicals (Alwazeer, Tan, and Örs 2020). Under low ORP the formation of biogenic amines (histamine, cadaverine, putrescine) is limited because it altered the cell permeability of decarboxylase-producing microbes (Riondet et al. 1999) or extends their lag phase (Bourel et al. 2003). RAP with 4% H2 in rainbow trout and horse mackerel has greatly reduced the production of biogenic amines, specifically heterocyclic, aromatic, and aliphatic di-amines (histamine, tyramine, putrescine, and cadaverine) (Sezer et al. 2022).

Vacuum Packaging

Vacuum packaging is an established technique involving the evacuation of air within a pack before sealing it hermetically. The chemical and biological deterioration of fish products like oxidation of fat, discolouration of pigment, and growth of aerobic spoilage microorganisms can be controlled using vacuum packaging. The packaging material used for the purpose should have excellent oxygen and light barrier properties preferably with O_2 transmission rates of less than 15 cm^3 m^{-2} day^{-1} atm^{-1}. Coextruded or laminated films including OPP/EVOH/PE, PA/PE, PET/PE, OPP/PVDC/PE, OPP/PVDC/OPP, and PVC/EVOH/PVC make up the majority of vacuum packing materials (Yamaguchi, 1990). A method called vacuum skin packaging was created to overcome the drawbacks of conventional vacuum packs in which a flexible plastic laminate is wrapped around a product forming second skin highlighting its natural shape, colour, and texture. Vacuum skin packaging reduces the number of processing stages required resulting in minimum handling during the manufacturing process. Pack integrity is increased and juice exudation is minimised since the bottom and top web films are sealed from the edge of the pack to the edge of the product and are ideally suited for freezing because the second skin prevents freeze burn and dehydration. Vacuum packaging doubles the shelf life of chilled and refrigerated products compared to conventional packaging. Cured fish products like smoked and sausage are called restructured products which are packed by shrink vacuum packing. However, vacuum packaging has some limitations like the growth of anaerobic pathogens, increase in drip loss due to an increase in pressure on muscle filaments, difficulty to use with crisp extruded products and discolouration of pigments on the surface of fish.

Controlled Atmosphere Packaging (CAP)

Controlled atmosphere packaging is a storage technique in which the gas composition inside the package is altered and monitored throughout the storage period to extend the shelf life. CAP product generally has a low concentration of oxygen (1%) and a high concentration of CO_2. CO_2 has an inhibitory effect and the presence of O_2 is required to prevent colour changes. Nitrogen is also used in combination because of the solubility of CO_2 in fish products. For this air exhaustion, a high vacuum chamber with a snorkel machine with programmable logic control (PLC) is used. In this device, a pouch is evacuated under pressure and further inflated under increasing external pressure to reduce tension while the pack is being gassed. Reliably achieving residual oxygen levels of 0.05% with cycle periods of the 30s. CAP products are kept in refrigerated condition and the film used in production is impermeable to gaseous exchange. The storage condition of CAP products is vital in maintaining the quality of the product. White Fish Authority (now Sea Fish Industry Authority and Torry Research Centre has evaluated the use of CAP in fishery products and found that a difference of 2°C could have a significant difference in the final product during storage. Use of O_2 scavengers and CO_2 emitters are also used to improve product quality. Monitoring of gaseous composition is possible in large storage facilities and so CAP application is limited in the case of retail packs. The capital cost, installation and management of CAP machinery and products have impacted their use in storage.

Smart Packaging

Traditional packaging systems are limited to four functions containment, protection and preservation, communication and convenience. Thus, conventional packaging concepts will not provide any details regarding the product's quality and safety. The term "smart packaging" refers to product packaging solutions that include embedded sensor technology for use with food, medication, and many other products. Active and intelligent packaging are both examples of smart packaging. It can keep track of internal and external product changes (intelligent) and intervene in the process (active) to extend the shelf life.

Active Packaging (AP)

Active packaging technology has elements that may discharge or absorb substances from or into the preserved food product or environment to maintain quality and extend shelf life (Arvanitoyannis and Stratakos 2012). AP has already been recognized as a potential improvement in food preservation, particularly for fisheries items. With active packaging, the goal is to prevent the degradation of seafood caused by microbial spoilage and oxidation of lipids (Maisanaba et al. 2017). AP films can be produced based on two methods whether the active chemicals are integrated into the matrix or coated onto the surface of the polymer. The constituents that are widely utilized in AP are Carbon dioxide emitters, oxygen scavengers, ethylene scavengers, Moisture regulators, flavour and odour absorbers/releasers, Antioxidant release and antimicrobial packaging (Nagarajarao, 2016). The technique for incorporating the active ingredient in packaging includes surface modification, coating, and incorporating directly into the package which can alter and enhance product quality (Laorenza et al. 2022). The major concern is the instability of the active agents used in the packaging. Bifunctional AP systems, combine oxygen scavengers and CO_2 emitters such that as oxygen is absorbed, it is replaced by CO_2, which improves the atmospheric conditions inside the package.

Carbon Dioxide Emitters and Absorbers

The use of CO_2 in MAP products is a standardized technology but the solubility of CO_2 in the food product leads to pack collapse which can be rectified by the use of CO_2 emitters. A higher concentration of CO_2 (10-80%) is used to inhibit the growth of microbes on the surface of the food by extending the lag phase so that the generation time of spoiling microbes will be increased thus extending the shelf life of fish and fishery products. Gram^{-ve} bacteria such as pseudomonas are sensitive to CO_2. CO_2 emitting substances include $NaHCO_3$, ascorbic acid, $FeCO_3$, citric acid, etc. When a weak organic acid (such as citric or ascorbic acid) and a carbonate or bicarbonate salt (such as sodium bicarbonate) interact chemically in the presence of water, carbon dioxide is produced as a byproduct (Cocolin, Dolci, and Rantsiou 2008). According to reports, using carbon dioxide emitters is more beneficial than MAP because it can control the G/P ratio (Volume of gas/volume of product) and the technique is simple and the product is pure gas (Hansen et al. 2009). Sachets that act as carbon dioxide absorbers contain either calcium hydroxide/iron oxide and sodium hydroxide, or potassium hydroxide, calcium oxide and silica gel which absorbs the CO_2 to prevent the bursting of the package during storage by respiring food products. They are widely used in poultry, fishery and meat products

Oxygen Scavengers

As a technique of increasing product quality and shelf life, oxygen-absorbing systems are an alternative to vacuum and gas flushing technologies. In the late 1970s, Mitsubishi Gas Chemical Company became the first company to commercialise oxygen absorbers under the trade name Ageless. An increase in O_2 content minimizes the shelf life by oxidation of pigments (myoglobin, carotenoids), rancidity of fats and loss of nutrients by oxidation of vitamin C, and vitamin E. Oxygen scavenger can be introduced into the package as sachet, film, card, and closure liner which absorbs the O_2 in the package throughout the storage period and maintains the concentration at $\leq 0.01\%$ O_2 to prevent oxidative changes. Technologies for scavenging oxygen currently in use make use of one or more of the following ideas: a) ascorbic acid oxidation b) iron powder oxidation c) unsaturated fatty acid oxidation d) photosensitive dye oxidation e) enzymic oxidation (i.e., through the use of glucose oxidase and alcohol oxidase).

Ethylene Scavengers

Ethylene is a volatile unsaturated hydrocarbon and a hormone produced in plants which stimulate physiological functions growth control, maturation, ripening, and accelerating senescence (Zhu et al. 2019). Because of the double bond, gaseous ethylene is a highly reactive chemical that can be altered or broken down in a variety of ways (Chowdhury et al. 2017). Its increase causes discolouration of green vegetables and develops bitter flavours in the fresh produce during storage so the removal of ethylene results in better quality and organoleptic characteristics. The methods for ethylene reduction can be divided into three categories: reduced ethylene content (via MAP using different gases), use of perforated packaging material (to permeate the gases inside and outside the package), and ethylene removal (by scavengers). Ethylene removal is carried out by scavengers (absorption by a chemical reaction between two substances) and absorption (physical adsorption). Ethylene scavenging systems make use of Potassium permanganate, which oxidizes ethylene to acetate and ethanol. They are available in sachets and polymeric film. According to Álvarez-Hernández et al. (2018), natural clay, activated carbon, Palladium, carbon fibre, silica gel, and zeolite are the common physical adsorbents.

Moisture Regulators

Fish has a water content of 70-80% in their body which constitutes its perishable nature due to microbial spoilage and autolysis. Hence humidity and high vapour pressure will increase due to the dripping fluid of fish products in the package leading to quality deterioration. In the case of dried fishery products, the moisture content will directly affect the texture and quality. Desiccants like silica gel, molecular sieves, Calcium oxide, and natural clays (montmorillonite) packaged in sachets can be utilised for dried fish applications. Relative humidity regulators most commonly used are sachets and liquid absorbers forms including blankets, pads, and sheets which acts as desiccants that can absorb moisture thus reducing water activity, reducing degradation of food products related to moisture and inhibiting microbes. These sachets may additionally include iron powder for O_2 scavenging or activated carbon for odour adsorption for dual function (Rooney, 2005). The blankets/sheets are made of many layers between which an absorbent polymer like starch, carboxymethyl cellulose (CMC), or polyacrylate salts are used and they can absorb up to 500 times their weight in water because of their affinity to water which is used during air freight transportation. Moisture absorber pads are used for packed meats, fish and poultry products to absorb the exudate from the tissue. Another method of reducing humidity in the package is the use of humectants (carbohydrate and propylene glycol) between two layers of water-permeable plastic film (polyvinyl alcohol (PVA)) to decrease the water content in the vapour phase thus extending shelf life.

Flavour and Odour Absorber/Releaser

The odour and flavour of the food product give the primary idea about the condition of food quality. Odour and flavour removal has both advantages and disadvantages because some of the flavours are beneficial and give the consumer the aesthetic feel when opening the product but in the latter case, undesirable odour leads to the conclusion that the food is spoiled even though the quality is maintained. The packaging material should act as a barrier to avoid the loss of desirable flavour from the pack. The inclusion of flavour and odour absorber/releaser into the packaging has two functions: to remove the odours from food products developed during storage and the packaging material will release some compounds into the package which is to be removed. In seafood products due to the presence of autolytic enzymes the protein molecules are broken down to produce sulfurous and amine compounds giving undesirable odour to the consumer. Further oxidative rancidity of lipids in fatty fish release aldehydes and ketones causing off-flavour from the product which should be removed. Different substances used for the purpose are cellulose triacetate, acetylated paper, citric acid, ferrous salt/ascorbate and activated carbon/clays/ zeolites in dairy products, cereals, poultry, fruits and fish

Antioxidant Release Packaging

The presence of O_2 in the food package is inevitable and it can degrade the quality of fish products in various ways. Fish lipids are high in unsaturated fatty acids prone to oxidation releasing aldehydes and ketones which give off odours and flavours. Myoglobin in heme proteins is responsible for the red colour of the product when oxidised forms metmyoglobin causing discolouration to the product and reducing its value. Fish is an important source of animal protein and its oxidation is caused either by reactive oxygen species (ROS) or by secondary oxidation products released during lipid oxidation like malonaldehyde. All the changes due to oxidation can be prevented by antioxidants in the packaging material and their release into the food system. Both synthetic and natural antioxidants are used in the packaging. Synthetic antioxidant includes butylated hydroxytoluene (BHT) and butylated hydroxy anisole (BHA) but their use in food systems is of concern because of their nature to accumulate in human adipose tissue causing health effects. To seek safety natural antioxidants like phytic acid, phenolic compounds from seaweeds, chitosan, polyphenols, vitamin C and tocopherol are being replaced in the packaging sector. The effect of antioxidants depends on their size and release into the packaging system (Kuai et al. 2021). The large size of compounds causes slow release and extends the period of preservation. Antioxidants used in the packaging system act by arresting the ROS and adsorbing the free radicals to stop the chain reaction.

Antimicrobial Packaging (AMP)

Recent advancement in packaging is antimicrobial/antibacterial packaging which can inhibit the growth of spoilage and harmful microbes growing on the surface of the food product. AMP provides a dual role in protecting food from the external environment and preserving the quality without degradation. AMP is the complementary process of the conventional preserving technique of fish products like drying, freezing, and thermal processing used to prevent microbial spoilage. The quality of the product is retained and can be applied to fresh fish as it does not alter its nature. Post-harvest handling of fish products causes contamination during all stages of processing which can be controlled by AMP. The use of antimicrobial agents in packaging is done in two ways: the migration of substances from the package onto food products and covalently immobilised agents in the package which will be activated in presence of food moisture. These are not directly used in food preservation because their interaction with food particles will reduce their effect and microbes in food obtained by contamination will cause spoilage on the surface initially which can be prevented by AMP. The antimicrobial agents are coated, coextruded, incorporated or modified along with packaging material for their slow release into the food surface so that the effect stays for a longer period, delaying the spoilage by extending the lag phase or growth phase of microbes to ensure good quality of the product. The antimicrobial classes used in packaging include bacteriocins (Nisin, lactisin), alcohol (ethylene vinyl alcohol copolymer), acid anhydride (sorbic anhydride), plant extract (clove, mustard, rosemary) polysaccharides (chitosan), organic acids (acetic acid, propionic acid), chelators, and enzymes (Lysozyme, Glucose oxidase)(Yildirim, 2011). But due to the laws governing the application of antimicrobial agents in food for human consumption and owing to the high cost of the packaging technology its application is limited. Although natural antimicrobial agents classified as GRAS or permitted as food ingredients are therefore potential for commercial applications.

Intelligent Packaging (IP)

Kerry et al. (2006) reported that intelligent packaging is mostly used to deliver information on the quality of the packaged commodity and monitor its status, such as meat and communicate to facilitate decision making on the quality. IP differs from AP in that it aims to notify the users in the food supply chains (such as producers, merchants, and consumers) about the food's quality rather than directly extending the shelf life of food. IP can perform intelligent actions like detecting, tracing, sensing, recording and communicating certain types of information and to do this activity hardware like gas detectors, freshness and/or ripening indicators, time temperature indicators, barcodes and radio frequency identification (RFID) systems are used (Kerry, O'Grady, and Hogan 2006). Fish products are packed with some headspace which changes over time in their composition and IP is capable of identifying and reporting the quality of food. The indicators used should be activated readily, exhibit a change, irreversible and ideally correlate with the quality of food (Yam, Takhistov, and Miltz 2005).

Freshness Indicators (FI)

Freshness indicators are used to know the quality by giving direct results from the chemical and microbial changes and are efficient intelligent packaging system which gives scientific information on the freshness of the product throughout the supply chain. FI provide the results without disturbing the package and destruction of food in real time compared to chemical methods which are time consuming. Unlike the traditional packaging system, FI offers protection from external sources and conveys the freshness information to the consumers. The biochemical change leads to loss of freshness and different techniques are used in their detection which forms the basis for FI. FI is available in two forms: indicator label and indicator film. The colour change on the FI label indicates the release of spoilage factors from chemical reactions and metabolites from microbial spoilage. The chemical compounds and metabolites that cause the colour change include CO_2, H_2S, NH_3, diacetyl and biogenic amines. CO_2 is produced during microbial metabolism and reduces the pH of the product during storage which indicates the beginning of spoilage. The reduction in pH causes a colour change in the FI label. H_2S are volatile compounds formed from cysteine in fish products by spoilage organisms which when combined with myoglobin of fish muscle form green pigment sulphmyocin that is used in agarose immobilised myoglobin-based FI and the colour change is not affected by CO_2 presence. The change in concentration of organic acids such as n-butyrate, acetic acid, lactic acid and D-lactate formed as metabolites from the breakdown of complex compounds indicate the spoilage of fish products. Total volatile-based nitrogen (TVBN) compounds and biogenic amines (histamine, cadaverine, putrescine) are produced from the degradation of protein in fish muscle by microbial action during the storage period indicating fish products deteriorate. FI can be used to access the freshness of the product effectively.

Time-temperature Indicators (TTI)

The first and most widely used version of IP technology is TTI. It is based on the fact that at high temperatures food deteriorates rapidly due to biochemical and microbial reactions. TTI are available in the form of small labels or stickers that records the time–temperature history of a perishable product like fish which is exposed to different temperatures spanning from processing, packaging, storage and distribution. The changes caused by temperature include physical (melting and diffusion of substances into the product), chemical(oxidation), enzymatic, electrochemical, and microbiological often exhibited as mechanical deformation and colour development (Taoukis and Labuza 2003). There are numerous diffusions, enzymatic, and polymer-based TTIs that are now on the market.

Leakage Indicators

Leakage indicators/gas detectors in the packaging system are used to monitor the changes occurring in the headspace of the can or package giving information on the packaging container's structural integrity throughout storage, handling and distribution. The presence and concentration of $CO_2$and O_2 can be used as an indicator of seal strength, to check the capacity of O_2 scavengers, to access different types of packaging system (vacuum, MAP, CAP) and gives information on the quality of the product. Most of the indicators show colourimetric change due to chemical (redox reactions, pH change, or luminescent dyes) and enzymatic reactions in the food package which are available in the form of tablets, labels or laminated film. Other forms like printed or immobilised layers inside the package are also used, since they are in contact with food, they should be non-soluble and non-toxic. Generally, redox dye (methylene blue, 2,6-dichloroindophenol) along with a base (NaOH/KOH) to maintain the pH and a reducing compound is used as an O_2 indicator which appears pink in absence of O_2 and turns blue in presence of oxygen (Hong and Park 2000). CO_2 indicators function by fast colour reaction consisting of calcium hydroxide (CO_2 absorber) and a redox dye (methyl red) incorporated in the film which changes colour in presence of CO_2. Their high sensitivity to low concentration of O_2 ($<1\%$) and reversible changes in the colour of redox dye upon oxidation and reduction has to be improved in packaging. Smart printing and ink technology advancements will also enable optical systems to automatically read gas indicators using optical devices. The recent development of such ink consists of a UV-absorbing semiconductor (such as TiO_2), redox dye, an electron donor, and a polymer. A blue oxygen indicator film is created using intelligent ink and printed on a variety of substrates; when activated by UV light, the film turns colourless. The reduced methylene blue reoxidizes back to its original blue colour when exposed to oxygen (Lee, Sheridan, and Mills 2005).

Biosensor

Biosensors are compact analytical devices able to detect chemical and microbiological changes in fish products and transmit signals to retailers and consumers during storage, transportation and distribution. A biosensor consists of a bioreceptor that includes, antigens, microbe, nucleic acids, hormones, enzymes or pH responsive dyes which is specific to a target analyte and a transducer that converts the biological response to an electrical signal. Depending on the type of transducer used in the packaging it is classified as an optical, thermal, piezoelectric or electrochemical biosensor (Tang et al. 2011). Intelligent packaging systems having biosensors are gaining importance because of their specificity, portability, reliability, and simplicity. However, an ideal biosensor used in packaging must be able to detect pathogens on line and in real time even in trace amounts for zero tolerance bacteria for consumer safety. Toxin Guard and Food Sentinel System are two examples of biosensors that are used commercially. A visual sensor called Toxin Guard uses antibody-antigen interactions to detect the presence of pathogenic germs. The bacterial toxin is linked to the antibodies and fixed on a thin layer of the film when a pathogenic bacterium is present, which causes the colour shift (Bodenhamer 2002; Bodenhamer, Jackowski, and Davies 2004). SIRA Technology USA has developed a Food Sentinel System which makes use of biosensor-barcode to identify pathogens in food packaging. This technique attaches a specific-pathogen antibody to a portion of the barcode that forms a membrane; when contaminating bacteria are present, this antibody will form a localised dark bar, making the barcode illegible when scanned (Ayala and Park 2000).

Smart Package Devices

It refers to small, inexpensive labels or tags that are attached to primary packaging (such as bottles, trays, and pouches) or more frequently secondary packaging (such as shipping containers) to aid in communication across the supply chain and enable implementation of the necessary actions to achieve desired improvements in food quality and safety. They act as data carriers that are used to transmit and store data, such as barcode labels and radio frequency identification [RFID] tags.

Barcodes

Barcodes inscribe the information about the product in form of bars which are optically read by a machine. Barcodes are less expensive and efficient as data carriers in food packaging. The UPC (Universal Product Code) barcode is a linear symbology introduced in the 1970s to represent 12 digits of data. It stores information about the manufacturer's identification numbers and item numbers. The advantages of 2D barcodes include their compact size, scalability, large data storage, good information accuracy, and great durability. If the information is sensitive, specific encryption technologies can be utilised along with the mistake corrector that is built into 2D barcodes. The PDF 417 (Portable Data File) is a 2-dimensional symbol that fits up to 1.1 kilobytes of data in the space occupied by a UPC barcode (Yam, Takhistov, and Miltz 2005). They can encrypt links to websites and other forms of data that mobile phones can capture. Several standardised 2D barcodes exist, including the DataMatrix and QR codes, as well as the Aztec Code and Nex Code. Recently, Reduced Space Symbology (RSS) has been adopted (Uniform Code Council 2004) and is well suited for product traceability and point-of-sale product identification. The entire 14-digit Global Trade Item Number (GTIN) is encoded by the RSS-14 Stacked Omni-directional barcode, which can be utilised for loose produce goods. The RSS Expanded Barcode in stacked format can store up to 74 alphanumeric characters and is appropriate for variable measure products (such as meat and seafood sold by weight), where a larger data capacity is needed to store additional data like the packed date, batch/lot number, and package weight.

Radio Frequency Identification (RFID)

Radio Frequency Identification (RFID) tags are used to store and wirelessly transmit data about the product. RFID is an electronic tag and an advanced version of an information carrier. This information is used for automatic traceability and product identification to prevent fraudulence, improve the economy, and optimize the retail chain conditions. RFID make use of radiofrequency electromagnetic field waves emitted by a reader to capture information from the RFID tag that is passed on to the computer for decision making. The RFID tag has a tiny microprocessor and a tiny antenna attached to it. RFID systems work across a variety of frequency bands, including low frequency (125 kHz), high frequency (13.56MHz), ultra-high frequency (400-900MHz), and super-high frequency (5.8GHz) (Gregor-Svetec 2018). RFID alone cannot give information about the kinetic change of a product but can be combined with sensors to be used in intelligent packaging. RFID has a large storage capacity and the tag need not be in the line of sight of the reader for data transmission thus saving time. The RFID tag in food products aids in the management of the entire supply chain from production to consumer and plays an important role in product recall (Kumar and Budin 2006). The three types of RFID tags are active, passive, and semi passive. The energy released by the reader powers passive tags, which lack batteries. The reading distance is only about 5 meters. Active tags function at a distance of up to about 50 meters and have their own battery to power the microchip's circuitry and transmit signals to the reader. Semi passive tags are unable to produce a return signal and have a battery to power their microchip (Dobrucka and Cierpiszewski 2014)

Conclusion

The different packaging systems aid in the storage of fishery products during their storage, transport and distribution. Type of packaging system depends on the product. A smart packaging technique increases shelf life and promotes consumer confidence in the product's quality. Integration of smart devices also promotes sustainability and supply chain traceability.

References

Álvarez-Hernández, Marianela Hazel, Francisco Artés-Hernández, Felipe Ávalos-Belmontes, Marco Antonio Castillo-Campohermoso, Juan Carlos Contreras-Esquivel, Janeth Margarita Ventura-Sobrevilla, and Ginés Benito Martínez-Hernández. 2018. "Current Scenario of Adsorbent Materials Used in Ethylene Scavenging Systems to Extend Fruit and Vegetable Postharvest Life." Food and Bioprocess Technology 11 (3): 511–25.

Alwazeer, Duried, Kadir Tan, and Betül Örs. 2020. "Reducing Atmosphere Packaging as a Novel Alternative Technique for Extending Shelf Life of Fresh Cheese." Journal of Food Science and Technology 57 (8): 3013–23.

Arvanitoyannis, Ioannis S., and Alexandros Ch. Stratakos. 2012. "Application of Modified Atmosphere Packaging and Active/Smart Technologies to Red Meat and Poultry: A Review." Food and Bioprocess Technology 5 (5): 1423–46.

Ayala, C. E., and D. L. Park. 2000. "New Bar Code Will Help Monitor Food Safety." Louisiana Agriculture 43 (2): 8–8.

Babic Milijasevic, J, M Milijasevic, and V Djordjevic. 2019. "Modified Atmosphere Packaging of Fish – an Impact on Shelf Life." IOP Conference Series: Earth and Environmental Science 333 (1): 012028.

Bodenhamer, William T. 2002. Method and apparatus for selective biological material detection. United States US6379908B1, filed April 17, 2000, and issued April 30, 2002.

Bodenhamer, William T., George Jackowski, and Eric Davies. 2004. Surface binding of the immunoglobulin to a flexible polymer using a water soluble varnish matrix. World Intellectual Property Organization WO2003093784A3, filed January 17, 2002, and issued March 25, 2004.

Bourel, G., S. Henini, C. Diviès, and D. Garmyn. 2003. "The Response of Leuconostoc mesenteroides to Low External Oxidoreduction Potential Generated by Hydrogen Gas." Journal of Applied Microbiology 94 (2): 280–88.

Chowdhury, Pritom, Madhurjya Gogoi, Sangeeta Borchetia, and Tanoy Bandyopadhyay. 2017. "Nanotechnology Applications and Intellectual Property Rights in Agriculture." Environmental Chemistry Letters 15 (3): 413–19.

Cocolin, Luca, Paola Dolci, and Kalliopi Rantsiou. 2008. "Molecular Methods for Identification of Microorganisms in Traditional Meat Products." In Meat Biotechnology, 91–127.

Dobrucka, R., and R. Cierpiszewski. 2014. "Active and Intelligent Packaging Food - Research and Development - a Review." Polish Journal of Food and Nutrition Sciences 64 (1).

Gregor-Svetec, Diana. 2018. "Chapter 8 - Intelligent Packaging." In Nanomaterials for Food Packaging, edited by Miguel Ângelo Parente Ribeiro Cerqueira, Jose Maria Lagaron, Lorenzo Miguel Pastrana Castro, and António Augusto Martins de Oliveira Soares Vicente, 203–47. Micro and Nano Technologies. Elsevier.

Hansen, Anlaug Adland, Turid Mørkøre, Knut Rudi, Marit Rødbotten, Frøydis Bjerke, and Thomas Eie. 2009. "Quality Changes of Pre rigor Filleted Atlantic Salmon (Salmo Salar L.) Packaged in Modified Atmosphere Using CO_2 Emitter, Traditional MAP, and Vacuum." Journal of Food Science 74 (6): M242-249.

Hong, Seok-In, and Wan-Soo Park. 2000. "Use of Color Indicators as an Active Packaging System for Evaluating Kimchi Fermentation." Journal of Food Engineering 46 (1): 67–72.

Kerry, J. P., M. N. O'Grady, and S. A. Hogan. 2006. "Past, Current and Potential Utilisation of Active and Intelligent Packaging Systems for Meat and Muscle-Based Products: A Review." Meat Science 74 (1): 113–30.

Kuai, Lingyun, Fei Liu, Bor-Sen Chiou, Roberto J. Avena-Bustillos, Tara H. McHugh, and Fang Zhong. 2021. "Controlled Release of Antioxidants from Active Food Packaging: A Review." Food Hydrocolloids 120 (November): 106992.

Kumar, Sameer, and Erin M. Budin. 2006. "Prevention and Management of Product Recall in the Processed Food Industry: A Case Study Based on an Exporter's Perspective." Technovation 26 (5): 739–50.

Laorenza, Yeyen, Vanee Chonhenchob, Nattinee Bumbudsanpharoke, Weerachet Jittanit, Sudathip Sae-tan, Chitsiri Rachtanapun, Wasaporn Pretescille Chanput, et al. 2022. "Polymeric Packaging Applications for Seafood Products: Packaging-Deterioration Relevance, Technology and Trends." Polymers 14 (18): 3706.

Lee, Soo-Keun, Martin Sheridan, and Andrew Mills. 2005. "Novel UV-Activated Colorimetric Oxygen Indicator." Chemistry of Materials 17 (10): 2744–51.

Maisanaba, S., M. Llana-Ruiz-Cabello, D. Gutiérrez-Praena, S. Pichardo, M. Puerto, A. I. Prieto, A. Jos, and A. M. Cameán. 2017. "New Advances in Active Packaging Incorporated with Essential Oils or Their Main Components for Food Preservation." Food Reviews International 33 (5): 447–515.

Nagarajarao, Ravishankar Chandragiri. 2016. "Recent Advances in Processing and Packaging of Fishery Products: A Review." Aquatic Procedia, 2nd International Symposium on Aquatic Products Processing and Health, ISAPPROSH 2015, 7 (August): 201–13.

Riondet, Christophe, Rémy Cachon, Yves Waché, Gérard Alcaraz, and Charles Diviès. 1999. "Changes in the Proton-Motive Force in Escherichia Coli in Response to External Oxidoreduction Potential." European Journal of Biochemistry 262 (2): 595–99.

Rooney, Michael L. 2005. "5 - Introduction to Active Food Packaging Technologies." In Innovations in Food Packaging, edited by Jung H. Han, 63–79. Food Science and Technology. London: Academic Press.

Sezer, Yasemin Çelebi, Menekşe Bulut, Gökhan Boran, and Duried Alwazeer. 2022. "The Effects of Hydrogen Incorporation in Modified Atmosphere Packaging on the Formation of Biogenic Amines in Cold Stored Rainbow Trout and Horse Mackerel." Journal of Food Composition and Analysis 112 (September): 104688.

Soccol, Marcilene C. Heidmann, and Marília Oetterer. 2003. "Use of Modified Atmosphere in Seafood Preservation." Brazilian Archives of Biology and Technology 46 (December): 569–80.

Speranza, B., M.R. Corbo, A. Conte, M. Sinigaglia, and M.A. Del Nobile. 2009. "Microbiological and Sensorial Quality Assessment of Ready-to-Cook Seafood Products Packaged under Modified Atmosphere." Journal of Food Science 74 (9): M473–78.

Tang, Xiaofeng, Yang Liu, Haoqing Hou, and Tianyan You. 2011. "A Nonenzymatic Sensor for Xanthine Based on Electrospun Carbon Nanofibers Modified Electrode." Talanta 83 (5): 1410–14.

Taoukis, P. S., and T. P. Labuza. 2003. "Time-Temperature Indicators (TTIs)." In Novel Food Packaging Techniques, 103–26. Elsevier Ltd.

Yam, Kit L., Paul T. Takhistov, and Joseph Miltz. 2005. "Intelligent Packaging: Concepts and Applications." Journal of Food Science 70 (1): R1–10.

Yamaguchi, Naohiko. 1990. "Vacuum Packaging." In Food Packaging, 279–92. Elsevier.

Yildirim, S. 2011. "Protective Cultures, Antimicrobial Metabolites and Bacterio-phages for Food and Beverage Biopreservation."

Section 4
Fisheries Economics Extension and Statistics

15

Climate Change Vulnerability in Fisheries: Impacts and Adaptation Strategies

V. Suvetha

ICAR-Central Institute of Fisheries Education, Mumbai, Maharashtra

Climate Change - An overview

The most studied topic in recent years was climate change, due to its increasing importance. Climate change refers to long-term shifts in temperatures and weather patterns. These shifts may be natural, such as through variations in the solar cycle or due to anthropogenic activities, since the 1800s, human activities have been the main driver of climate change, primarily due to burning fossil fuels like coal, oil and gas. The process of raising the earth's temperature is known as global warming. Climate change is one result of global warming in different places on the planet. Many people think that climate change means only global warming or an increase in temperatures. But the rise in the temperature is only the beginning of the story. Because in the Earth system everything is connected to one another, changes in one area or in one aspect of the system can influence changes in other areas. The consequences of climate change include flooding, intense droughts, water scarcity, wildfires, rising sea levels, melting polar ice, catastrophic storms and declining biodiversity. Adapting to climate change consequences protects people, businesses, livelihoods, infrastructure and natural ecosystems. It covers current impacts and those likely in the future.

Climate change and fisheries are interconnected, the latter influenced directly or indirectly by the former. Climate change such as rising ocean temperatures, ocean acidification and ocean deoxygenation are radically altering marine aquatic ecosystems, while freshwater ecosystems are being impacted by changes in water temperature, water flow, and fish habitat loss. The context of each fishery is affected by these impacts differently. The availability and trade of fish and fishery products are predicted to undergo major changes as a result of climate change. Peak catch season is becoming harder to anticipate because of climate change, which also reduces the fishing frequency and offshore fishing time. Due to the effects of climate change, fishermen are catching fewer fish, which results in a limited supply of fish.

Climate Change on Coastal Regions

The adaptive capacity of all areas should be improved but must be prioritized now for the most vulnerable population with the least resources to cope with climate change hazards. Coastal regions are adversely affected by many external stressors like cyclones, sea level rise, etc. Around the world, coastal communities are the most sensitive and vulnerable community to climate change. Climate change affects the marine sector by raising sea surface temperatures, increasing the frequency and severity of extreme weather, altering the patterns of precipitation and freshwater runoff brought on by the El Nino and La Nina phenomena, and causing shifts in ocean water circulation patterns and sea-level rise. Coastal areas are more vulnerable to developmental pressures as well as to the impacts of climatic variability, exposing both human populations and ecosystems to the impacts of climate change.

Fishers have been observing the negative impacts of climate change on their fishing livelihoods. The places located in the immediate vicinity of the sea and low in elevation, are particularly vulnerable to sea level rise. A study conducted in the coastal countries around the world revealed that the world's least developed nations are the most vulnerable, and small island developing states make up seven of the 10 most vulnerable nations. The top half of the vulnerability index contains more than 87% of the least developed nations. No such trends are apparent from the exposure or sensitivity indices, which indicates that the main cause of this is the enormous difference in each country's potential for adaptation. The UNDP states that "99% of the casualties from climate change will be in developing countries". The high vulnerability of coastal communities is mainly due to the high resource dependency (depending on the sea) of the community.

Vulnerability

A location, industry or group's level of vulnerability depends on its ecological and socioeconomic attributes. It is also dynamic since its exposure, sensitivity and ability for adaptation change with time, to various stimuli and ecological, social, economic, political, and technological factors. According to this viewpoint, vulnerability assessments call for various techniques at various spatial and temporal scales, in various geographical locations, and for various policy objectives. Due to the complexity of coastal zone dynamics and the long term effects of changing climatic conditions, new, comprehensive, integrated assessment and management techniques at all scales such as national, state and district levels are needed for coastal policy and management.

The phrase "vulnerability" has gained popularity recently. It is connected to social and/or natural hazards: Sea-level rise, coastal flooding, floods, storm surges, global warming, etc. are examples of natural hazards, whereas social hazards include poverty, economic dependency, population, unemployment, etc. The inclination or predisposition to suffer from a hazard is known as vulnerability. It includes a range of ideas and components, such as exposure, sensitivity or susceptibility to harm and a lack of adaptability.

The Intergovernmental Panel on Climate Change (IPCC, 2007) defines "vulnerability" as "the degree to which a system is susceptible to, and unable to cope with, adverse effects of climate change". The IPCC definition of vulnerability – decomposed as a function of exposure, sensitivity and adaptive capacity – tends to be used as a starting point for most analyses, either in its original form or specifically tailored to fisheries and aquaculture. Vulnerability to hazards caused by climate change is a function of the following three key elements: Exposure(E) to the physical effects of climate change, the natural resource system's level of inherent sensitivity(S) or the country's dependence on the sector's social and economic returns, and the degree to which adaptive capacity (AC) enables these potential impacts to be mitigated.

$$\text{Vulnerability} = (\text{Exposure} + \text{Sensitivity}) - \text{Adaptive Capacity}$$

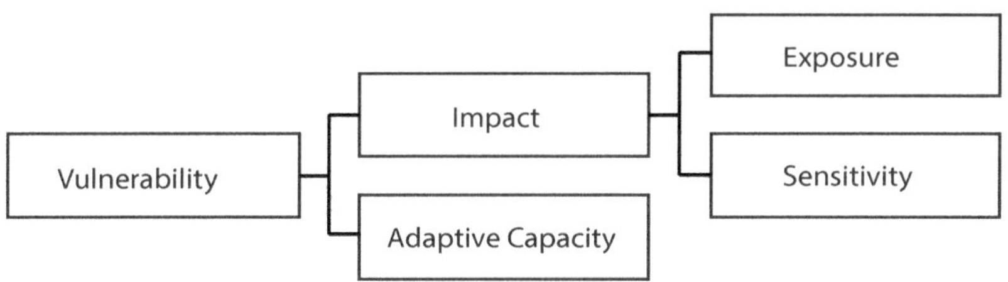

Figure 1 : Vulnerability Concept and its Components

a) Exposure

Exposure is external to the system or community; it is the degree to which an individual or a community or a system to exposed to external stressors like climate change. Think of a seaside town that is only 10 km from the ocean, cyclones, rising sea levels, and saltwater intrusion in aquifers would all affect this settlement. Mainly the geographical location of the community/village influences the exposure.

b) Sensitivity

The degree to which the system is affected by the risks to which they are exposed is known as **sensitivity**. Sensitivity is the degree/extent to which a system is impacted by climate variability or change, either adversely or beneficially. A systems exposure and sensitivity are combinedly known as the **'Impact'.** The "climate impact" on the system is the result of exposure to external stressors and sensitivity to climate change impacts.

Climate impact = Exposure (E) + Sensitivity (S)

Adaptive Capacity

The term "adaptive capacity" describes an individual's, a family's, or a community's potential to become resilient and cope with climatic threats. It's like a defensive mechanism of a system to survive and revive from any vulnerability. Access to financial, technological, educational, and community resources affects one's ability to adapt. Accessibility and Adaptive capacity are positively correlated. The community which has high accessibility to available resources usually has a high ability to cope with the adverse effects of climate change vulnerability. Sensitivity and adaptive capacity internal to the system.

Vulnerability (V) = Impact – Adaptive Capacity (AC)

Exposure and Sensitivity are positively related to vulnerability; it means high sensitivity and exposure increase vulnerability. Adaptive capacity has a negative relationship with vulnerability. High Adaptive capacity reduces the vulnerability of a system. High exposure to climatic stressors, high system sensitivity to climatic risks, and inadequate adaptive capacity to cope with the impacts of changing climate are altogether considered as the high vulnerability of a system or community.

Impacts of Climate Change on Fisheries/Coastal Communities

The fish season becomes unpredictable – One of the main impacts of climate change related to capture fisheries is change/alteration in the fishing season.

Decreased fishing frequency & Reduced fishing time – Climate change adversely affects the marine ecosystem and reduced the fish stocks in the ocean. Because of this reduction in fish stocks, the catch also decreased and the catch per unit effort increased. This scenario is likely to get even worsen. Due to the less availability of fish in the sea, the fishing frequency has decreased.

- Sea level rise
- Changes in ocean currents
- Ocean acidification
- Damage to productive assets like boats and nets
- Changing the distribution of fish stocks and altering the structure of ecosystems

- Cyclones and hurricanes are strongly influenced by sea surface temperatures
- Distribution, abundance, seasonality and fisheries production
- Decreased productivity - Primary production & secondary production
- Changes in stratification, mixing, and nutrients in lakes and marine upwellings
- Coastal profile changes, loss of harbours and homes
- Increased exposure of coastal areas to storm damage

Adaptation Strategies

In order to lessen a system's vulnerability to climate change, proactive adaptation seeks to minimise risk and/or improve the system's resilience. Nicholls and Klein (2005) identified five proactive adaptation goals especially for coastal zones: enhancing societal awareness and preparedness, increasing the robustness of infrastructure layouts and long-term investments; enhancing the adaptability of vulnerable natural systems; and increasing the flexibility of vulnerable managed systems. Coastal adaptation is indeed a complex process and requires more understanding.

For coastal zones, there is another classification of three fundamental adaptation strategies that are frequently used (Smit et al., 2006; Nicholls and Klein, 2005).

1. Protect - to lower the risk of the event by minimizing the likelihood of its occurrence, in other terms, it is the method of reducing the exposure of the community to a particular vulnerability. But it is not possible in real life, because the exposure is primarily due to the geographical location. In the fisheries context, the fishers are likely to reside in the coastal region, which has more exposure to adverse climatic events. To a certain extent, this strategy can be used by increasing the robustness of the coastal region – building dykes and beach nourishment.
2. Accommodate - to increase society's capacity/ability to cope with the effects of the event; the most sensible way of dealing with vulnerability is to increase the adaptive capacity of the coastal communities. This can be achieved by various means as more variables are contributing to increasing the ability of a coastal community to cope with the vulnerability. In simple terms, it is to increase the flexibility of the community to cope with the hardships faced by them.
3. Retreat - to control/reduce the risk of the event by reducing its potential effects, one or other way around it is the way to reduce the sensitivity of the coastal region to climate change vulnerability. It can be achieved

through improved awareness and preparedness about climate change and its impacts.

Other Adaptation Strategies

i. **Sustainable management**: Marine protected areas, ecological restoration and spatial planning - This is to protect the ecological balance in the marine ecosystem. Such as Wetland restoration programmes to increase the adaptive capacity of the dependent communities.

ii. **Accessibility** - Increasing accessibility of the resources to the most vulnerable coastal communities, results in the reduction of vulnerability. Resources include social as well as economic resources. Availability and accessibility to the market, health facilities, community infrastructure, transport facilities, technical knowledge and information, etc., can make the community more resilient and aware of the impacts of climate change.

iii. **Migration of people** - it may not be an affordable solution, but in extreme events it can be followed. Moving people settlement from a location that is more prone to climatic variabilities to a location which is comparatively less exposed and suitable for the community. For example, migrating local fishers to a location which is far from their fishing ground may not be a good option for them.

iv. **Community-based adaptation** - Not all the communities are same in their adaptive capacity and vulnerability. There is a need to develop community-specific or regional-specific coping mechanisms to effectively limit the adverse effect of climate change. Coastal regions are dynamic in nature, within the coastal region there is variability in its physical and geophysical characteristics.

v. **Climate change policies and long-term action plans** - Sensitivity to any climate change hazards cannot be reduced in a short duration. It requires long-term action plans to improve the sensitiveness of the community. Sensitivity is influenced by variables like population density, net sown area, educational status, economic dependency ratio, etc., which are not increased or decreased by short-term plans.

vi. **Alternative and diversified livelihoods** - One of the main impacts of climate change on fisheries is depleting fish stocks. Depending on one livelihood option which is climate sensitive makes the community more vulnerable to changing climate. Alternative and diversified livelihood options like fish farming, livestock farming, cattle breeding, seaweed culture, crab fattening, etc., should be adapted to improve adaptability.

vii. **Socio-economic stability** such as insurance, welfare schemes, credits and minimum wage – Socio-economic factors play a crucial role in vulnerability reduction. Socially and economically disadvantaged people are more vulnerable to any hazard, by improving socio-economic factors like education, income, household facilities, housing conditions, water availability, etc., vulnerability can be reduced to a great extent. Many comparative studies on climate change have emphasised that climatic changes will increase the vulnerability of the poor and marginalised, who have always been most at risk from natural disasters.

viii. **International research and collaboration programmes** - In any sector research and development play a major role in risk reduction by bringing the problems to light, mapping the risk-prone areas and predicting the future impacts of the problem. Collaborative research and developmental programmes can assess the climate change impacts in a particular community and suggest appropriate coping strategies. There are lots of research going on around the world about climate change and its impacts.

References

H, V., Karmakar, S., Ghosh, S., Murtugudde, R., 2020. A comprehensive India-wide social vulnerability analysis: highlighting its influence on hydro-climatic risk. Environ. Res. Lett. 15, 014005. https://doi.org/10.1088/1748-9326/ab6499

IPCC, 2001. TAR Climate Change 2001: Impacts, Adaptation, and Vulnerability — IPCC. URL https://www.ipcc.ch/report/ar3/wg2/

IPCC, 2007. AR4 Climate Change 2007: Impacts, Adaptation, and Vulnerability — IPCC. URL https://www.ipcc.ch/report/ar4/wg2/

IPCC, 2014. AR5 Climate Change 2014: Impacts, Adaptation, and Vulnerability - IPCC. URL https://www.ipcc.ch/report/ar5/wg2/

IPCC, 2019. Special Report on the Ocean and Cryosphere in a Changing Climate. URL https://www.ipcc.ch/srocc/

Klein, R., Nicholls, R., 1999. Assessment of Coastal Vulnerability to Climate Change. AMBIO J. Hum. Environ. 28, 182–187.

Krishnan, P., Ananthan, P.S., Purvaja, R., Joyson Joe Jeevamani, J., Amali Infantina, J., Srinivasa Rao, C., Anand, A., Mahendra, R.S., Sekar, I., Kareemulla, K., Biswas, A., Kalpana Sastry, R., Ramesh, R., 2019. Framework for mapping the drivers of coastal vulnerability and spatial decision making for climate-change adaptation: A case study from Maharashtra, India. Ambio 48, 192–212. https://doi.org/10.1007/s13280-018-1061-8

Maiti, S., Jha, S.K., Garai, S., Nag, A., Chakravarty, R., Kadian, K.S., Chandel, B.S., Datta, K.K., Upadhyay, R.C., 2015. Assessment of social vulnerability to climate change in the eastern coast of India. Clim. Change 131, 287–306. https://doi.org/10.1007/s10584-015-1379-1

Nelson, M.C., Ingram, S.E., Dugmore, A.J., Streeter, R., Peeples, M.A., McGovern, T.H., Hegmon, M., Arneborg, J., Kintigh, K.W., Brewington, S., Spielmann, K.A., Simpson, I.A., Strawhacker, C., Comeau, L.E.L., Torvinen, A., Madsen, C.K., Hambrecht, G., Smiarowski, K., 2016. Climate challenges, vulnerabilities, and food security. Proc. Natl. Acad. Sci. U. S. A. 113, 298–303. https://doi.org/10.1073/pnas.1506494113

Nicholls, R.J., Klein, R.J.T., 2005. Climate change and coastal management on Europe's coast, in: Vermaat, J., Salomons, W., Bouwer, L., Turner, K. (Eds.), Managing European Coasts, Environmental Science. Springer, Berlin, Heidelberg, pp. 199–226. https://doi.org/10.1007/3-540-27150-3_11

Ramachandran, A., Praveen, D., Radhapriya, P., Divya, S.K., Remya, K., Palanivelu, K., 2016. Vulnerability and adaptation assessment a way forward for sustainable sectoral development in the purview of climate variability and change: insights from the coast of Tamil Nadu, India. Int. J. Glob. Warm. 10, 307–331. https://doi.org/10.1504/IJGW.2016.077896

Satapathy, S., Porseche, I., 2014. A framework for climate change vulnerability assessment: Satapathy, S., (Ed.). GIZ, New Delhi.

Smit, B., Wandel, J., 2006. Adaptation, adaptive capacity and vulnerability. Glob. Environ. Change, Resilience, Vulnerability, and Adaptation: A Cross-Cutting Theme of the International Human Dimensions Programme on Global Environmental Change 16, 282–292. https://doi.org/10.1016/j.gloenvcha.2006.03.008.

16

Fisheries Technologies and Its Adoption

Ankush Kamble, Jayapratha T, Rujan J and and Shivaji Argade

Fisheries Economics, Extension and Statistics Division, ICAR – CIFE Mumbai, Maharashtra

Introduction

Technology is the application of knowledge and/or equipment in undertaking and accomplishing certain activity. Fisheries technology is a systematically organized knowledge and materials applicable to production problems to help to boost the present level of fish productivity and or extend the existing range of production. Fisheries technologies are wide range of importance; Technological innovation in aquaculture is leading an evolution in aquaculture practices, thereby reducing losses and increasing efficiency. Technology in aquaculture affects many areas of fishery, such as seed production, feed production and management, pest and diseases management, production, harvesting, marketing, processing, etc. Induced breeding technology in Indian Major Carps transformed the growth of aquaculture sector from traditional to intensive aquaculture practices and led to success of modern aquaculture industry. Mechanization in fishing increased the harvest with less time and minimal human drudgery. Fish processing technologies improved product shelf-life, reduced fish loss and added value to the products. Cold storage technologies in fish chain increased availability of fish in off season, on time delivery, availability in remote areas. Similarly digital tools are increasing the adoption of fisheries knowledge as well as awareness of fish producers and consumers. Analytical tools are helping to improve the decision-making power of fish producers, consumers, researchers, policy makers, etc. In brief, technical progress have laid to more efficient and economical fishing operations, reduction of the physical labour required per unit of output and improved access to resources.

Scope of technologies in fishery: Technologies have wide scope in fishery sector ranging from production to marketing up to consumption. It has scope in;

- Increase in fish productivity
- Cutting down cost of production
- Decrease in use of water (Ex. Bioflock)
- Increase efficiency of feed (Ex. Nano feed)
- Controlled pest and diseases
- Reducing deterioration of natural ecosystems
- Providing chemical frees sewage water in rivers and reservoirs
- Producing less pollutant fish by proper sewage treatment
- Increasing fish worker safety by mechanization
- Greater fish production and marketing efficiencies
- Optimization of prices and improving production and marketing efficiency
- Providing quality and safety fish food
- Reducing environmental and ecological impact
- Cutting down maintenance costs of plant
- Providing on time, reliable market information to fish folks
- Making better production and marketing decisions, etc.

Importance of technology transfer in fishery sector: Technology transfer is an important part of the technological innovation process, promoting scientific and technological research and the associated skills and procedures to wider society and the marketplace. Thus, process of transferring innovation from research institutions/universities to organizations /persons that transform knowledge into new products which benefit to society (Ravi & Janodia, 2022). Transfer of technology is needed for improving the production potential and productivity of culture and capture fishery. Transfer of technology in fishery is needed for improving decision making of fish farmers, their production, income and standard of living. Further it requires to empower small scale fishermen, fish retailers, fish processors as well as fish consumers.

Methods to transfer technologies in fishery sector: Mainly three methods are prevailing in technology transfer in fishery sector. The transfer of technology in fishery has three approaches like top-down approach, feedback approach and participatory approach. The technology transfer models are discussed below;

a) **Top-down-model:** This is the conventional technology transfer model being used in transferring fisheries technologies from lab to land. This is one-way process where fisheries technologies developed by scientists/

researchers are transferred to final users through extension agencies. The fish farmers act as passive recipients of technologies and they do not have direct contact with scientists who developed technology. Here extension plays the role to persuade the fish farmers to adopt new technologies. Therefore, inherent problem of this model is that it does not involve fish farmers in identifying the constraints and adapting the research to local conditions. This model has failed in areas where the fish farming system is more complex and in case of technologies not refined according to locality.

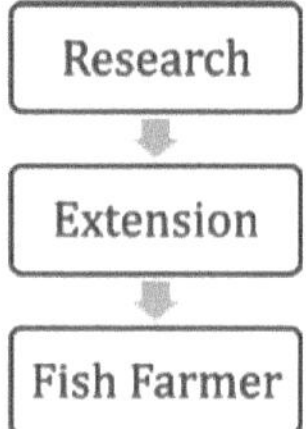

Figure 1 : The top-down fisheries technology transfer model

b) **Feedback model:** In this model an attempt has been made to overcome the glitches of top-down model. The response about adoption of new fisheries technologies is collected through continuous interaction of extension agencies and research scientists. However, this feedback is considered to be frail as the users remain passive recipients of technology and the feedback mechanism exclusively rests with the extension service (Stoop, 1988). This model deliberate researchers, extension personnel and social scientists to have a holistic understanding of farmers' problems.

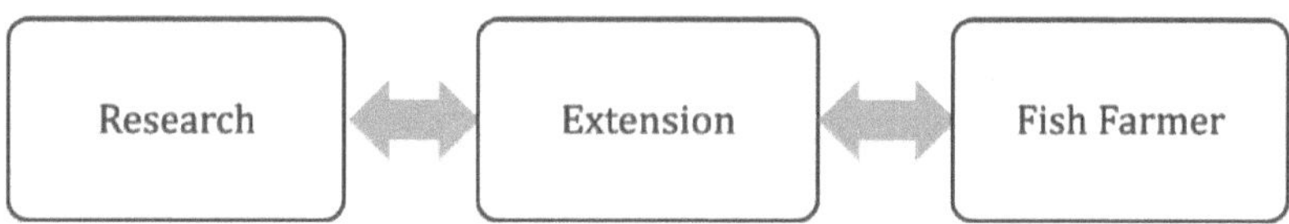

Figure 2 : The feedback fisheries technology transfer model

c) **Farmer- back to- farmer model:** It is also called fish farmer participatory model. This is a best alternative model to above two models. In this model fish farmers must be incorporated as fully active members of the problem-solving team and research study must begin with the fish farmer and also end with them only. This model (Figure 3, based on Rhoades and Booth, 1982) is designed to improve two-way communication. This model involves diagnosis to define problems; multidisciplinary team research to develop best solutions; on-farm / experiment station trials and adaptation of proposed technologies to farmer's conditions, and fish farmer evaluation and adaptation of the technology and monitoring of its adoption (Stoop, 1988). Compared with the modified feedback model, the degree of fish farmer participation and integration between on-station and on-farm research is high in this model.

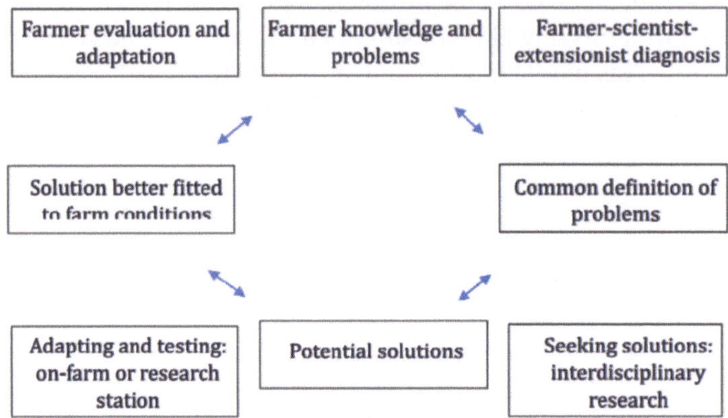

Figure 3: 'Fish farmer-back-to-fish farmer' technology invention and transfer process

Different technologies in fisheries: The technologies in fisheries can be broadly classified into production technologies, marketing technologies and new age technologies.

Fish production technologies: The technologies have direct impact on fish production, production cost, and net profit may be considered as a production technology. Some of the technologies help in boosting fish production and profit are;

- The fish seed induced breeding technology in Indian Major Carp
- Different types of fish hatcheries like Chinese hatchery (eco- hatchery), glass jar hatchery, transparent polythene jar hatchery, galvanized iron jar hatchery, etc.
- Efishery sensor-based fish feeding technology which uses sensors to detect the hunger level of the fish and feed them accordingly
- Wet type fish feed production line: This production line produces floating fish feed, sinking fish feed, pet feed, livestock feed, etc.
- Different types of aerators like surface aerators, self-priming submersible aerators, rotary aerator, water aerator, etc.
- Fish insect / parasite net to protect fish from different insect / parasite infestations

- Different fish growth stimulators, hormones

Fish marketing technologies: The technologies improve the producer and consumer welfare and increase the marketing efficiency may be recorded as a fish marketing technology. These are;

- Different fish processing methods like high pressure processing, low pressure processing, irradiation, pulsed light technology, microwave processing, etc.
- Different packaging technologies like modified atmosphere packaging, active and intelligent packaging, etc.
- Open system and close system transportation
- Cold storage technologies

New age technologies in fish: Further some technologies may be considered as a new age technology that have impact on both producer and consumer welfare. These are new data processing technologies in fisheries include: big data, block chain, smart weighing at sea, Radio-frequency identification (RFID), smartphones for monitoring, artificial intelligence, drones, on-board cameras, augmented reality (AR), virtual reality (VR), blockchain, etc., have made their way into the aquaculture industry. Further, Recirculating Aquaculture Systems (RAS) are becoming more common with water recycling and waste reused as fertilizer for agriculture. RAS can reduce the carbon footprint of seafood by up to 50%, and fish in these systems can be grown in a controlled and traceable environment without the use of hormones or antibiotics. Further

- ***Satellite – based vessel monitoring system*** *to track stamped location, course and speed of vessels*
- ***Very small aperture terminal (VSAT)*** *system to provide* independent communication network connecting to a large number of geographically dispersed sites
- ***Global positioning system (GPS) to*** find the aquaculture potential and threats sites in streams and locating these points on map.
- ***Remote* sensing technology** for mapping sea surface temperature, potential fishing sources in seas and oceans, seasonal variation, extent of floods, etc.
- ***Potential fishery zone forecasting (PFZ)*** to obtain maximum thermal gradient information

Organizations / Projects Involved in Technologies Transferred

The All-India co-ordinated Research Project (AICRP) on carp culture was started in 1971 for demonstrating composite fish culture technology in different agro-climatic zones in India (IASRI Online).

- The ***Krishi Vigyan Kendra (KVK)*** were started in 1974 as grassroot level organization to impart need-based and skill–oriented vocational training in agriculture and allied fields including fisheries through work experience.
- The ***Operational Research Project (ORP)*** to disseminate technologies in a cluster of fisheries villages for fast dissemination of technical knowledge was launched in 1974-75.
- The ***Lab to Land Project (LLP)*** was launched in 1979 with the objective to transfer technology from research laboratory to fish farmers' fields.
- To increase the adoption of fish culture technologies ***Fish Farmers Development Agency (FFDA)*** was begun in 1974-75
- The ***International Development Research Centre (IDRC)* – Rural Aquaculture Project** was launched in 1975 to conduct demonstrations on various aspects of aquaculture like carp culture and carp seed production, by providing technology and all necessary inputs.
- The ***Brackish Water Fish Farmers' Development Agency (BFDA)*** was established in 1985-1990 with an objective to promote shrimp fish farming in brackish waters.
- In order to assess and refine technologies in the light of bio-physical and socio-economic conditions of farmers an, ***Institute Village Linkage Programme (IVLP)*** was launched in 1996.
- The ***Jai Vigyan Mission*** was started in 2000 for ensuring food and nutritional security through increased fish productivity in backward, hilly and tribal areas in Orissa, Assam and Chhattisgarh.
- The ***World bank - aided inland fisheries project*** was established in 1979. It provided credit assistance for construction of modern fish- seed hatcheries. In Total 63 Hatcheries had come up with World Bank Assistance in Uttar Pradesh, West Bengal, Bihar, Madhya Pradesh And Odisha
- The ***agriculture technology information centres (ATICs)*** project was started with the objective of 'single window system' delivery of products, information and services to farmers and entrepreneurs. Presently there are 44 ATICs of which 3 are operating in specialized fisheries institute.
- The ***Trickle-Down System (TDS)*** in Bangladesh to develop self – reliance and awareness in the minds of fish farmers about aquaculture by repeated training, demonstration and close supervision by field extension personnel.

Constraints to transfer of aquaculture technologies: Generating appropriate technologies is not the end in technological advancement process. It should be reached to and adopted by the end users for their benefits. However, the technology adoption in fishery has several constraints. The important constraints to technology adoption are economic factors, complexity of new technology, and farmers' attitudes towards risk and change (Guerin & Guerin, 1994). Beside theses social, cultural, environmental, political constraints are also restrict the effective adoption of technologies in fisheries sector. The situational constraints that prevent a fish folks to adopt a technology are listed below:

- **Technological constraints:** The major obstacle in the adoption of aquaculture technologies is its incompatibility. Most of the new technologies are incompatible with fishers' current farming practices and its present environment. Uncertainty regarding the qualities of the new innovation and its complexity in adoption as well as inability of the technology to perceive fishers' problems are the major technological constraints curb the adoption of new technologies in aquaculture.
- **Economical constraints:** The adoption cost and return from new technology; availability and cost of complimentary inputs necessary for technology adoption; availability of market and price to the product generated from technology; consumer preference towards product; capital requirement of technology; availability and cost of post-harvest chains viz., transportations, storage, packaging, processing, etc. are the major economical constraints in technology adoption in fishery sector.
- **Inputs constraints:** Feed constitutes around 50-60% of the total variable cost in farming (Ayyappan, 2006). Low-cost technology using appropriate combination of locally available low –priced feed ingredients need to be evolved and promoted.
- **Marketing constraints:** Market for its speedy disposal is a pre-requisite for fish farming and transportation also poses a problem for fish farmers. In the absence of organized market, fish farmers sell their produce to middlemen. Due to lack of refrigeration / preservation facilitates at the village level, major portion of fish catch is sold as fresh (Ayyappan, 2006).
- **Institutional constraints:** The lack of awareness of available innovations, asymmetric information, non-availability of quality inputs, indifference of the fisheries department towards technology adoption, static and rigid rules and regulation in inputs use, poor linkages with research extension system, inappropriate training facilities etc., are some institutional constraints in technology adoption in aquaculture.
- **Social constraints:** The transfer of aquaculture technologies is greatly affected and hampered by the social constrains. The *social structure* in a

system can act as barrier to the transfer of aquaculture technologies. Loss of status and victim of censure leads an individual to throwaway a particular fish technology. Further, *social system* norms relating to technology, *roles* & *status* of individual within the system, and relationship of one's system with other social system, may not permit an individual decision-maker to adopt a particular technology.

- **Personnel constraints:** The technology may be treated as inappropriate or irrelevant by the individual due to his ignorance about details of the technology and/or financial inability. Further individual technical competency to bring about the adoption of new knowledge and skill is important in technology adoption. Lack of expertise in dealing with disease and mortality. Technical competency to bring about the adoption of new knowledge and skill is being emphasized. Informal education, group discussion, community video, method demonstration all help in reinforcing the skill for improved technology (Ayyappan, 2006).
- **Natural constraints:** The natural / climatic factors like quality water, floods, cyclone, hurricane, drought, high temperature, low temperature, high rainfall low rainfall, etc., affects technology adoption in fish farming.
- **Biological constraints:** The incidence of pest and disease, parasites restrict the efficient adoption of new technology.
- **Political constraints:** Undesirable intervention by local leaders during selection of appropriate fishers for technology demonstrations is very serious problem in technology adoption.
- **Other constraints:** The other constraints like lack of skilled manpower necessary for technology application, lack of proper equipment, sharecropping: discouragement of tenant in adoption of new technology because no guarantee of getting land next year, etc.

Conclusion

The benefits of modern technology adoption in aquaculture cannot be exhausted, there is increased fish productivity, reduced impact on natural ecosystems, increased fish worker safety, decreased use of water resources, and increased efficiency of feeds among many others. However, in order to achieve these goals, fishers need to understand the concept of modern fish farming and the use of technology. And researchers need to develop and promote low-cost technology using appropriate combination of locally available inputs.

References

Ayyappan, S. 2006. Hand Book of Fisheries and Aquaculture. Directorate of Knowledge Management in Agriculture, New Delhi.

Guerin, L.J. and T.F. Guerin. 1994. Constraints to the adoption of innovations in agricultural research and environmental management: a review. Australian Journal of Experimental Agriculture. 34:549-571.

IASRI Online. http://ecoursesonline.iasri.res.in/mod/page/view.php?id=91200, accessed on 12.10.2022

Ravi, R. and M.D. Janodia. 2022. Factors affecting technology transfer and commercialization of university research in India: A cross-sectional study. Journal of the Knowledge Economy. 13:787-803.

Rhoades, R. and R. Booth. 1982. Farmer-back-to-farmer: a model for generating acceptable technology. CIP [International Potato Research Centre] Social Sciences Department. Working Paper No. 1982-1.

Stoop, W.A. 1988. NARS linkages in technology generation and technology transfer. ISNAR. Working Paper No. 11.

17

Management Implications for Hokersar Wetland: A Ramsar Site in Jammu and Kashmir

[1*]Regu Atufa, [2]Nadeem Qadri, [1]Saba N Reshi and [3]Sadaf Gul

[1]Fisheries Economics, Extension and Statistics Division, ICAR-Central Institute of Fisheries Education, Mumbai, Maharashta

[2]Amicus Curiae, High Court of Jammu & Kashmir, Srinagar

[3]Rinchen Shah Center for West Himalayan Cultures-Islamic University of Science and Technology, Awantipora, Jammu & Kashmir

Abstract

A review synthesis on Hokersar wetland, a Ramsar site in Jammu and Kashmir, India was carried out with the aim to highlight the current issues/problems, suggest a way forward for its wise use, and to preserve its conservation functions of serving as one of the important ecological site. This chapter presents a quick snapshot about the present status of the Hokersar and guides the policy makers to prioritize the important unexplored areas of research that can be taken further to evoke the comprehensive management decisions for its conservation.

Keywords: Hokersar, Wetland, Ramsar site, Migratory Birds, Wildlife Reserve, Livelihood, Conservation, Governance and Management

Introduction

Hokersar wetland (34°06' N latitude, 74°05' E longitude), a Ramsar site and a protected wildlife reserve is located in the northwest Himalayan bio geographic province of Kashmir, back of the snow-draped Pir Panchal. The wetland shelters about two million migratory water-fowl during winter that migrate from Siberia and the Central Asian region. The wetland is fed by two inlet streams Doodhganga (from east) and Sukhnag Nalla (from west). It is the only site in Kashmir with remaining reed beds and serves as a migration route for specifically 68 waterfowl species from Siberia, China, Central Asia, and Northern Europe, including the Large Egret, Great Crested Grebe, Little Cormorant, Common Shelduck, Tufted Duck, and endangered White-eyed Pochard (Ramsar, 2021). The Hokersar wetland is a designated bird sanctuary. The waterbirds fly to Kashmir Valley via the Central Asian Flyway. They begin to arrive in September–October and -leave by May. Over 500,000 waterbirds were recorded in the Hokersar Wetland in 2000–01. It is an important source of food, spawning ground and nursery for fishes, besides offering feeding and breeding ground to a variety of water birds. Typha, Phragmites, Eleocharis, Trapa, and Nymphoides species inhabit the marshy vegetation complexes, which range from shallow water to open water aquatic flora. Despite water withdrawals since 1999, sustainable exploitation of fish, fodder, and fuel is significant. Recent housing developments, littered garbage, and other potential threats Potential threats include recent housing facilities, littered garbage, and demand for increasing tourist facilities.

Literature Synthesis

(Joshi et al., 2002) studied temporal mapping of the wetland using the data sets for the autumn and spring seasons to assess the land cove/land use dynamics. The increase in the settlement has been observed proportionate to the rate of fragmentation in the wetland. This study has created an information base, which will help to design conservation schemes for long term maintenance of the wetland. Khan,2004 evaluated Hokersar's environmental contamination by assessing the allocation and loading of bacterial communities in water. Bacterial load variability was clearly visible at the Doodhganga inlet and Zainakoot locations. Considerable seasonal fluctuations were evident in bacterial densities, highest values were observed in summer and lowest in winter. Annual averages showed that dense aquatic vegetation zone harboured least number of bacteria (159/100 mL) suggesting increased die-off of the bacterial population. All the samples were found contaminated to grossly-contaminated. The nature of qualitative composition of bacterial species including prevalence of Escherichia coli and species of Yersinia, Staphylococcus, Pseudomonas, Shigella indicate probable contamination of waters due to ingress of sewage and fecal matter posing health hazards.

(Mir et al., 2009) reveals that macrophytic species composition in Hokersar has witnessed great variations. The wetland area, which was once spread across 13.26 square kilometres, has now been dramatically decreased to 5.6 square kilometres. The paper stresses the Hokersar Wetland's rapidly deteriorating environmental status and the pressing need for its Ecological restoration, with a focus on macrophytes.(Foziah, 2009) elucidated to compare the water-bird population and extent of human use in Hokersar for conservation land use-land cover characteristics of wetland landscapes and also studied the attitude of residents and concluded that all respondents (100%) knew about the siltation and nutrient problems and perceived removal of excessive weeds and desilitation as the best option of management.

(Romshoo et al., 2011) looked at changes in land use/land cover and wetland area over the last four decades. Topographic maps of 1969 and the remote sensing data for 1992, 2001, 2005 and 2008 was used for determining the spatial-temporal dynamics of the wetland. The analysis of data over the last four decades discovered substantial changes in the Hokersar wetland and surrounding uplands. The wetland area has decreased from 1875.04 Ha in 1969 to 1300 Ha in 2008, with a marked decline in open waters in the wetland. Based on the findings of this study, the main causes of wetland depletion are farmer encroachment, sediment load brought by the Doodhganga River, and the intensification of willow plantations in the wetland. The marshy lands, which serve as migratory bird habitat, have shrunk from 754 ha in 1969 to 610 ha in 2008. In addition, the increasing development of settlements around the wetland over the past few decades has adversely affected its varied aquatic flora and fauna. Apart from these causes, change in climatic conditions in the study area is also responsible for decreasing the water level and water spread in the wetland. These changes in the composition and structure of the wetland have affected its functionality and are manifest in the deterioration of water quality and changes in the aquatic vegetation composition. (Jan et al., 2014) investigated the relative effect of some physicochemical parameters of water on the occurrence of water-borne fungi in a high altitudinal wetland from March 2008 to February 2009. The results reflect that fungal communities are more influenced by the seasonal variation.

(Ahmad et al., 2014) aimed to assess the heavy metal sequestration capability of one of the most common wetland plant species Phragmites australis in Hokersar wetland. The accumulation of the different elements was in order of Al > Mn > Ba > Zn > Cu> Pb > Mo > Co > Cr > Cd > Ni. Translocation factor, i.e. ratio of shoot to root metal concentration revealed that metals were largely retained in the roots of P. australis, thus reducing the supply of metals to avifauna and preventing their bio-accumulation. (Mushtaq et al., 2019) documented the diversity of two fish species in Hokersar and found that overall contribution of the common carp was recorded as 68.4% by biomass and 69.1% by number than Schizothoracids, S. niger. (Habib, 2014) studied avifaunal diversity of Hokersar wetland during 2012-13. A total of 58 species were recorded belonging to 27 families from 7 different habitats. Family Anatidae shows maximum species diversity. Among the recorded species(41%) were resident, summer migrant (28%), winter migrant (21%), and local migrant (10%). Open water habitat supports most of the winter migrant birds and plantation supports most of the summer migrant species. Most of the species belong to insectivorous category (45%) and most of the waterfowl species were herbivorous (14%).

(Romshoo & Rashid, 2014) The spatiotemporal analysis of data from 1969 to 2008 was used to monitor the spatiotemporal changes in Hokersar wetland. The wetland area has shrunk from 18.75 km2 in 1969 to 13 km2 in 2008, owing to a drastic reduction in the wetland's water depth. Marshy lands, migratory bird habitat, have shrunk from 16.3 km2 in 1969 to 5.62 km2 in 2008, and have been colonised by a variety of other land cover types. Significant changes in the forest cover (88.33–55.78 km2), settlement (4.63–15.35 km2), and water bodies (1.75–0.51 km2) were observed in the catchment.

(Yousuf et al., 2015) concluded that Hokersar is moderately eutrophic particularly due to various human activity occurring in their vicinity. Complete dryness of wetlands needs to be avoided since it leads to easy human access and interference. Hence constant supply of water and relieved of excessive silt and nutrients (by treatment plants) is the need of hour. (J. A. Shah et al., 2017) studied the seasonal dynamics of various physico-chemical characteristics of water during 2014 and found that the wetland has still the capacity to absorb major nutrients especially the nitrogen and phosphorus that drain from the immediate catchment into the Queen wetland of Kashmir. (Lolu et al., 2019) highlights the carbon sequestration efficiency of twelve abundant macrophytes in Hokersar Wetland. This study will be helpful to refine our understanding about the role of macrophytes in carbon sequestration of wetlands and could also be explored in earning carbon credits in the global carbon market.

(M. A. Shah et al., 2020) found that the most abundant birds were the Purple Moorhen and Indian Moorhen, while the most abundant macrophytes were Typha, Phragmites, Trapa, Ceratophyllum, and Nymphoides. They conclude that holistic knowledge of links between various life forms inhabiting the wetland is pivotal for evolving appropriate conservation strategies for relatively rarer elements of wetland biodiversity and ecosystem functioning. (Jamal & Ahmad, 2020) while examining the spatiotemporal dynamics of LULC in wetland ecosystems Hokersar found horticulture +4884.3 ha, built-up +6071.96 ha, agriculture −11,605.43 ha, and water body −106.01 ha in Hokersar and Anchar Lake. Anthropogenic activities threatening the wetland ecosystems may include urban development, agriculture and horticulture practices, impacts of insecticides and pesticides, fertilizers, climate change and invasion of alien species. Study also suggested that a thorough and more exhaustive study must be taken up to examine the inter-linkages between wetland ecosystems and population growth and its associated activities at micro- and macro-levels.

Issues/ Problems in Hokersar Wetland

Hokersar wetland has historically been used for a variety of livelihood purposes. People who live on the outskirts of a wetland grow paddy crop, vegetables, rare poultry, cattle, fish, and engage in other activities such as wood collection, mulch collection, and reed collection. Wetlands of Kashmir provide overwintering resort to millions of water birds. It has decreased in size and exhausted due to the anthropogenic activities and encroachments. The wetland is bordered by urban habitations on the northern side along the Srinagar-Baramulla highway. All kinds of waste generated by the people are dumped into the Wetland. Encroachments cause mud flats and grasslands to be converted into agricultural lands, dividing the habitat into small sections and affecting the population of birds in the wetland. Further, throwing of domestic waste into the wetland has resulted in eutrophication and finally in excessive weed growth.

Literature review and the scanning of news reports in recent times, certain problems in Hokersar stand out. They are namely untenable changing Land use patterns, increasing siltation, shrinkage of the lake area, increasing frequency of floods and droughts, pollution, conversion of vast catchment into agriculture land, and changing diversity of waterfowl and migratory birds.

Key Learnings

There is a dearth of the literature available on conservation aspects of Hokersar Wetland and so far the focus of studies is on assessment and ecosystem valuation may provide an ample scope to develop actionable interventions. The constant efforts of a Non Profit Organisation, Environmental Policy Group (EPG) with more than 100 member have called for an eco-restoration of Hokersar Wetland and as a result of which the Hokersar has gained attention lately and the works relating to restoring its water holding capacity, main water regulators on inlet and outlet are being executed with great fervour by the Department of Jal Shakti.

Suggestions

1. *For Eutrophication:* Construction of Sewage treatment plants. Municipal Corporations can be given the responsibility to perform the job.
2. *To improve fish catch:* Quota of fish stocks can be increased. DoF is an important player to give directions and execute decision for fisheries development.
3. *To reduce Ecological degradation:* Local people should be brought into consideration. NGO's can be encouraged to come up as there are no such

4. NGO's who work on ground particularly with fishers. Media, both online and offline can be used as a tool for change in a positive direction.
5. *Integrated fish farming model to augment the livelihood:* As for subsistence it is expensive to afford a cage but through SHG's or Public Private Partnership (PPP) basis it can be advised to people and to non-bonafide license holders.
6. *Exchange / rotation / accountability of departments engaged*: it is suggested to increase the transparency, equity and will also enable to utilize the convergent thinking. Training of staff needs to be done on other important aspects like capacity building of stakeholders, providing informal education, market led extension services rather than watch & ward, compensation & license recovery
7. *Recruiting Professional Staff*: for strengthening the data base by instructing to record credible information
8. *Local Participation and awareness:* for willingness of local masses to conserve environment: about the gravity of situation and involving the locals in planning, monitoring, implementation and evaluation of any such program.
9. *Enforcement and accountability* for Conservation measures taken - Jammu & Kashmir Wildlife Protection Act (1978) and other legislations.
10. Need to prioritize and rethink various proposals so far suggested: like cutting of weeds, dredging, raising of bunds, diversion of Doodganga flood channel to reduce siltation and erection of a perimeter fence.

Way Forward

On Hokersar Ramsar site, the studies have focused on hydrology ((Ahmad et al., 2014; Bano et al., 2018; Jan et al., 2014; M. A. Shah et al., 2020; Yousuf et al., 2015), avifauna and macrophytic diversity ((Bano et al., 2018; Foziah, 2009; Habib, 2014; Lolu et al., 2019; Mir et al., 2009; G. M. Shah et al., 2000; G. Mustafa Shah, 1984), Land Use-Change (Jamal & Ahmad, 2020; Joshi et al., 2002; Romshoo et al., 2011; Romshoo & Rashid, 2014), biology of fishes ((Mushtaq et al., 2019). However, the conceptual framework that will holistically aid in management and wise use of Hokersar wetland resource demand robust base line studies at present level e.g Socio-economic, Governance, Alternative Livelihood, Gender Analysis, Total Economic Valuation studies based on stakeholders assessment and preliminary studies is the need of hour for the effective action plan which can be drawn as a roadmap for its wise use and participatory management, monitoring and evaluation.

References

Ahmad, S. S., Reshi, Z. A., Shah, M. A., Rashid, I., Ara, R., & Andrabi, S. M. (2014). Phytoremediation potential of Phragmites australis in Hokersar wetland-a Ramsar site of Kashmir Himalaya. International Journal of Phytoremediation, 16(12), 1183–1191.

Bano, H., Lone, F. A., Bhat, J. I., Rather, R. A., Malik, S., & Bhat, M. A. (2018). Hokersar Wet Land of Kashmir: Its utility and factors responsible for its degradation. Plant Arch, 18, 1905–1910.

Foziah, H. (2009). A Study on Waterfowl Population and Human Use of Hokersar and Hygam Wetlands of Kashmir Valley for Conservation Planning [PhD Thesis]. Saurashtra University.

Habib, M. (2014). Bird community structure and factors affecting the avifauna of Hokersar wetland Kashmir. Int. J. Curr. Res, 6, 7397–7403.

Jamal, S., & Ahmad, W. S. (2020). Assessing land use land cover dynamics of wetland ecosystems using Landsat satellite data. SN Applied Sciences, 2(11), 1–24.

Jan, D., Mir, T. A., Kamilli, A. N., Pandit, A. K., & Aijaz, S. (2014). Relationship between fungal community and physico-chemical characteristics in the Hokersar Wetland, Kashmir Himalayas. African Journal of Microbiology Research, 8(4), 368–374.

Joshi, P. K., Rashid, H., & Roy, P. S. (2002). Landscape dynamics in Hokersar Wetland, Jammu & Kashmir—An application of geospatial approach. Journal of the Indian Society of Remote Sensing, 30(1), 1–5.

Khan, M. A. (2004). Kashmir Himalayan waterfowl habitat, Hokersar: Species composition and threat perceptions. In Environmental contamination and bioreclamation (pp. 188–193). APH Publishing Corporation.

Lolu, A. J., Ahluwalia, A. S., Sidhu, M. C., & Reshi, Z. A. (2019). Carbon sequestration potential of macrophytes and seasonal carbon input assessment into the Hokersar Wetland, Kashmir. Wetlands, 39(3), 453–472.

Mir, A. A., Mahajan, D. M., & Saptarishi, P. G. (2009). Composition and distribution of macrophytes in Hokersar-A wetland of international importance in Kashmir Himalaya. Int. J. Climate Change: Impacts and Responses, 1(4), 23–35.

Mushtaq, S. A., Balkhi, M. H., Abubakr, A., Kumar, A., Shah, T. H., Ahmed, B., Shah, F., Talia, S., Qadri, S., & Farooq, I. (2019). Current status of fish fauna, catch composition of Hokersar Wetland, Kashmir.

Romshoo, S. A., & Rashid, I. (2014). Assessing the impacts of changing land cover and climate on Hokersar wetland in Indian Himalayas. Arabian Journal of Geosciences, 7(1), 143–160.

Romshoo, S. A., Ali, N., & Rashid, I. (2011). Geoinformatics for characterizing and understanding the spatio-temporal dynamics (1969 to 2008) of Hokersar wetland in Kashmir Himalayas. International Journal of Physical Sciences, 6(5), 1026–1038.

Shah, G. M., Qadri, M. Y., & Jan, U. (2000). Species composition and population dynamics of birds of Hokersar wetland, Kashmir. Environment, Biodiversity and Conservation (Ed. Khan MA) Pp, 113–131.

Shah, G. Mustafa. (1984). Birds of Hokersar: Food, feeding and breeding biology of resident and non-resident birds [PhD Thesis]. PhD Thesis, Univ of Kashmir, Srinagar.

Shah, J. A., Pandit, A. K., & Shah, G. M. (2017). Dynamics of physico-chemical limnology of a shallow wetland in Kashmir Himalaya (India). Sustainable Water Resources Management, 3(4), 465–477.

Shah, M. A., Bashir, S., & Kamili, A. N. (2020). Response of Avian Community to Vegetation Pattern and Hydrological Variability in Hokersar Wetland, a Kashmir Himalayan Ramsar Site. European Journal of Molecular & Clinical Medicine, 7(11), 1229–1238.

Yousuf, T., Yousuf, A. R., & Mushtaq, B. (2015). Comparative account on physico-chemical parameters of two wetlands of Kashmir, Valley. Intern J Recent Sci Res, 6(2), 2876–2882.

18

Management Options for Sustainable Exploitation of Fishes on the Coast of Gulf of Mannar of Tamil Nadu, India

N. Neethiselvan, S. Archana, J. Amala Shajeeva and D. Sundareswari

Department of Fishing Technology and Fisheries Engineering, Fisheries College and Research Institute, Thoothukudi, Tamilnadu

Abstract

The Gulf of Mannar has drawn the attention of conservationists even before the initiation of the Man and Biosphere (MAB) program by UNESCO in 1971. With its rich biodiversity of about 4223 species of various flora and fauna, part of this Gulf of Mannar between Rameswaram and Tuticorin covering 21 islands and the surrounding shallow coastal waters has been declared a Marine National Park in 1986 by the Government of Tamil Nadu and later the first Marine Biosphere Reserve of India in 1989 by the Government of India. The Gulf of Mannar is one of the best regions in the Indian subcontinent in fish biodiversity richness. The residents of coastal villages in the Gulf of Mannar pursue fishing as a primary occupation since agricultural activities have proved to be unproductive. Though the marine fishery resources are renewable overexploitation leads to a drastic reduction in the fish catch and the individual fisherman's economic share is not profitable. Hence, proper resource management options need to be evolved to ensure minimum returns to the fisherman by considering the livelihood of the coastal community.

Introduction

Fishermen of Gulf of Mannar coast were largely depending on traditional fishing gears, and they were mainly exploiting the coastal fishery resources before 1960's. After the introduction of trawl gears and mechanized fishing boats during 1960's fishermen were more comfortable with plenty of fish catch, especially the high-valued demersal shrimps. The fishing effort was also increased at a tremendous pace in this region with the addition of stern trawlers. Due to this increment in the fishing effort, the fishery resources and their life-supporting systems such as coral reefs and sea grass beds are seriously disturbed. The multi-stage exploitation of fishery resources especially the high-value item such as shrimps, by traditional fishermen in the coastal waters and mechanized vessels in the off-coast areas lead to the sudden depletion of stocks. The non-selective nature of trawl gears and shore seines also caused serious damage to the marine biodiversity, especially the trawl gears also known as bulldozers of sea affecting the ecosystems and biodiversity of the Gulf of Mannar. The declining catch in the Gulf of Mannar coast may be attributed not only to the declining stock due to overfishing but also due to sharing of catch by the increased number of fishing vessels which has led to economic overfishing.

Study of the Research Area

The Gulf of Mannar, the first Marine Biosphere Reserve (GOMMBR) in South and South East Asia, running down from Rameswaram to Kanyakumari in Tamil Nadu, India is located between Longitudes from 78008 E to 79030 E and along Latitudes from 8035 N to 9025 N. This Marine Biosphere Reserve encompasses a chain of 21 Islands (2 Islands already submerged) and adjoining coral reefs off the coasts of the Ramanathapuram and the Tuticorin districts forming the core zone; the Marine National Park and the buffer zone includes the surrounding seascape and a 10 km strip of the coastal landscape covering a total area of 10,500 Km2 , in the Ramanathapuram, Tuticorin, Tirunelveli and Kanyakumari Districts with a long coastline of 364.9 Km.

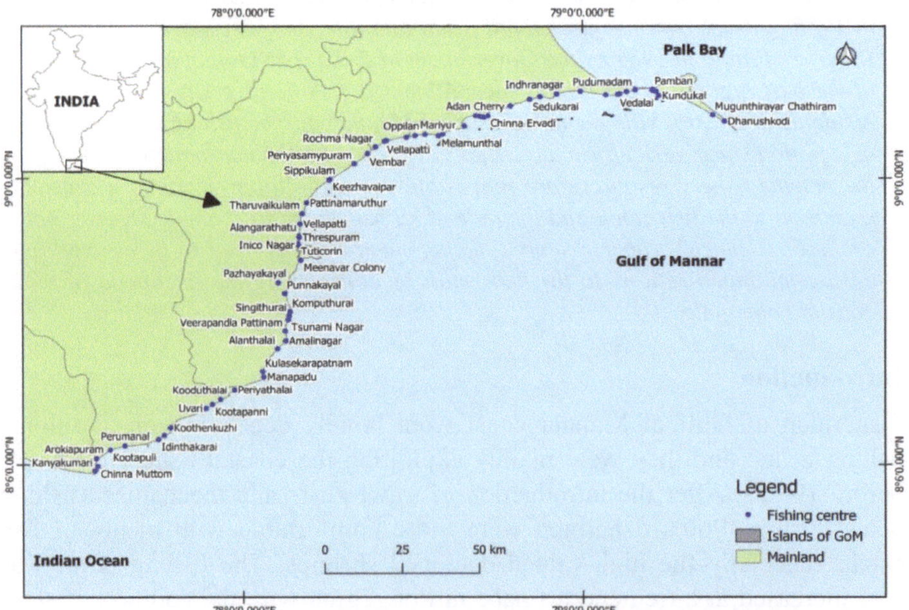

Figure 1 : Map showing the the study area of Gulf of Mannar

Table 1: Management option for sustainable fishing in Kanyakumari District

 **- Less Exploited**  **- Optimally Exploited**  **- Over Exploited**

S.No	**Species**	**Vernacular Name**	**Current status**	**Management option for sustainable Exploitation**
1	*Abalistes stellatus*	Killathi		1% of fishing effort may be increased from the current level.
2	*Ablennes hians*	Vaalamural		1% of fishing effort may be increased from the current leve.l
3	*Amblygaster sirm*	Keeri meen Chalai		20% of fishing effort may be increased from the current level.
4	*Caranx sexfasciatus*	Paarai		4% of fishing effort may be increased from the current level.
5	*Chirocentrus dorab*	Vaalai Meen		4% of fishing effort may be increased from the current level.
6	*Decapterus russelli*	Olugu salai		Needs immediate attention by fishery managers. 18% of fishing effort needs to be reduced from the current level.
7	*Dussumieria acuta*	Thondon		18% of fishing effort may be increased from the current level.
8	*Epinephealus fuscocuttatus*	Kalava		6% of fishing effort may be increased from the current level.
9	*Katsuwonus pelamis*	Vari Choori		3% of fishing effort may be increased from the current level.
10	*Leiognathus lineolatus*	Kaaral		Needs immediate attention by fishery managers. 12% of fishing effort needs to be reduced from the current level.
11	*Lethrinus lentjan*	Vilameen		No further increment in the current level of fishing effort.
12	*Lutjanus fulviflamma*	Manjakeeli		No further increment in the current level of fishing effort.
13	*Nemipterus furcosus*	Lomio		No further increment in the current level of fishing effort.
14	*Parupeneus indicus*	Nagarai		No further increment in the current level of fishing effort.

15	*Pellona ditchela*	Venganai		17% of fishing effort may be increased from the current level.
16	*Rastrelliger kanagurta*	Ayilai / Kumla		Needs immediate attention by fishery managers. 23% of fishing effort needs to be reduced from the current level.
17	*Sardinella fimbriata*	Soodai		No further increment in the current level of fishing effort.
18	*Sardinella gibbosa*	Chalai		Needs immediate attention by fishery managers. 25% of fishing effort needs to be reduced from the current level.
19	*Scarus rivulatus*	Kileemeen		8% of fishing effort may be increased from the current level.
20	*Scolopsis bimaculatus*	Pramuttan		30% of fishing effort may be increased from the current level.
21	*Scomberomorus commerson*	Seela		No further increment in the current level of fishing effort.
22	*Sepia pharoanis*	Kanava		5% of fishing effort may be increased from the current level
23	*Sphyraena jello*	Ooli		10% of fishing effort may be increased from the current level
24	*Thryssa setirostris*	Kola		20% of fishing effort may be increased from the current level
25	*Upeneus vittatus*	Sennagarai		6% of fishing effort may be increased from the current level

Kanyakumari District

Of the 25 species studied. 4 species namely *Decapterus russelli, Leiognathus lineolatus, Rastrelliger kanagurta* and *Sardinella gibbosa* were found over exploited. The fishing gears largley their species may be reduced to the twise of 12-25% for sustainable fishing. 5 species namely *Lethrinus lentjan, Lutjanus fulviflamma, Nemipterus furcosus, Parupeneus indicus* and *Scomberomorus commerson* were found optimum exploited. No further increment in the current level of fishing effort. 16 species namely *Abalistes stellatus, Ablennes hians, Amblygaster sirm, Caranx sexfasciatus, Chirocentrus dorab, Dussumieria acuta, Epinephealus fuscocuttatus, Katsuwonus pelamis, Pellona ditchela, Sardinella fimbriata, Scarus rivulatus, Scolopsis bimaculatus, Sepia pharoanis, Sphyraena jello, Thryssa setirostris* and *Upeneus vittatus* were found less exploited. The fishing effort may be increased from the current level. (Mentioned above Table :1)

Table 2 : Management option for sustainable fishing in Tirunelveli District

- Less Exploited **- Optimally Exploited** **- Over Exploited**

S. No	Species	Vernacular Name	Current status	Management options for sustainable Exploitation
1	*Amblygaster clupeoides*	Keeri meen Chalai	😁	4% of fishing effort may be increased from the current level.
2	*Amblygaster sirm*	Keeri meen Chalai	😊	No further increment in the current level of fishing effort.
3	*Caranx sexfasciatus*	Paarai	😊	No further increment in the current level of fishing effort.
4	*Chirocentrus dorab*	Vaalai meen	😊	No further increment in the current level of fishing effort.
5	*Decapterus russelli*	Olugu salai	😡	Needs immediate attention by fishery managers. 10% of fishing effort needs to be reduced from the current level.
6	*Dussumieria acuta*	Thondon	😁	8% of fishing effort may be increased from the current level.
7	*Epinephealus fuscocuttatus*	Kalava	😊	No further increment in the current level of fishing effort.
8	*Gerres filamentosus*	Oodagam	😁	5% of fishing effort may be increased from the current level.
9	*Leiognathus lineolatus*	Kaaral	😡	Needs immediate attention by fishery managers. 17% of fishing effort needs to be reduced from the current level.
10	*Lethrinus lentjan*	Vilameen	😊	No further increment in the current level of fishing effort.
11	*Lutjanus fulviflamma*	Manjakeeli	😁	7% of fishing effort may be increased from the current level.
12	*Nemipterus furcosus*	Lomio	😊	No further increment in the current level of fishing effort.
13	*Parupeneus indicus*	Nagarai	😊	No further increment in the current level of fishing effort.
14	*Pellona ditchela*	Venganai	😊	No further increment in the current level of fishing effort.
15	*Rastrelliger kanagurta*	Ayilai / Kumla	😡	Needs immediate attention by fishery managers. 20% of fishing effort needs to be reduced from the current level.

16	*Sardinella fimbriata*	Soodai		No further increment in the current level of fishing effort.
17	*Sardinella gibbose*	Chalai		Needs immediate attention by fishery managers. 16% of fishing effort needs to be reduced from the current level.
18	*Scarus rivulatus*	Kileemeen		10% of fishing effort may be increased from the current level.
19	*Scolopsis bimaculatus*	Pramuttan		20% of fishing effort may be increased from the current level.
20	*Scomberoides tala*	Katta		No further increment in the current level of fishing effort.
21	*Scomberomorus commerson*	Seela		No further increment in the current level of fishing effort.
22	*Siganus canaliculatus*	Ora		No further increment in the current level of fishing effort.
23	*Sphyraena jello*	Ooli		Needs immediate attention by fishery managers. 17% of fishing effort needs to be reduced from the current level.
24	*Thryssa setirostris*	Kola		25% of fishing effort may be increased from the current level.
25	*Tylosurus crocoidilus*	Kalinga mural		6% of fishing effort may be increased from the current level.
26	*Upeneus vittatus*	Sennagarai		No further increment in the current level of fishing effort.

Tirunelveli District

Collectively 26 species studied and of these 5 species namely *Decapterus russelli, Leiognathus lineolatus, Rastrelliger kanagurta Sardinella gibbosa* and *Sphyreana jello* were found over exploited. The fishing gears largley their species may be reduced for sustainable fishing. 12 species namely Amblygaster sirm, Caranx sexfasciatus, Chirocentrus dorab, Epinephealus fuscocuttatus, Lethrinus lentjan, Nemipterus furcosus, Parupeneus indicus, Pellona ditchela, Sardinella fimbriata, Scomberoides tala and Scomberomorus commerson and Siganus canaliculatus were found optimum exploited. No fishing effort may be increased from the current level.9 species namely *Amblygaster clupeoides, Dussumieria acuta, Gerres filamentosus, Lutjanus fulviflamma, Scarus rivulatus, Scolopsis bimaculatus, Thryssa setirostris, Tylosurus crocoidilus* and *Upeneus vittatus* were found less exploited. The fishing effort may be increased from the current level. (Mentioned above Table :2)

Table 3: Management option for sustainable fishing in Tuticorin District

 **- Less Exploited**  **- Optimally Exploited**  **- Over Exploited**

S.No	Species	Vernacular Name	Current Status	Management options for sustainable Exploitation
1	*Ablennes hians*	Vaalamural		Needs immediate attention by fishery managers. 15% of fishing effort needs to be reduced from the current level.
2	*Amblygaster sirm*	Keeri meen Chalai		Needs immediate attention by fishery managers. 15% of fishing effort needs to be reduced from the current level.
3	*Auxis thazard*	Urulai Shurai		No further increment in the current level of fishing effort.
4	*Caranx sexfasciatus*	Paarai		No further increment in the current level of fishing effort.
5	*Cephalopolis sonnerati*	Thamban		No further increment in the current level of fishing effort.
6	*Chirocentrus dorab*	Vaalai meen		3% of fishing effort may be increased from the current level.
7	*Coryphaena hippurus*	Ilas (or)Avilus		No further increment in the current level of fishing effort.
8	*Decapterus russelli*	Olugu salai		Needs immediate attention by fishery managers. 25% of fishing effort needs to be reduced from the current level.
9	*Dussumieria acuta*	Thondon		Needs immediate attention by fishery managers. 10% of fishing effort needs to be reduced from the current level.
10	*Epinephealus fuscocuttatus*	Kalava		No further increment in the current level of fishing effort.
11	*Euthynnus affinis*	Kaka Choorai (or) Choorai		Needs immediate attention by fishery managers. 15% of fishing effort needs to be reduced from the current level.
12	*Gerres filamentosus*	Oodagam		No further increment in the current level of fishing effort.
13	*Hemiramphus far*	Kattamural		6% of fishing effort may be increased from the current level.

14	*Katsuwonus pelamis*	Vari Choori		Needs immediate attention by fishery managers. 15% of fishing effort needs to be reduced from the current level.
15	*Leiognathus lineolatus*	Kaaral		Needs immediate attention by fishery managers. 20% of fishing effort needs to be reduced from the current level.
16	*Lethrinus lentjan*	Vilameen		Needs immediate attention by fishery managers. 20% of fishing effort needs to be reduced from the current level.
17	*Lutjanus fulviflamma*	Manjakeeli		No further increment in the current level of fishing effort.
18	*Nemipterus furcosus*	Lomio		7% of fishing effort may be increased from the current level.
19	*Parupeneus indicus*	Nagarai		Needs immediate attention by fishery managers. 25% of fishing effort needs to be reduced from the current level.
20	*Pellona ditchela*	Venganai		5% of fishing effort may be increased from the current level.
21	*Plectrorhinchus schotaf*	Kombu-ora		10% of fishing effort may be increased from the current level.
22	*Rastrelliger kanagurta*	Ayilai / Kumla		Needs immediate attention by fishery managers. 25% of fishing effort needs to be reduced from the current level
23	*Sardinella fimbriata*	Soodai		Needs immediate attention by fishery managers. 50% of fishing effort needs to be reduced from the current level
24	*Sardinella gibbose*	Chalai		Needs immediate attention by fishery managers. 40% of fishing effort needs to be reduced from the current level
25	*Scarus rivulatus*	Kileemeen		7% of fishing effort may be increased from the current level.
26	*Scolopsis bimaculatus*	Pramuttan		Needs immediate attention by fishery managers. 10% of fishing effort needs to be reduced from the current level

27	*Scomberoides tala*	Katta		20% of fishing effort may be increased from the current level.
28	*Scomberomorus commerson*	Seela		10% of fishing effort may be increased from the current level.
29	*Selaroides leptolepis*	Kocchaam paarai		No further increment in the current level of fishing effort.
30	*Sepia pharonis*	Kanava		Needs immediate attention by fishery managers. 20% of fishing effort needs to be reduced from the current level.
31	*Siganus canaliculatus*	Ora		20% of fishing effort may be increased from the current level.
32	*Sphyraena jello*	Ooli		Needs immediate attention by fishery managers. 20% of fishing effort needs to be reduced from the current level.
33	*Stoliphorus indicus*	Nethili		Needs immediate attention by fishery managers. 15% of fishing effort needs to be reduced from the current level.
34	*Strongylura leiura*	Ooshe murral		Needs immediate attention by fishery managers. 15% of fishing effort needs to be reduced from the current level.
35	*Thunnus albacares*	Kerai		Needs immediate attention by fishery managers. 15% of fishing effort needs to be reduced from the current level.
36	*Tylosurus crocodilus*	Kalinga mural		No further increment in the current level of fishing effort.
37	*Upeneus vittatus*	Sennagarai		8% of fishing effort may be increased from the current level.

Tuticorin District

Collectively 37 species were studied and of these 18 species namely *Ablennes hians, Amblygaster sirm, Decapterus russelli, Dussumieria acuta, Euthynnus affinis, Katsuwonus pelamis, Leiognathus lineolatus, Lethrinus lentjan, Parupeneus indicus, Rastrelliger kanagurta, Sardinella fimbriata, Sardinella gibbosa, Scolopsis bimaculatus, Sepia pharonis, Sphyraena jello, Stoliphorus indicus, Strongylura leiura* and *Thunnus albacares* were found overexploited. The fishing gears largely their species may be reduced for sustainable fishing. 8 species namely *Auxis thazard, Caranx sexfasciatus, Cephalopolis sonnerati, Epinephealus fuscocuttatus, Gerres filamentosus, Lutjanus fulviflamma* and *Selaroides leptolepis* were found optimum exploited. No fishing effort increased from the current level. 11 species namely *Chirocentrus dorab, Coryphaena hippurus, Hemiramphus far, Nemipterus furcosus, Pellona ditchela, Plectrorhinchus schotaf, Scarus rivulatus, Scomberoides tala, Scomberomorus commerson* and *Siganus canaliculatus* were found less exploited. The fishing effort may be increased from the current level.

Table 4 : Management option for sustainable fishing in Ramanathapuram District

😁 **- Less Exploited** 😊 **- Optimally Exploited** 😡 **- Over Exploited**

S.No	**Species**	**Vernacular Name**	**Current status**	**Management options for sustainable Exploitation**
1	*Ablennes hians*	Vaalamural	😁	3% of fishing effort may be increased from the current level.
2	Amblygaster sirm	Keeri meen Chalai	😊	No further increment in the current level of fishing effort.
3	*Caranx sexfasciatus*	Paarai	😊	No further increment in the current level of fishing effort.
4	*Chirocentrus dorab*	Vaalai meen	😁	5% of fishing effort may be increased from the current level.
5	*Decaptreus russelli*	Olugu salai	😡	Needs immediate attention by fishery managers.20% of fishing effort needs to be reduced from the current level.
6	*Dussumieria acuta*	Thondon	😊	No further increment in the current level of fishing effort.
7	*Epinephelus fuscoguttatus*	Kalava	😁	5% of fishing effort may be increased from the current level.
8	*Leiognathus lineolatus*	Kaaral	😡	Needs immediate attention by fishery managers. 20% of fishing effort needs to be reduced from the current level.
9	*Lethrinus lentjan*	Vilameen	😁	10% of fishing effort may be increased from the current level.
10	*Lutjanus fulviflamma*	Manjakeeli	😁	10% of fishing effort may be increased from the current level.
11	*Nemipterus furcosus*	Lomio	😁	5% of fishing effort may be increased from the current level.

12	*Parupeneus indicus*	Nagarai		No further increment in the current level of fishing effort.
13	*Pellona ditchella*	Venganai		3% of fishing effort may be increased from the current level
14	*Rastrelliger kanagurta*	Ayilai / Kumla		Needs immediate attention by fishery managers. 15% of fishing effort needs to be reduced from the current level.
15	*Sardinella fimbriata*	Soodai		Needs immediate attention by fishery managers. 15% of fishing effort needs to be reduced from the current level.
16	*Sardinella gibbosa*	Chalai		Needs immediate attention by fishery managers. 20% of fishing effort needs to be reduced from the current level.
17	*Scarus rivulatus*	Kileemeen		6% of fishing effort may be increased from the current level.
18	*Scolopsis bimaculatus*	Pramuttan		No further increment in the current level of fishing effort.
19	*Scomberoides tala*	Katta		No further increment in the current level of fishing effort.
20	*Scomberomorus commerson*	Seela		No further increment in the current level of fishing effort.
21	*Selaroides leptolepis*	Kocchaam paarai		5% of fishing effort may be increased from the current level.
22	*Siganus canaliculatus*	Ora		8% of fishing effort may be increased from the current level.
23	*Sphyraena jello*	Ooli		Needs immediate attention by fishery managers. 20% of fishing effort needs to be reduced from the current level.
24	*Stoliphorus indicus*	Nethili		No further increment in the current level of fishing effort.
25	*Thryssa setristoris*	Kola		No further increment in the current level of fishing effort.
26	*Upeneus vittatus*	Sennagarai		5% of fishing effort may be increased from the current level.

Ramanathapuram District

Collectively 26 species were studied and of these 6 species namely, *Decaptreus russelli, Leiognathus lineolatus, Rastrelliger kanagurta, Sardinella fimbriata* and *Sardinella gibbosa* were found over exploited. 9 species namely *Amblygaster sirm, Caranx sexfasciatus, Dussumieria acuta, Parupeneus indicus, Scolopsis bimaculatus, Scomberoides tala, Scomberomorus commerson* and *Stoliphorus indicus* were found optimum exploited. No fishing effort may be increased from the current level. 11 species namely *Ablennes hians, Chirocentrus dorab, Epinephelus fuscoguttatus, Lethrinus lentjan, Lutjanus fulviflamma, Nemipterus furcosus, Pellona ditchella, Scarus rivulatus, Selaroides leptolepis* and *Siganus canaliculatus* were found less exploited. The fishing effort may be increased from the current level. (Table:4)

Conclusion

With respect of the exploitation status, Tuticorin coast of Gulf of Mannar show the maximum number of species (18) over exploited, followed by Ramanathapuram (6), Tirunelveli (5) and Kanyakumari (4). The fishery resources of Tuticorin require immediate attention with suitable management resoures. As for as Ramanathapuram, Tirunelveli and Kanyakumari coasts of Gulf of Mannar show the maximum number of less exploited fish species and the current level of effort may be continued without any increment.

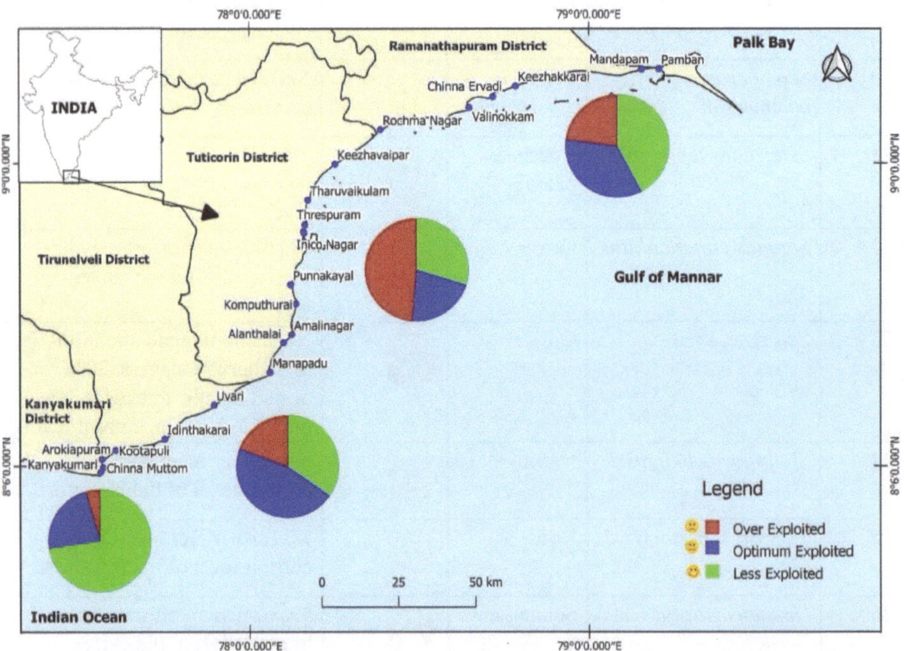

Figure 2 : Status of exploitation of fish species along Gulf of Mannar

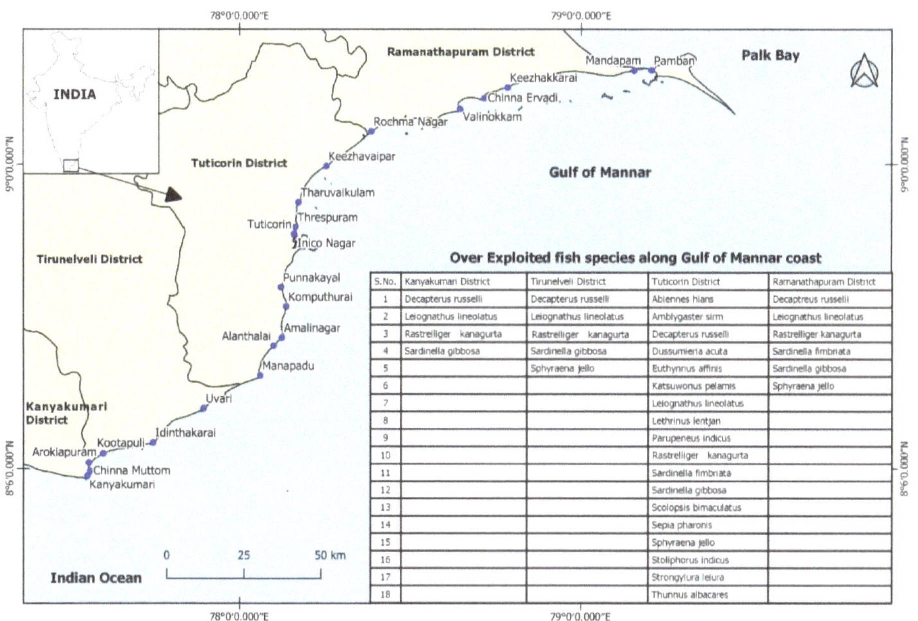

Over Exploited fish species along Gulf of Mannar coast

S.No.	Kanyakumari District	Tirunelveli District	Tuticorin District	Ramanathapuram District
1	Decapterus russelli	Decapterus russelli	Ablennes hians	Decaptreus russelli
2	Leiognathus lineolatus	Leiognathus lineolatus	Amblygaster sirm	Leiognathus lineolatus
3	Rastrelliger kanagurta	Rastrelliger kanagurta	Decapterus russelli	Rastrelliger kanagurta
4	Sardinella gibbosa	Sardinella gibbosa	Dussumieria acuta	Sardinella fimbriata
5		Sphyraena jello	Euthynnus affinis	Sardinella gibbosa
6			Katsuwonus pelamis	Sphyraena jello
7			Leiognathus lineolatus	
8			Lethrinus lentjan	
9			Parupeneus indicus	
10			Rastrelliger kanagurta	
11			Sardinella fimbriata	
12			Sardinella gibbosa	
13			Scolopsis bimaculatus	
14			Sepia pharonis	
15			Sphyraena jello	
16			Stoliphorus indicus	
17			Strongylura leiura	
18			Thunnus albacares	

Figure 3 : District wise status of over exploited fish species along Gulf of Mannar

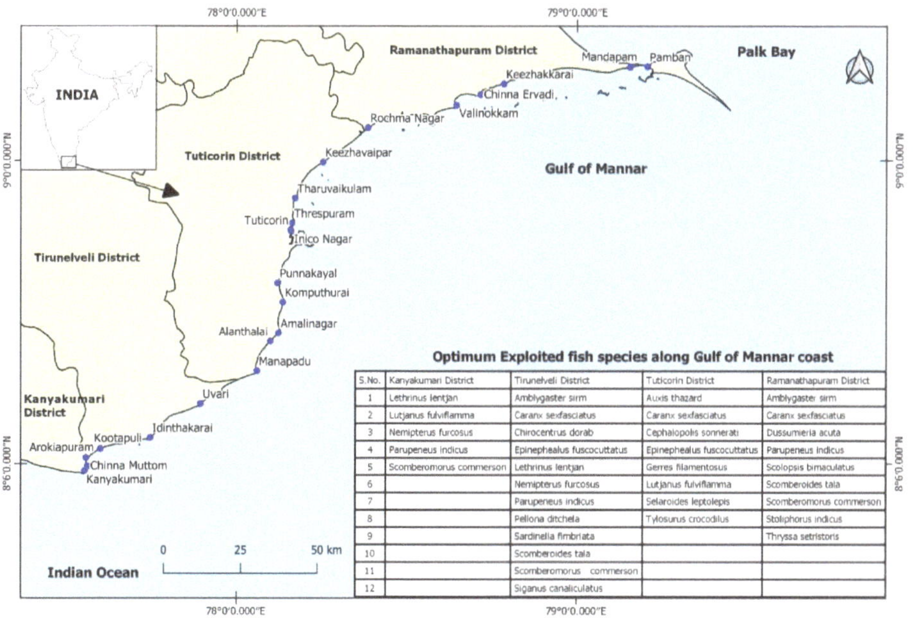

Optimum Exploited fish species along Gulf of Mannar coast

S.No.	Kanyakumari District	Tirunelveli District	Tuticorin District	Ramanathapuram District
1	Lethrinus lentjan	Amblygaster sirm	Auxis thazard	Amblygaster sirm
2	Lutjanus fulviflamma	Caranx sexfasciatus	Caranx sexfasciatus	Caranx sexfasciatus
3	Nemipterus furcosus	Chirocentrus dorab	Cephalopolis sonnerati	Dussumieria acuta
4	Parupeneus indicus	Epinephealus fuscocuttatus	Epinephealus fuscocuttatus	Parupeneus indicus
5	Scomberomorus commerson	Lethrinus lentjan	Gerres filamentosus	Scolopsis bimaculatus
6		Nemipterus furcosus	Lutjanus fulviflamma	Scomberoides tala
7		Parupeneus indicus	Selaroides leptolepis	Scomberomorus commerson
8		Pellona ditchela	Tylosurus crocodilus	Stoliphorus indicus
9		Sardinella fimbriata		Thryssa setristoris
10		Scomberoides tala		
11		Scomberomorus commerson		
12		Siganus canaliculatus		

Figure 4: District wise status of optimum exploited fish species along Gulf of Mannar

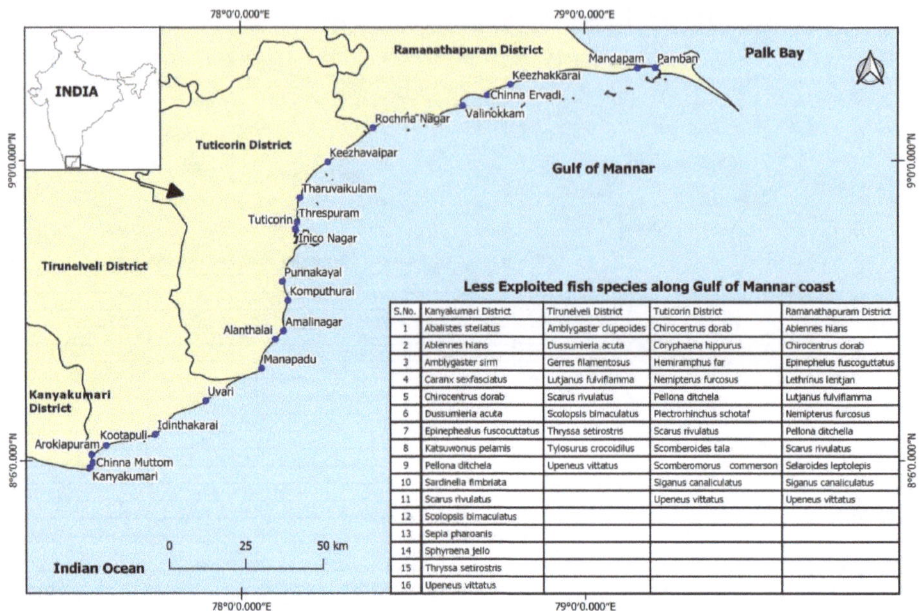

S.No.	Kanyakumari District	Tirunelveli District	Tuticorin District	Ramanathapuram District
1	Abalistes stellatus	Amblygaster clupeoides	Chirocentrus dorab	Ablennes hians
2	Ablennes hians	Dussumieria acuta	Coryphaena hippurus	Chirocentrus dorab
3	Amblygaster sirm	Gerres filamentosus	Hemiramphus far	Epinephelus fuscoguttatus
4	Caranx sexfasciatus	Lutjanus fulviflamma	Nemipterus furcosus	Lethrinus lentjan
5	Chirocentrus dorab	Scarus rivulatus	Pellona ditchela	Lutjanus fulviflamma
6	Dussumieria acuta	Scolopsis bimaculatus	Plectrorhinchus schotaf	Nemipterus furcosus
7	Epinephealus fuscocuttatus	Thryssa setirostris	Scarus rivulatus	Pellona ditchella
8	Katsuwonus pelamis	Tylosurus crocoidilus	Scomberoides tala	Scarus rivulatus
9	Pellona ditchela	Upeneus vittatus	Scomberomorus commerson	Selaroides leptolepis
10	Sardinella fimbriata		Siganus canaliculatus	Siganus canaliculatus
11	Scarus rivulatus		Upeneus vittatus	Upeneus vittatus
12	Scolopsis bimaculatus			
13	Sepia pharoanis			
14	Sphyraena jello			
15	Thryssa setirostris			
16	Upeneus vittatus			

Figure 5: District wise status of less exploited fish species along Gulf of Mannar

19

Role of NGOs in Fisheries Development

J. Rujan, Ankush L. Kamble and P. Seenivasan

Fisheries Economics, Extension and Statistics Division, ICAR-CIFE Mumbai, Maharashtra

Abstract

The fisheries sector has been acknowledged as a significant source of income and employment because it encourages the growth of numerous ancillary industries, provides cheap and wholesome food, and provides a means of subsistence for a sizeable portion of the nation's economically disadvantaged population. NGOs play a significant role in the socioeconomic advancement of fishermen within the fisheries sector. These effectively nudge the intended fish farmers/producers to improve aquaculture practices. In this chapter, we describe NGOs in terms of their function within the fisheries sector.

Introduction

NGOs (Non-Governmental Organizations) are organisations with professional staff that work to lessen human suffering and advance developing nations. The number of Non-Governmental Organizations (NGOs) in India is close to 3.4 million, and they work in a range of sectors, from disaster relief to advocacy for underserved and neglected groups. It is a nonprofit organisation that operates without interference from any governmental body. It is task-oriented and organised by people with common interests in a local, national, or international setting. It performs a variety of services and charitable tasks. Particularly, some are structured around particular concerns, such as health, the environment, or human rights. These make up a sizeable portion of civil society and bring about quick social change. They have developed into a vital and vocal platform for civil society participation in public affairs (Desai, 2014.).

Features of NGOs

- *Voluntary associations*: NGOs are voluntary associations that are created by people having a standard interest.
- *Autonomous*: Free from government interference. They are governed by their own policies and guidelines.
- *Service Motive:* These don't seem to be profit-making business organizations. Shows lots of concern in welfare aspects like education of youngsters, protection of animals, wildlife, environment, improving the status of women, etc.
- *Own funds*: These are created and maintained by their own funds. They often collect contributions from the general public. Some are financed by private business organizations and international authorities.

Sector-wise NGOs in India

The NGO-DARPAN is a portal that gives VOs (Voluntary Organizations) and NGOs a platform to interface with government ministries/departments, government authorities, and other organisations. Later, it is suggested that it apply to all central government agencies, departments, and ministries. There are currently 154461 NGOs are registered in NGO-DARPAN website, and they are affiliated to 53 ministries and departments, including 33 online ones. Though fisheries has separate sector NGOs in Animal Husbandry, Dairying and Fisheries other sector NGOs also involving a fisheries development activities (figure 1). A healthy partnership between Voluntary organizations (VOs)/NGOs and the Government of India was to be created and promoted through the NGO-DARPAN, an initiative of the Prime Minister's Office. NITI Aayog currently oversees the management of the portal. The number of NGOs currently operating in various sectors is listed below (table 1).

Table 1: Sector-specific list of VOs / NGOs registered with NGO-DARPAN

Sectors	Signed up NGOs	Percentage
Animal Husbandry, Dairying & Fisheries	26737	2.38
Aged/Elderly	27502	2.44
Agriculture	37338	3.32
Art & Culture	40521	3.60
Biotechnology	13830	1.23
Children	56735	5.04
Civic Issues	21794	1.94
Differently Abled	24945	2.22
Disaster Management	23678	2.10

Sectors	**Signed up NGOs**	**Percentage**
Dalit Upliftment	20790	1.85
Drinking Water	30729	2.73
Education & Literacy	89406	7.94
Environment & Forests	41521	3.69
Food Processing	22203	1.97
Health & Family Welfare	60242	5.35
HIV/AIDS	25738	2.29
Housing	12442	1.11
Human Rights	27948	2.48
Information & Communication Technology	23206	2.06
Legal Awareness & Aid	23093	2.05
Labour & Employment	27055	2.40
Land Resources	11205	1.00
Micro Finance (SHGs)	18878	1.68
Minority Issues	18764	1.67
Micro Small & Medium Enterprises	19376	1.72
New & Renewable Energy	12646	1.12
Nutrition	18908	1.68
Panchayati Raj	16822	1.49
Prisoner's Issues	5909	0.53
Right To Information & Advocacy	13103	1.16
Rural Development & Poverty Alleviation	38931	3.46
Scientific & Industrial Research	9210	0.82
Skill Development	13236	1.18
Science & Technology	16413	1.46
Sports	20970	1.86
Tribal Affairs	18018	1.60
Tourism	10701	0.95
Urban Development & Poverty Alleviation	20015	1.78
Vocational Training	41063	3.65
Water Resources	17246	1.53
Women's Development & Empowerment	53576	4.76
Youth Affairs	27789	2.47
Any Other	25159	2.24
Total	**1125391**	**100.00**

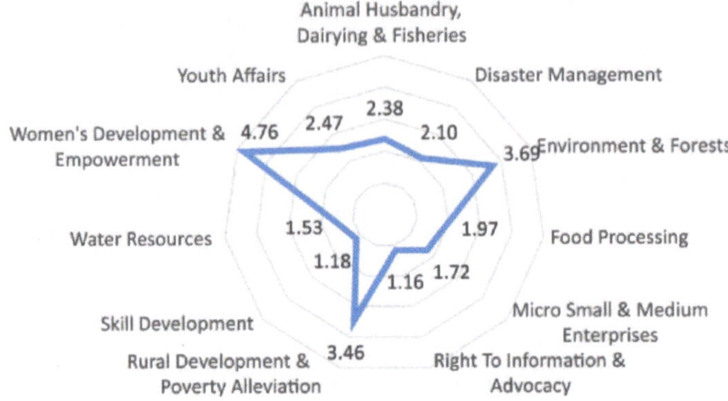

Figure 1: Fisheries-related NGOs in India

Categories of NGOs

The World Bank identifies two broad groups of NGOs,

- Operational NGOs, which target the planning and implementation of development projects, e.g., Organic agriculture in rural development, funding support for one-acre farmers (small hold farmers) for improving production, etc.,
- Advocacy NGOs, which defend or promote a specific cause and seek to influence public policy, e.g., human rights, social justice, environmental movements, etc.,
- NGOs' role in fisheries
- Planning routine training programs.
- Publishing a standard scientific international journals.
- Aiding budding scientists financially and technologically to carry out excellent research.
- Addressing significant fisheries challenges through publications, symposia, and seminars.
- NGOs work to change policy in the seafood sector by educating customers about sustainability issues, frequently through awareness campaigns and product guides.
- The Sustainable Fisheries Partnership (SFP), a global non-governmental organisation (NGO), was established to work with the private sector to advance sustainable fisheries by advancing commercial fishing and aquaculture techniques.
- The goal of SFP is to involve and stimulate the world's seafood supply chains in lowering the negative effects of fishing and aquaculture on the environment and restoring depleted fish supplies.

- NGOs aid in sustaining community development by enhancing their ability (Langran, 2002).
- It serves as a global organization that stands for traditional fishing communities whose means of subsistence are directly impacted by the sustainable management of fisheries resources.
- Promotes the recognition and upholding of small-scale and indigenous fishers' rights and the distinctive culture of fishing communities through advocacy and awareness campaigns.
- Enables organisations of small-scale fishermen to have an impact on national and international laws that affect their access, use, and control of resources for fishing as well as the sustainability of fisheries.
- Motivating and mobilizing people to be self-reliant and to participate in development activities (Nikkhah & Redzuan, 2010).

Role of NGOs in Transfer of Technology in Fisheries

- NGOs complement the government agencies in extension services.
- NGOs can engage in technology transfer through demonstrations, awareness campaigns, exhibitions, training programmes run by their own employees or staff from research organisations, and the regular dissemination of information through the media and publications.
- They can also serve as a link between the research system and client system and can thus provide vital feedback on field problems and field performance of the innovations.
- NGOs now provide services such as laboratory analysis works (water & soil quality analysis, seed & feed quality testing & disease diagnostics), and input supplies (certified feed and seed) and helps the clients to avail institutional credit (micro finance to fisher's) for the adoption of some innovations.
- NGOs play the multiple roles of technology generation, technology diffusion system, and supporting system for adoption in the field in the transfer of technology process. (Mondal, et al., 2015).

Fisheries-related NGOs in the World

1. Aquaculture Stewardship Council (Utrecht, Netherlands)

Through its certification programme, it acknowledges and encourages sustainable fish farming techniques. It is responsible for a set of standards addressing the key social and environmental impacts of fish farming.

2. Fisheries Innovation (Scotland)

It strives to bring the fish business, scientists, and government officials together to conduct research, promote knowledge transfer, and encourage innovation in Scotland's marine fisheries. The main focus is on education and capacity-building programs, the development of improved fisheries management models, and the assessment and development of new fishing gear and techniques. In 2017, FIS launched a Fisheries Innovation Award Competition, encouraging innovative ideas and approaches that could inspire new research and development projects.

3. Institute for Fisheries Resources

It works to carry out the fishery research and conservation needs of working fishermen and fisherwomen. It carries an establishment of alliances among fish workers, government agencies, and concerned citizens, to protect fish populations and restore aquatic habitats. It fought for Pacific Salmon restoration, toxins regulation, and agricultural reform. It provides advocacy work on common resource protection, conservation issues, and informs policy debates at the regional, national, and international levels.

4. International Collective in Support of Fish Workers (Rome)

It strives to create fisheries that are both socially equitable environmentally sound. Mainly focusing on advocating for policies for recognizing the rights of small-scale and traditional fishing communities, the role of women in fishing and fishing communities, the impact of climate change on fisheries resources, and the impact of trade on food and livelihood security in fishing communities.

5. International Seafood Sustainability Foundation (McLean, The United States)

It facilitates long-term conservation and sustainable use of global tuna stocks, reducing by-catch and promoting tuna ecosystem health. It works to reduce illegal, unreported, and unregulated tuna fishing by advocating for stronger RFMOs and strengthening MCS.

6. Marine Conservation Alliance (Juneau, Alaska)

Ecosystem-based fishery management, research on vital fish habitats and deep-sea coral, as well as improved risk control and catch limits regulations, are all MCA initiatives.

7. Marine Fish Conservation Network (The United States)

It is mainly promoting policies that improve opportunities for fishing communities, provide stronger protections for fish habitats, and secure more effective monitoring of fishing vessels.

8. Marine Stewardship Council (London, United Kingdom)

It is an international NGO that uses an eco-label and fishery certification programme to recognise and reward sustainable fishing methods. It assures that fisheries are properly managed to maintain healthy fish stocks and maritime environments.

9. Monterey Bay Aquarium (California, The United States)

It offers suggestions for sustainable seafood that are supported by research to customers, chefs, and companies who want to make informed decisions about their seafood purchases.

10. National Oceanic and Atmospheric Administration (The United States)

It is in coordination with federal, state, local, tribal, and international authorities, NOAA regulates fisheries and marine sanctuaries as well as protects threatened and endangered marine species. It provides factual information to guide consumers toward sustainable seafood purchases.

11. Natural Resources Defence Council (New York, The United States)

With the aid of its Smart Seafood Buying Guide, consumers may increase the variety of seafood they consume, stay away from

12. Northwest Atlantic Marine Alliance (The United States)

It is an organisation run by fishermen that works to preserve marine biodiversity and advance economic, social, and environmental justice. The main function is advocating for policy and market strategies that advance the rights of small and medium-scale fishermen.

13. Oceana (The United States)

It is the largest international advocacy organization focused solely on ocean conservation, through targeted policy campaigns. It mainly involved signing petitions on buying ocean-friendly products and reducing plastic waste.

14. Sailors for the Sea (Japan)

It promotes sustainable seafood farming and fishing.

15. Sustainable Fisheries Partnership (The United States)

It is an NGO that works to reduce the negative effects of fishing and fish farming on the environment and to stimulate the development of depleted fish supplies. It fills a specific gap between industry and the marine conservation community, operating through two main principles: information and improvement.

16. World Forum of Fish Harvesters and Fish Workers (Kampala, Uganda)

It serves as an international entity that represents the interests of traditional fishing communities, whose means of subsistence directly depend on the sustainable management of fisheries resources, and it gives small-scale fishers' organisations the authority to influence national and international laws that affect their rights to access, use, and control resources and the sustainability of fisheries.

Fisheries related NGOs in India

India has 154461 NGOs spread throughout 28 states and 9 union territories (table 2). The state with the greatest number (22348) from that is Uttar Pradesh (14.47%), followed by Maharashtra (13.81%), and West Bengal (7.41%) (figure 2). In the meantime, Sikkim has the fewest NGOs (143) (figure 3). Delhi has the most NGOs (11929) while Lakshadweep has the fewest of the Union territories (6). Below, a few significant NGOs involved in fisheries development are discussed.

Table 2 : State & UT wise NGOs list in India

States	Numbers	Percentage
Andaman & Nicobar Islands	174	0.11
Andhra Pradesh	5506	3.56
Arunachal Pradesh	513	0.33
Assam	2533	1.64
Bihar	5386	3.49
Chandigarh	241	0.16
Chhattisgarh	2043	1.32
Dadra & Nagar Haveli	32	0.02
Daman & Diu	18	0.01
Delhi	11929	7.72

Goa	293	0.19
Gujarat	7314	4.74
Haryana	3607	2.34
Himachal Pradesh	894	0.58
Jammu & Kashmir	1870	1.21
Jharkhand	2851	1.85
Karnataka	8869	5.74
Kerala	4335	2.81
Ladakh	176	0.11
Lakshadweep	6	0.004
Madhya Pradesh	7122	4.61
Maharashtra	21338	13.81
Manipur	2628	1.70
Meghalaya	328	0.21
Mizoram	275	0.18
Nagaland	474	0.31
Orissa	4470	2.89
Puducherry	327	0.21
Punjab	1940	1.26
Rajasthan	5868	3.80
Sikkim	143	0.09
Tamil Nadu	10632	6.88
Telangana	3643	2.36
Tripura	529	0.34
Uttar Pradesh	22348	14.47
Uttarakhand	2264	1.47
West Bengal	11542	7.47
Total	**154461**	**100**

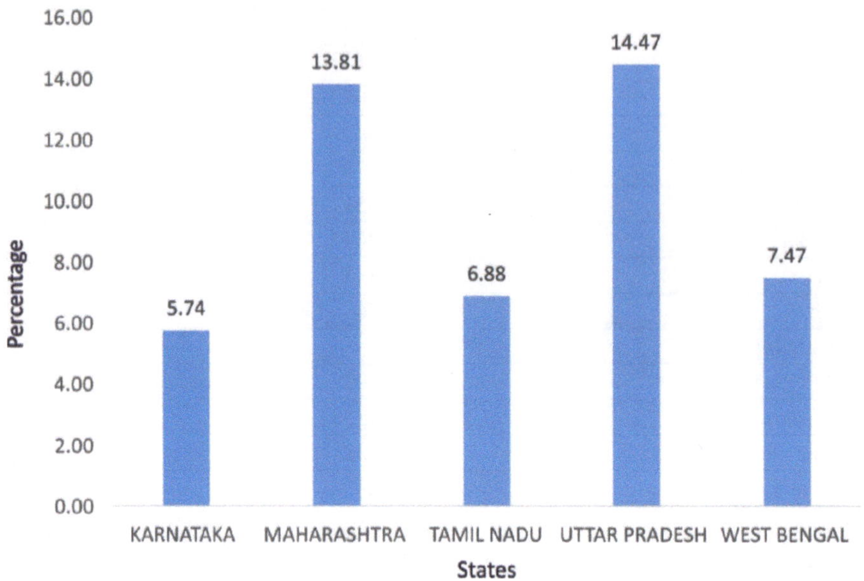

Figure 2 : Contribution of top 5 States in total numbers of NGOs in India

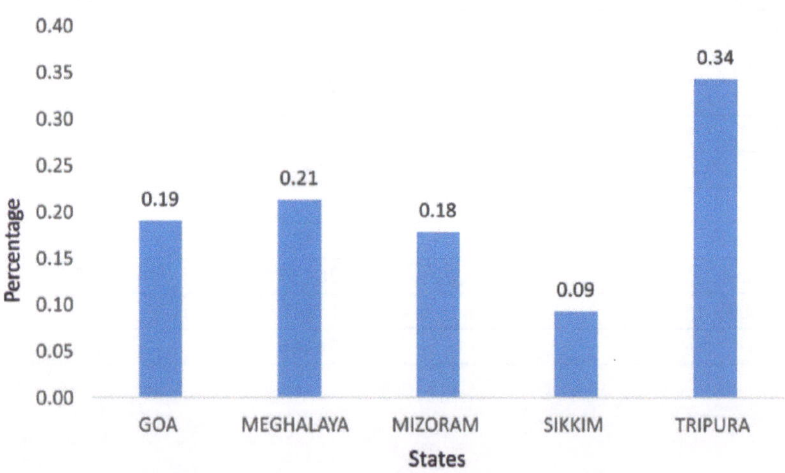

Figure 3 : Contribution of least 5 States in total numbers of NGOs in India

1. Watershed Organization Trust (WOTR) – Maharashtra

Watershed Organization Trust (WOTR) is a well-known Maharashtra-based organization that works on a variety of rural development projects. Agriculture, allied sector growth, climate change adaptation, watershed, natural resource management, social development, training, and capacity building are among the topics addressed.

2. Universal Versatile Society (UVS) - Maharashtra

UVS' initiatives are aimed at reducing farmer suicides by offering basic education, creating awareness, imparting employable skills, developing infrastructure, and growing small businesses, all of which lead to long-term revenue generation and employment prospects.

3. Dilasa Sanstha - Maharashtra

Dilasa Sanstha, founded in 1994, collaborates with a number of smaller NGOs in Maharashtra's Vidarbha and Marathwada districts. It promotes creative and long-term irrigation technologies to help tribal people and small Indian farmers expand their agriculture and allied activities. Dilasa's low-cost technological solutions assist farmers in improving irrigation practices and practicing mixed agriculture.

4. Islanders Sangathan Manch - Andaman and Nicobar Islands

Islanders Sangathan Manch is based in the Indian territory of Andaman & Nicobar Islands. Established in the year 2005, Islanders Sangathan Manch works in the area of Aged and elderly, Animal Husbandry, Dairying & Fisheries. The organization works towards the promotion of sustainable development.

5. Al Ihsan Enterprises - Lakshadweep

A non-governmental organisation (NGO) operating in Lakshadweep, India, is called Al Ihsan Enterprises. Al Ihsan Enterprises was founded in 1998 and focuses on animal husbandry, dairying and fishing, animal welfare, art and culture, children's welfare, Dalit welfare, ecotourism, education, and literacy. The non-profit organisation promotes sustainable development.

6. Lakshadweep Marine Research and Conservation Centre - Lakshadweep

Lakshadweep Marine Research and Conservation Centre was established in the year 2005. It works in the area of Aged and elderly, Animal Husbandry, Dairying & Fisheries, Education & Literacy, Environment, Food and Agriculture.

7. The M S Swaminathan Research Foundation (MSSRF) – Tamil Nadu

The M S Swaminathan Research Foundation (MSSRF) was established in 1988 as a not-for-profit trust. MSSRF was envisioned and founded by Professor M S Swaminathan with proceeds from the First World Food Prize that he received in 1987. The Foundation aims to accelerate the use of modern science and technology for agricultural and rural development to improve the lives and livelihoods of communities. Since its small start, the Foundation has changed the lives of over 6,00,000 agricultural families and the daily livelihoods of 100,000 farmers and fishermen. This impact has been felt over the years in many different ways.

8. South Indian Federation of Fishermen Societies (SIFFS) - Kerala

The South Indian Federation of Fishermen Societies (SIFFS) is an NGO that works in the marine fisheries industry. SIFFS is the leading group for small-scale artisanal fish workers. It has a three-tier organizational structure. With over 9104 member fishermen, organized through 153 primary societies in 8 districts of Southern peninsular India, SIFFS over the last two decades has kept its focus on strengthening the artisanal fisheries. SIFFS was initially founded primarily as a fish marketing organisation, but it now offers both members and non-member fish workers a variety of services. More than 65,000 fish workers, including non-members are availing of these services.

9. Centre for Agriculture and Rural Development (CARD) - Maharashtra

The Centre for Agriculture and Rural Development is a premier NGO that was founded and incorporated in 2000 under the Societies Registration Act of 1860. It is an integral part of India's attempts to develop through agricultural transformation. CARD is committed to reaching out to all parts of rural society, especially the farming community.

Conclusion

NGOs are organisations that work on a variety of social and developmental concerns and are non-profit, voluntary, and independent of the government. Although Animal Husbandry, Dairying and Fisheries sector has 26737 NGOs out of 154461, due to the significance of the fisheries development, some of the other sector NGOs also involving in fisheries activities, which are successful at inspiring the target fish farmers and producers to improve their aquaculture techniques. It has been successful in enhancing fish producers' access to markets, technology, credit, fish seed, and water resources. As a result, they have been very successful at reaching aqua farmers with an extension message. They are primarily responsible for encouraging and supervising small-scale farmers or fish farmers. Some NGOs support farmers who have ponds that provide a living for aqua farmers who are poor and in need.

References

Bihar Animal Sciences University (2020): https://www.basu.org.in/wp-content/uploads/2020/06/4.-Role-of-NGOs-and-SHGs-in-fisheries-Copy.pdf

Desai, V., 2014. 8 The role of non-governmental organizations (NGOs). The companion to development studies, pp.590-594.

Eva Perroni., (2017): https://foodtank.com/news/2017/10/sustainable-fisheries-list/

Langran LV 2002. Empowerment and the Limits of Change: NGOs and Health Decentralization in the Philippine, Department of Political Science. Ph.D. Thesis, Toronto: University of Toronto.

Mondal, D., Chowdhury, S. and Basu, D., 2015. Role of non-governmental organization in disaster management. Research Journal of Agricultural Sciences, 6, pp.1485-1489.

NGO-DARPAN: https://ngodarpan.gov.in/index.php/home/sectorwise

Nikkhah, H.A. and Redzuan, M.R.B., 2010. The role of NGOs in promoting empowerment for sustainable community development. Journal of Human Ecology, 30(2), pp.85-92.

Petersson, M.T., 2020. Transparency in global fisheries governance: The role of non-governmental organizations. Marine Policy, p.104128.

Streeten, P., 1997. Non-governmental Organizations and Development. Annals of the American Academy of Political and Social Science, 554, pp.193-210

Willetts, P., 2002. What is a non-governmental organization? Conventions, treaties and other responses to global issues, 2(11), pp.229-248.

Section 5
Fish Pathology, Genetics Biotechnology and Nutrition

20

Organ On A Chip

Rajendiran Rajeshkannan

Department of Fish Biotechnology, Fish Genetics and Biotechnology Division
ICAR-Central Institute of Fisheries Education (ICAR-CIFE)
Mumbai, Maharashtra

Abstract

Organs-on-chips (OoCs), often referred to as microphysiological devices or "tissue chips", have garnered a great deal of attention in recent decades due to their potential to be instructive at different phases of drug discovery. These cutting-edge tools might offer insights into the healthy functional status and disease pathophysiology of various organs, in addition to more precisely forecasting the efficacy and safety of experimental therapeutics in humans and fishes. As a result, they are probably going to supplement current in vivo animal studies and preclinical cell culture techniques in the short term and, in some situations, replace them in the long haul. The OoC sector has dramatically advanced in the last ten years in terms of the complexity of biology and engineering, the proof of physiological relevance, and the variety of applications. To fully achieve the potential of OoCs for core and therapeutic applications, insight from a variety of pharmacology, biomedical, and engineering domains will be required. These advancements have also highlighted new obstacles and opportunities. This chapter covered microfabrication methods and contemporary uses of this fast-developing technology in pharmacology and pathophysiology research on several human organs as well as zebrafish.

Keywords: Microfluidic system, OOC, Zebrafish-On-a-Chip, Drug Discovery and Tissue Chips.

Introduction

Organ-On-a-Chip (OOC) is a developing trans-disciplinary technique that shares similarities with tissue engineering and lab-on-a-chip technologies. It has benefited from recent advancements in microtechnology (particularly microfluidics), cell biology, physiology, and tissue engineering and is motivated by the need for reliable, low-cost in vitro drug screening models. It can be characterized as a microfluidic-based perfusion device that facilitates an in vitro cell (co-)culture and tries to reproduce certain structure(s), function(s), and important elements of human or animal metabolism of a particular tissue or organ in healthy and pathological physiology. Important features of live organs are such as extracellular microenvironments, spatiotemporal cell-cell interactions, and physiologically important tissue microarchitecture. Due to their optical qualities, which allow for live-cell imaging, polydimethylsiloxane (PDMS) and glass constitute the most frequent fabrication materials used in their construction. With the ability to control important variables including concentration gradients, shear stress, cell patterning, tissue-boundaries, and tissue-organ interactions, it can simulate the organ's physiological environment.

In order to investigate the physiological functions that characterize the interaction between organs, the immune system, and exogenic (for example, pharmaceutical and nutraceutical) stimuli in health and disease states, the OOC technology aims to develop efficient and reproducible comprehensive microphysiological models. OOCs of a number of different organs or tissues, including the liver, lung, gut, kidney, skin, bone, adipose, heart, brain/blood-brain barrier, vasculature, and illnesses such cancer, diabetes, infection, and thrombosis. The great majority of the OOC devices currently in use rely on straightforward microfluidic chips that are made up of either single or double-layered circuits with a permeable membrane. Before integrating with another functioning organ, it is essential to have a comprehensive understanding of the cell-cell interaction and general tissue function and structure of a single organ in order to create multi-organ-based models.

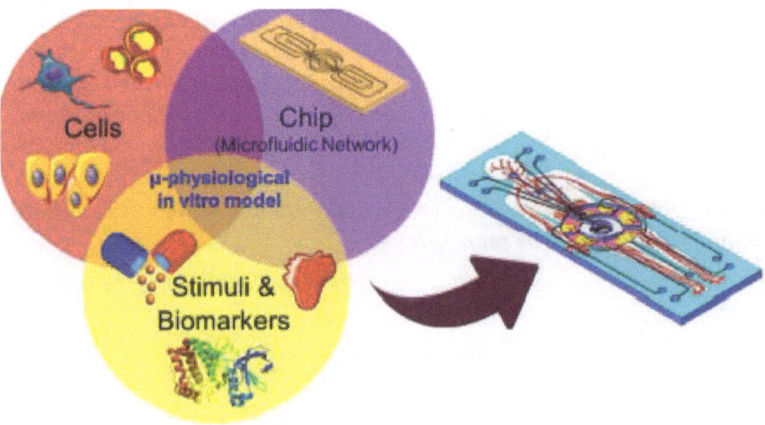

Figure 1 : Main components of Organ-On-a-chip (Ramadan & Zourob, 2020).

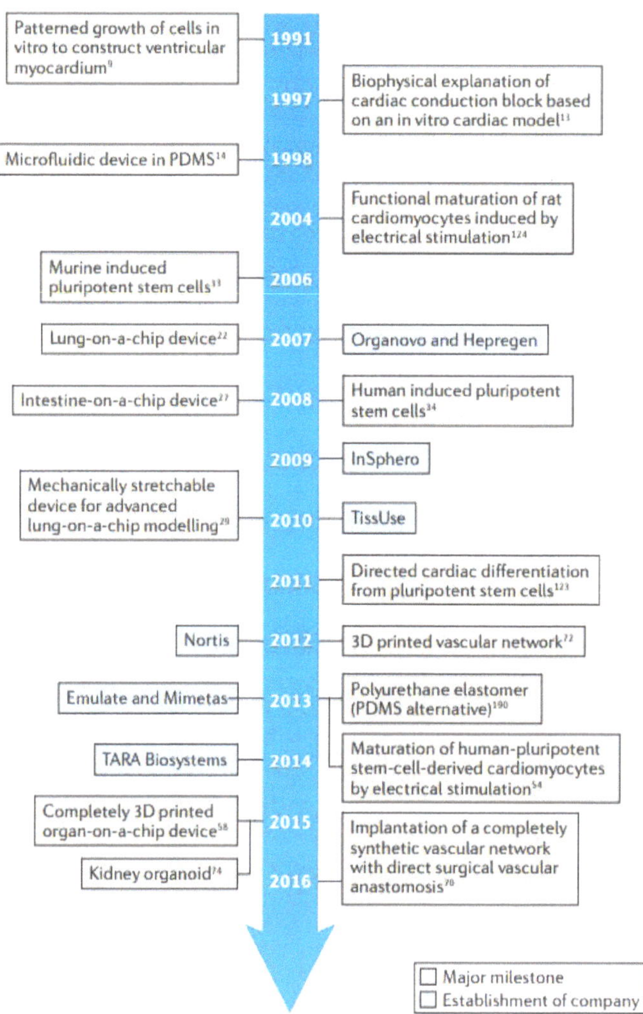

Figure 2 : Major milestones in organ-on-a-chip technology and establishment of organ-on-a-chip companies (Zhang et al., 2018).

Table 1 : Key Features of 2D and 3D Engineered Tissues (Low et al., 2020)

Parameter	Conventional 2D systems	3D systems	
		Organoid	**Organ-on-chip**
Production characteristics	Grown on rigid flat surfaces, often as a cellularly homogeneous monolayer	Embedded in hydrogels/suspended in 'hanging drops', and left to self-organize into multiple cell types	Multiple relevant cell types seeded into engineered chambers with perfusion and/or biomechanical forces included
Production complexity and speed	Generally straightforward and fast (minutes to days)	Generally straightforward, but slower (days to weeks) depending on cell sources	Variable complexity (depends on platform design), slower (days to weeks) depending on cell sources and required tissue maturation metrics
Level of control over cell architecture	High	Very low	High
Maturation of iPS cell-derived cells allowed by platform[a]	Immature	Improved but still highly immature	Platform designs can improve and encourage cell maturity
Resulting cell morphology	Unnatural, with limited ECM composition and contact with cells	Size and shape similar to in vivo case, allows relevant ECM interaction during cell proliferation	Size and shape similar to in vivo case, allows relevant ECM interaction throughout cell lifetime
Diffusion of signal factors and nutrients	Short distances possible	Ineffective transport to interior can cause cell death or immaturity	Allows precisely controlled temporal and spatial gradients
Vascularization or perfusion?	Not possible, generally perfusion via medium change	Depends on cell types but likely creates non-functional vessels; externally perfused; can include fluid flow across tissue surfaces	Yes — by microfluidic channels or design which can include/create endothelialized vessels
High-throughput feasibility?	Yes	Possibly, depending on tissue	Depends on platform design; generally low to medium throughput
On-platform assay and analysis difficulty	Low difficulty, easy access to cells and readouts	Tissue function analyses possible; cell separation not possible	Real-time tissue/organ function analyses possible
Variability and in vivo relevance of resulting tissues in manufactured platform	Low variability and relevance — simple, homogeneous cultures	Can show high variability and low relevance as there is little control over resulting cell subtypes and location	Can show low variability and high relevance — allows high levels of control over cell type and placement

ECM, extracellular matrix; iPS cell, induced pluripotent stem cell. [a] Immaturity of iPSC-derived cells is still a general issue.

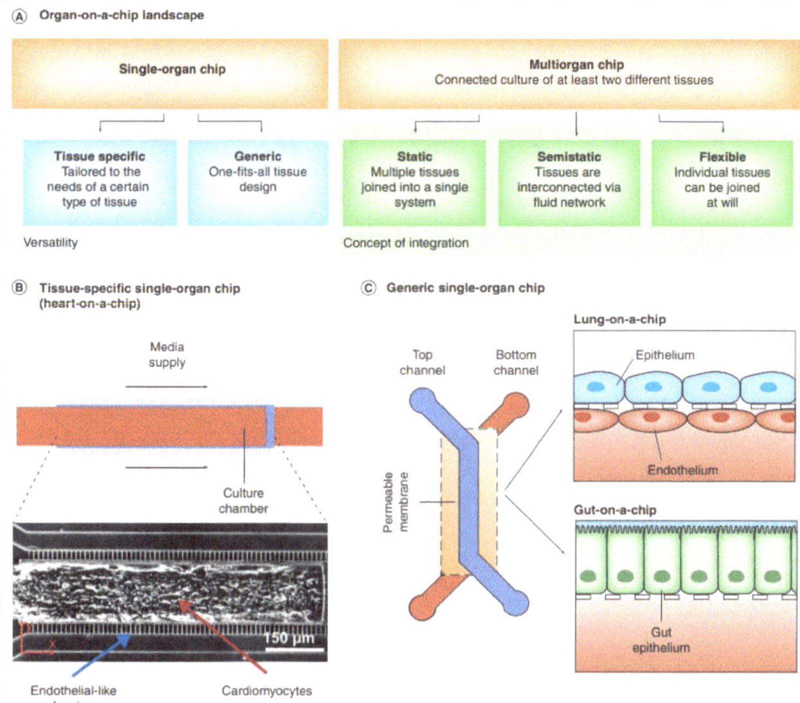

Figure 3 : The basic concept of organ-on-a-chip technology (Rogal et al., 2017)

(A) The two basic types of organ-on-a-chip systems: single-organ chips, which integrate just one kind of tissue or organ, and multi-organ chips, which incorporate at least two various kinds of tissue or organ chambers. Tissue-specific **(B)** and generic **(C)** single-organ chips are two subcategories of single-organ systems. A generic one-geometry-fits-all-tissues strategy enables quick commercialization, but the geometry of organ/tissue-specific chips is properly adjusted to the requirements of a particular type of tissue.

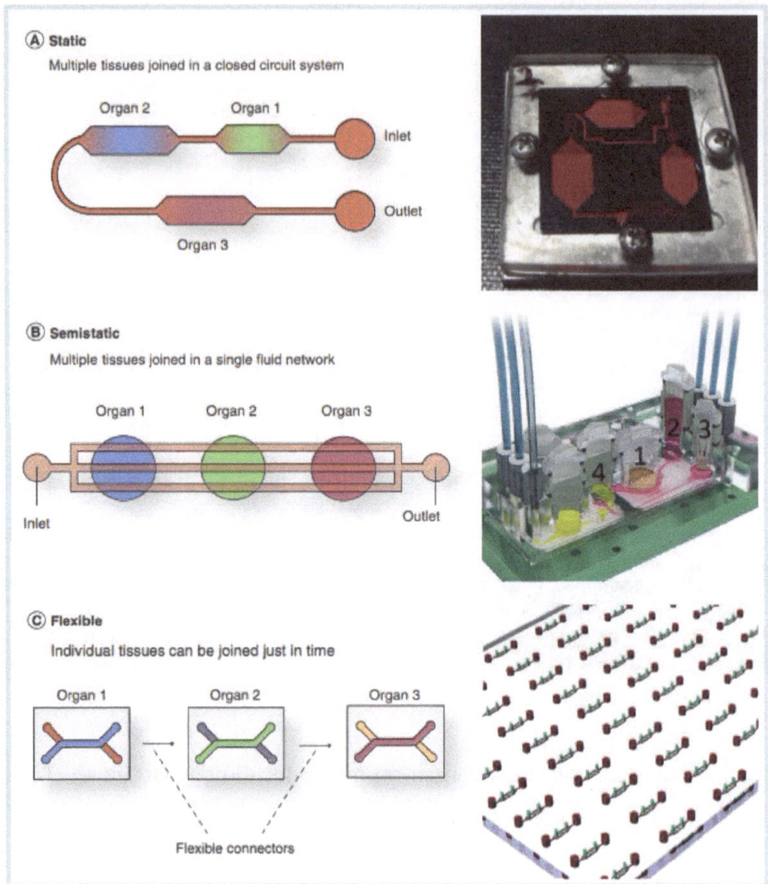

Figure 4 : Basic strategies for integrating with multi-organ devices (Rogal et al., 2017)

(A) Static systems: Several tissues are combined into a single, interconnected device. **(B)** Semistatic systems: Transwell®-based tissue implants connect tissues through a fluidic network. **(C)** Flexible systems: Flexible microconnectors are used to connect several platforms that are tailored to certain organs or tissues.

OOC Engineering

Requirements of the Microfluidic Platforms

OOC performs differently from other lab-on-chip (LOC) systems since that maintains a healthy cell co-culture in a synthetic environment for a considerable amount of time and simulates in vivo conditions in a tiny bioreactor. The OOC system's ideal use is to offer the equipment and setting necessary to research cellular arrangement in vitro that resembles its counterpart in vivo and to record the spatiotemporal behaviour of cells in response to exogenic stimuli and chemicals. The integration of process control, sensors, imaging systems, and other analytical components would be made possible by miniaturization (Ashammakhi et al., 2019). However, these could differ according to the relevant in vitro model, such as liver, skin, heart, etc.

A. Supply of Cell Nutrients and Waste Removal In A Stable Perfusion Circuit

The OOC's cellular system simulates how cells and tissues interact with blood and other circulating materials. A dependable fluidic circuit must be created in order to make this possible. It must provide constant fluid flow at the given flow rate, with the possibility of flow modification and medium dilution. The experimental circumstances are severely impacted by bubble creation in microfluidic channels because of their small capacity, hence the fluidic circuit must contain a device to remove or prevent bubbles. Cell population density changes during the culture (in most cases increases), and in the case of co-culture, various cell populations may have diverse proliferation profiles that necessitate adjusting the nutrient supply correspondingly. One of OOC's key differentiators is perfusion, which is achieved by keeping cells in a channel or chamber while changing the medium that carries nutrients to support cell growth and viability. OOC systems keep cells alive in culture for substantially longer periods of time by continuously feeding them with fresh media and clearing away cell trash.

Advantages of Perfusion

- By supplying nutrients continuously and removing waste, it is possible to maintain nutrient levels for ideal growth circumstances while removing cell waste products to prevent cell toxicity.
- The protein of interest is secreted by the cultured cells into the flowing media, where it can be sampled and examined in situ or off the chip.
- Avoiding exposing the medicine or stimulant under consideration to excessive waste that could lead to a departure from the intended conditions (e.G., Drift in the ph value).
- The OOC microfluidic chip's cell-to-liquid ratio can be optimized, which will utilize less culture media overall and lower the cost of culture.
- Permitting the cell membrane to experience physiologically meaningful shear stress. Particularly in vivo, endothelial cells undergo development and differentiation when subjected to flow-induced shear stress.
- Endothelial cells must be cultivated under constant supply for an extended period of time with a specified flow rate to simulate the effects of certain shear stress in vitro.

- Triggering liquid exchange in the segmented fluidic system and allowing time-dependent monitoring for downstream analyses.

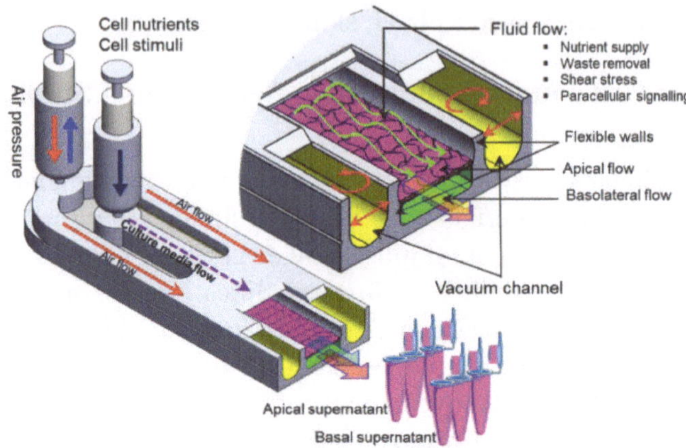

Figure 5 : A basic OOC platform with important features (Ramadan & Zourob, 2020)

B. Device Fabrication Materials

The soft lithography method is used by PDMS to produce the great majority of LOC and OOC devices. Due to the ease of fabrication, elasticity, gas permeability, biocompatibility, quick molding and good optical clarity of PDMS, the research community has been able to apply it in various applications and achieve high-impact results. However, PDMS used as an OOC material which may not be considerable material for mass production of LOC and OOC by its limitations including absorbing drugs, small biomolecules and other organic compounds and the lack of industrial standards. Recently, several various materials including polystyrene (PS) and polymethylmethacrylate (PMMA) are used to fabricate the microfluidic/LOC/OOC devices to replace and overcome the shortcoming of PDMS.

Table 2 : List of widely used materials in the fabrication of LOC and OOC devices (Ramadan & Zourob, 2020)

Materials	Biocompatibility	Optical Property (Transparency)	Mechanical Property (Elasticity)	Chemical Resistance	Manufacturability	Cost
PDMS	Good	Good	Good	poor	Lab based soft lithography No scale up production	High
Polymethyl-methacrylate (PMMA)	Good	Good	Rigid	Good	Microinjection molding, hot embossing, castimg, reactive ion etching, mechanical milling (CNC), Laser micromaching	Low
Polystyrene (PS)	Good	Good	Rigid	Poor	Injection molding, hot embossing	Low
Polyimide (PI)	Good	Poor	Poor	Good	Photosensitive lithography	High
Polycarbonate (PC)	Good	Poor	Rigid	Good	Injection molding, hot embossing	Low
Cyclic Olefin Copolymer (COC)	Good	Poor	Rigid	Good	Injection molding, hot embossing	Moderate

Principles of Microfabrication

Organ-on-a-chip devices highly depend on fabrication processes to accurately simulate in vivo microenvironments. Because it is possible to create tissue conditions at the microscale, organ-on-a-chip tools have been made using microfabrication techniques on a large scale (Jiang et al., 2013). Organ-on-a-chip devices have been created using a variety of methods, including replica molding, soft lithography, and microcontact printing. Polydimethylsiloxane (PDMS), illustrates its flexibility and biocompatibility. Because PDMS is optically clear, it can be used to take clear pictures of cultivated cells (Bhatia & Ingber, 2014).

In this design, a consistent photoresist coating is applied on the silicon wafer. By being exposed to UV light, the photoresist may be either crosslinked or uncrosslinked. Only the desired areas of the photoresist can be lighted when a photomask is used. The silicon wafer substrate now serves as a mold by transferring the pattern to them. As it bakes in the oven, the PDMS polymerizes after being put onto the patterned mold. After that, the PDMS platform is separated from the substrate. The procedure is known as replica molding since the substrate can be utilized for numerous fabrications. Soft lithography's subordinate technology is replica molding.

Fabricating PDMS tubes for moving fluids is another application for soft lithography. The only difference between the fabrication processes and the replica molding technique is that the PDMS system is subsequently attached to other substrates (usually glass slides or other complementary PDMS layers). On a microscale, the flow pattern of liquid in the microfluidic channels is laminar. Cell-flow interactions can be studied by manipulating laminar flow. Cell aggregation, heart tissue creation, and differentiation can all be studied using a microfluidic chip with regulated laminar flows. By cultivating two distinct kinds of cells on each side of the porous PDMS substrate, microfabrication techniques also enable the creation of porous membranes that imitate the vascular endothelium and parenchymal tissues (Wang et al., 2015).

The microcontact printing method allows for accurate regulation of cell patterning in culture. An elastomeric PDMS stamp is created and covered with a monolayer of proteins after a PDMS membrane has been constructed. The desired regions of the proteins could be transferred to the membrane substrate by stamping the PDMS stamp onto the substrate. Since cells only proliferate in protein-containing regions, the membrane's cells can also be designed by varying the pattern of printed proteins. The most often employed elements for organ-on-a-chip platforms designed to research in vivo interactions, such as cell-cell interactions, cell behaviours, and cell responses to chemical and mechanical stimulations, are microwells, fluidic channels, and porous membranes.

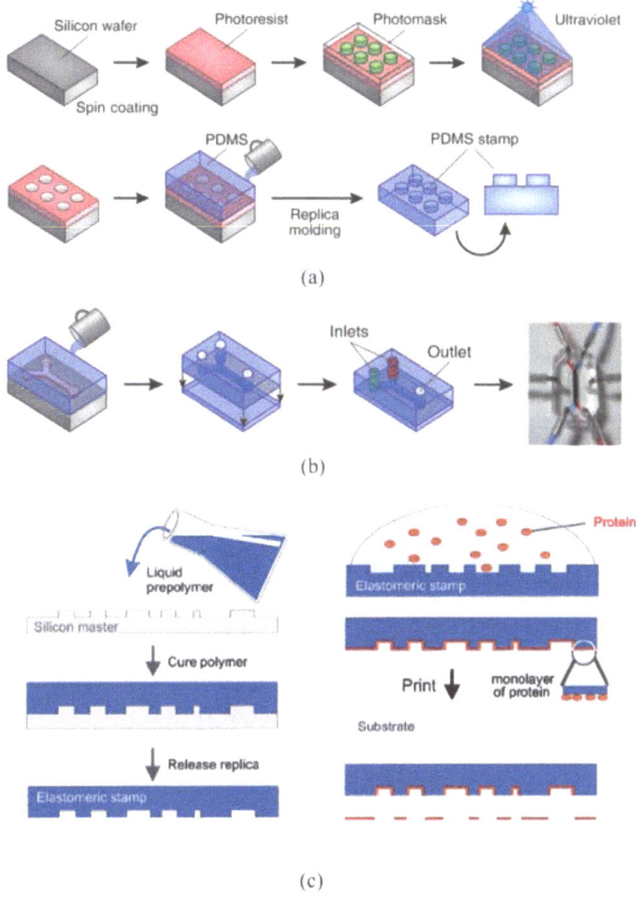

Figure 6 : Microfabrication techniques and applications (Wang et al., 2015)

(a) Replica molding is used to make PDMS stamps. **(b)** PDMS channel with one outlet and two inlets fabricated using soft lithography. **(c)** Proteins are microcontact printed to pattern cells.

C. Tissue Architecture: Cell–Cell and Tissue–Tissue Interface

OOC is an advanced type of cell culture design that guarantees precise cellular placement and in vivo-like cell polarization by giving cells a template on which to replicate a complicated assembly and imitate the actual tissue organization. Microfluidic systems are created to enable the organization of diverse cell types with the proper tissue architecture, allowing for the construction of an in vitro organotypic cellular structure with fluidic/bio/chemical exchange. Spheroids or multicellular structures are produced using three-dimensional

(3D) cell culture (Ronaldson-Bouchard & Vunjak-Novakovic, 2018). However, spheroidal tissue is not always present in the tissue topologies found in living organisms. The asymmetrical protein arrangement in the cell membrane and the development of cell-cell tight junctions (tjs), which separate the basolateral and apical membranes, are what cause polarity, an ineluctable structural characteristic of organs. It is a key component of hepatic, endothelial, and epithelial structure and function and can be created through cell culture or extracellular matrix (ECM) techniques (Materne et al., 2015; Lelievre et al., 2017).

The integration of numerous cell types and/or organs into a single fluidic system using microfluidic technology allows for organ-organ crosstalk while maintaining the functionality of each separate organ and simulates the function of vascular perfusion in vivo. Using a standard cell culture medium to sustain all the cell types and organs within the integrated system is an easy solution. This strategy is restricted to tissues that are phenotypically stable and mature, though. Multi-tissue structures may be housed in compartmentalized microfluidic systems that are divided by an endothelial barrier to imitate the tissue-blood interface in the body. Tissue-specific media may be retained in each compartment to promote and mature each tissue optimally while facilitating the crosstalk among the tissue components via vascular flow (Ramadan & Zourob, 2020).

The size of the hosted organ, the sequence in which the organs are connected, the orientation of the tissue, and the perfusion rate within each compartment are all important factors in the design of the compartmentalized fluidic system for OOC, which ensures a meaningful in vitro system. A correctly constructed OOC system could offer a greater comprehension of cellular metabolism and organ-organ interaction because cell metabolism changes from one organ to another and during cell maturation. The physiological parameters that must be reproduced in vitro serve as the primary design criteria for OOC. This might be accomplished by choosing the key cell models (from either cell lines or the induced pluripotent stem cell (ipsc) source), biochemical stimuli (e.g., Drugs and toxins), and physical stimuli (e.g., Hydrodynamic, mechanical, and electrical) to create a basic functional structure of a single organ or integrated organs.

Various Connection Strategies in OOC

Convection-Based Fluidic Transfer (Manual Pipetting or Through Tubing)

This straightforward connection technology allows for the integration of many separate organs and permits organ-organ communication via secreted factors with no requirement for a micro-fabricated channel to join the fluidic chambers. It is restricted to the use of organs that can be supplemented by the same culture media and lacks to simulate the natural flow between organs (Satoh et al., 2018).

Planar Oriented Gel or Porous Barriers-Based Connection

In this setup, cells are constructed in a planar compartmentalized microfluidic framework in a two-dimensional (2D) organization. In such designs, semi-porous vertical barriers serve as physical barriers separating the several compartments. Therefore, many cell types are cultivated in close proximity to one another while remaining fluidically and chemically coupled. The array of tiny pores that make up the semi-porous walls allow only liquid and not cell exchange via the neighbouring compartments. To manage the mass transfer between the various compartments, the size of these pores might be modified to meet unique requirements (such as the kind and size of the cells) (Liu et al., 2018).

For the purpose of enabling biomimetic vascular structures between various organs, the porous channels can be loaded with porous materials (for example, hydrogel). In turn, this would make it possible for cells in various nearby compartments to interact chemically and biologically (a process known as paracrine signaling). Internal or external pumps or valves can be used to regulate the direction of the inter-compartment flow, flow profile, and rate. Multi-cell type cultures can be formed in the appropriate sequence by taking use of the porous barrier structure, and the interactions between the cells/tissue within each compartment are determined by the course of the perfusion flow. The substance and concentration of chemicals within the lateral compartment can be precisely changed by using the perfusion micro-pores in conjunction with the micro-flow in the neighbouring compartment to establish chemical gradients as needed. Due to its simplicity in manufacture and usefulness, the horizontal arrangement of organs is a typical configuration for studying how particular organs and cells interact with the immune system (Parittotokkaporn et al., 2019).

Connection Through Vertically Oriented Porous Barriers

This arrangement allows for the co-culture of two or more cell types in close proximity to one another in two vertical orientations by stacking two or more fluidic compartments vertically and allowing a porous membrane to act as an interface between them. The two cellular architectures can be physically separated in this structure by a permeable membrane with optimal pore size and thickness. The porous membrane can also be utilized as a substrate for growing two different kinds of cells on either side. This structure would imitate the structure and function of significant human body parts when epithelial/endothelial/epidermal cells are cultivated on the upper surface of the membrane and the equivalent parenchymal tissue is grown on the lower side. Examples of organs and tissues that often regulate how the body interacts with medications, food, and the environment include the skin, small intestine, lung parenchyma, and blood vessels (Ramadan & Zourob, 2020).

A model of a particular organ, including the liver, adipose tissue, muscle, bone, etc., may be housed in the lower compartment. To assess the bioavailability of drugs, chemicals, and other molecules in the target organs and their metabolic profile, this structure can be utilized to recreate to transport (absorption and distribution) them (these bioactive substances) via the epithelium. Additionally, migrating immune cells can be hosted in the bottom compartment/channel to simulate the activation of the immunological system following the passage of foreign particles through the epithelia.

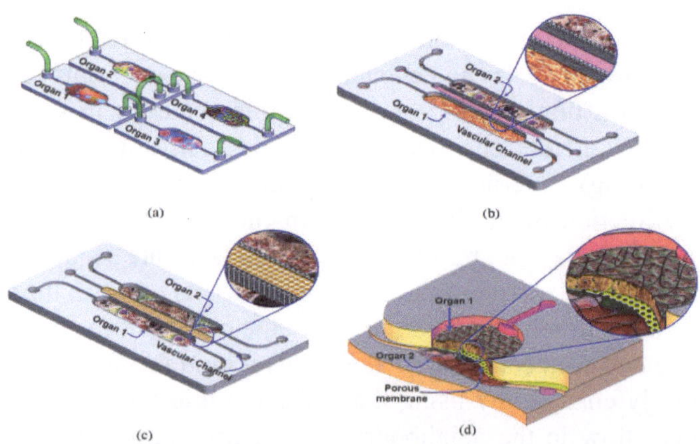

Figure 7 : In vitro models' designs of fluid circulation between cell culture chambers (Ramadan & Zourob, 2020)

(a) External tubing connects several microfluidic devices. **(b)** Porous thin solid barriers divide the fluidic chambers. **(c)** A narrow conduit constructed of porous materials divides the fluidic chambers. (e.g., gel). **(d)** A permeable membrane distinguishes two compartments that are stacked vertically.

D. Allometric Scaling

Allometric criteria must be used in a successful OOC design to account for the interaction between cells, cell metabolism, and exchange in the human body in order to produce suitable cell proportions in the system. To enable the predictive potential, two independent allometric scaling models namely, the metabolic scaling model and the cell number scaling model must be taken into consideration. For instance, Ucciferri et al. (2014) suggested that is to use connecting hepatocytes with endothelial cells to construct an in vitro model of biotransformation and distribution in the liver before adding any other cells. Since the endothelium in the model is scaled utilizing the surface area of the human circulatory system, the hepatocytes in the model are scaled using the basal metabolism. The allometric design method starts by taking into account the metabolic activity of a two-dimensional culture of human hepatocytes in a single system because cells are often coated in monolayers. Allometric scaling would allow us to precisely predict human pharmacokinetics and pharmacodynamics by combining *in vivo* and *in vitro* data.

E. Consistent Performance Over Long Term

In order to examine the time-dependent cellular behaviour that varies according to the kind of cells/tissue and the downstream analysis, the OOC platform is a compact bioreactor that houses the hybrid cellular structure for lengthy periods of time. During the relatively long lifespan of the OOC model, it is crucial to maintain steady cell viability and manage the phenotypic stability in their artificial microenvironment in order to obtain valid in vitro data (Mandenius, 2018). The three steps of the OOC system's life cycle are (i) developing the model's structure, (ii) characterizing the model's functionality, and (iii) validating and testing the model. When moving from the prototype to product development, these functions and processes should guide the engineering design of the OOC system.

Model Structure (Micro-Anatomy)

Controlling the environment and keeping an eye on cell development and phenotyping are crucial during the stage of creating the in vitro model's structure.

(i) The Microfluidic Dynamic Properties

A key element of the dynamic cell culture is the perfusion flow dynamically. The essential cell-liquid/nutrient interaction as well as the physiological information of humans must be taken into consideration while optimizing these factors, which include the shape and dimensions of the channels/chambers and the time-dependent pressure gradient. Furthermore, the flow/pump mode must be originally established in order to configure the kind of flow, such as constant or pulsatile flow. The drive frequency should be specified in the second scenario within a physiologically appropriate range (i.e., In the range of the human heartbeat of 60–180 beats per minute or a frequency of

1–3 Hz). The cellular microenvironment, which regulates the proliferation, differentiation, phenotypic, and migration of cells and tissues, is strained by mechanical shear stresses imposed by the fluid flow within the microfluidic channel (Kaarj & Yoon, 2019).

(ii) The Physicochemical Parameters (Temperature, pH, oxygen, CO_2)

The ideal micro-bioreactor would be able to manage the physical and chemical conditions of the cell culture while also supplying cells with dynamic profiles of nutrients, oxygen, and growth factors. Mammalian cells flourish at temperatures of about 37 °C, therefore maintaining suitable culture temperatures is essential for optimum cell proliferation. Additionally essential to cellular activity is the pH level. The buildup of metabolic acidic products by cultures that have proliferated too densely or for too long in that medium, the low oxygen levels, and contamination by quickly growing bacteria or fungi can all cause the pH of the cell culture to depart from the ideal ranges. The great majority of cell culture media comprise carbonate-based buffers, which, like those found in vivo, operate with high levels of CO_2 in the incubator to stabilize the pH of the cells.

In a small micro-bioreactor, Zhang et al. (2017) exhibited a combined pH and oxygen sensor. In contrast to the oxygen sensor, which uses the oxygen-quenchable luminous dye [ru(dpp)3]2 + cl2 - tris(4,7-diphenyl-1,10-phenanthroline) ruthenium (ii) chloride, the pH sensor monitors variations in the light absorption of phenol red in the culture medium to convert into a voltage change. Soucy et al. (2019) have presented a thorough overview of numerous instrumented in vitro microphysiological platforms. The quasi vivo® system was created by Kirkstall and comprises interlinked cell culture compartments and a peristaltic pump to maintain a constant supply of media over cells. The three separate culture chambers that make up the quasi vivo® system (QV500, QV600, and QV900) allow for the monitoring of several parameters throughout a cell culture experiment. The technology is also compatible with common cell culture inserts (transwell).

(iii) Mass Transfer at The Tissue–Fluid Interface

Adequate fluidic connections among organ modules (fluidic chambers) are necessary to imitate in vivo cross-organ interactions. Recirculating microfluidic systems' perfusion circuit makes it possible to simulate blood circulation and makes it easier for organs to communicate. Maintaining a suitable level of cell viability during the cell culture process is one of the fundamental issues in OOC systems, especially when using 3D cell culture or a multi-organ model. Therefore, it is essential that oxygen, growth factors, nutrients, and any other regulatory substances be efficiently delivered via the multi-tissue structure from the majority of culture media (internal mass transfer) (Salehi-Nik et al., 2013). Inter-tissue mass transfer is often influenced by the permeability and size of the microfluidic system, the diffusion rate through the biomaterial, the sequence or orientations of the tissue in the multi-organ system, and a conjunction of convection and diffusion mechanisms. The porosity, thickness, and structure of the substance (such as hydrogel) are the most crucial factors in upgrading the microfluidic design will lead to effective mass transfer. Due to the poor solubility of oxygen in the culture media and taking into account the oxygen diffusion distance in tissue (which generally ranges from 100 to 200 m) because oxygen transfer is another important parameter in the cell culture setup that needs to be monitored (Muschler et al., 2004).

Functional Characterization (Micro-Physiology)

In the downstream processes, the designed in vitro system must be evaluated by proving pertinent in vivo physiology and evaluating the analogy between both the in vitro model and its in vivo counterpart in order to develop confidence in it.

(i) Differentiation

In addition, highly specialized organ-specific tissues have been developed using induced pluripotent stem cells (ipscs). For pharmacokinetics screening/ drug absorption, an in vitro assay of the small intestine should, for instance, have the major cell types including enterocytes, paneth cells, and goblet cells with an appropriate in vivo ratio. It is challenging to fit all the essential features onto a single chip; the model's complexity varies based on the precise function desired. To verify the model's applicability during culture, it is crucial to monitor cell differentiation within the OOC system. For instance, it was found that the "liver" phenotype rapidly disappeared during culture, making the culture system unsuitable for research involving repeated doses (Ingber, 2018).

(ii) Nutrient and Metabolite Profile in Cell Culture (Glucose and Lactate)

A thorough understanding of a cell culture's physiological status can be obtained by measuring its metabolic activity. Large-scale bioreactors often employ bulky electrochemical detection systems and colorimetric assays to determine the amounts of glucose and lactate. Prill et al. (2014) showed that consistent tracking of glucose uptake and lactate release by a hepatic tumour cell line allowed for automated identification of drug-induced alterations in cell survival. By subjecting the cells to the complex I inhibitor rotenone, which lowers OXPHOS and thus increases other catabolic pathways such as anaerobic glycolysis, the approach investigates the mitochondrial apoptotic route with a help of microfluidically addressed electrochemical sensors. In a liver-on-chip device, Bavli et al. (2016) described real-time mitochondrial respiration monitoring employing tissue-embedded phosphorescent microprobes with two-frequency phase modulation.

(iii) Permeability of Epithelial Tissue Barriers

The precise alignment of cells that significantly express intercellular tight junctions (tjs) is a key property of epithelial tissue structure types, including the gastrointestinal tract epithelium, blood vessel endothelium, and skin epidermis. Intercellular tight junctions (tjs) are crucial in controlling how quickly nutrients and medicines are transported from the apical side of cells to the basolateral circulation. This barrier can be affected by particular stimuli and is not static. As an illustration, the pathophysiology of inflammatory bowel disease, ulcerative colitis, crohn's disease, and food allergies have all been connected to the increased porosity of intestinal epithelium. On the other hand, desirable bioactive molecules, which are often poorly absorbed, may have their bioavailability enhanced by a brief increase in paracellular transport (Kosinska & Andlauer, 2013). One of the often used techniques for assessing epithelial integrity involves utilizing two sets of electrodes attached to a volt-ohmmeter to measure the transepithelial electrical resistance (TEER) of barrier-forming cells cultured on porous membranes. One of the most precise and sensitive measurements of epithelial integrity and permeability is TEER through reflects the resistance to the flow of ions through the biological epithelial barrier. TEER can also be utilized to assess any biological responses of the underlying cells in the bottom chamber as well as the impact of various stimuli on barrier permeability. These quantitative data offer a useful tool to forecast how tissue exposure to chemical, biological, electrical, thermal, and mechanical stimuli may affect organ inflammation (such as skin and gut inflammation), toxicity, and immune regulation (Sun et al., 2010).

(iv) Response to Mechanical Stimulation

The lung-on-a-chip model is a biomimetic microsystem that simulates the dynamic microenvironment and microarchitecture of the human lung's alveolar-capillary unit. To promote differentiation and the development of the alveolar-epithelium barrier tissue, the alveolar epithelial cells were exposed to air. In order to simulate the breathing action of the lung, cyclic mechanical stretching of the adherent cell layers was induced by applying cyclic vacuum suction to two side hollow compartments next to the cell culture channels. Additionally, it has been demonstrated that mechanical strain increases nanoparticle uptake by the epithelial layer and induces their shipping into the underlying microvascular channel. The skin, heart, bone, stomach, and urothelium (Guenat and Berthiaume, 2018) are among the human body's other tissues that are subjected to mechanical stress in addition to the lung. The prospect of targeting mechano-transduction systems to diminish scar formation (Barnes et al., 2018) is made possible by recent studies on the function of mechanical forces in wound healing and repair.

(v) Migration

All stages of a cell's life require motility, which is a fundamental property of living cells. Understanding several physiological and pathological mechanisms in tissue in vitro models, such as organogenesis, cancer metastasis, and inflammation requires careful observation of cell migration. These tests' results may provide information that can be used to understand how a specific cell type migrates or reacts to chemical stimulation by migrating in that direction. Wound healing, transwell migration, and invasion assays are the most popular cell migration assays (Castellone et al., 2011). The in vivo settings are greatly simplified by the fact that the majority of microfluidic-based cell migration investigations have concentrated on single cell types rather than diverse cellular architectures. In order to investigate tissue-level cell migration for clinical applications, more advanced methods could be needed. The cancer metastasis serves as an illustration of OOC-based cell movement research. Cancer cells spread to distant organs during metastasis, which involves cell-cell communication and chemotaxis (Calvo & Sahai, 2011) and results in the growth of new tumours. To evaluate the impact of biochemical parameters from the contacting cells on carcinoma cell intravasation, Zervantonakis et al. (2012) created a microfluidic-based experiment to simulate the tumour-vascular interface. The study demonstrated that an increase in intravasation rates and endothelial barrier impairment results got from signalling with macrophages through tnfα release, as confirmed by live imaging. Additionally, it was discovered that endothelial barrier dysfunction is linked to both an increase in the number and speed of interactions between tumour cells and endothelial cells.

(vi) Immune Competence: (Response to Toxic or Pathogenic Substances)

A biologically active in vitro model must have the ability to communicate among parenchymal cells and immune cells, which can happen either through soluble substances or directly between cells at the interface between drug/

nutrient metabolism and immunological regulation. Inflammation is an immunological response that aids in the clearance of infectious and other foreign substances. It is controlled by soluble immune signaling molecules called cytokines and involves a series of immunological, physiological, and behavioural processes (Ramadan & Gijs, 2015). Additionally, this causes a range of cellular reactions in the tissues, such as increased microvessel permeability, anchoring of migrating cells to the vessels close to the injury site, circulation of different cell types, cell death, and the formation of new tissue and blood vessels (Ramadan & Gijs, 2015). Utilizing OOC systems including the small intestine (Ramadan and Jing, 2016), skin (Pupovac et al., 2018), and lung/airway (Esch et al., 2012), researchers looked at how different epithelium in vitro models responded to allergen exposure.

Testing, Validation, and Screening

In order to verify that the biological processes mimicked in vitro are reflective of the native tissues in vivo, that is essential to enable the use of this technique in industrial applications because the use of OOC tools for screening hinges on successful device validation. To strengthen confidence in the results of OOC systems and translate OOC technology, a direct comparison against the present industry standards as well as between various technologies may be required (Zhang et al., 2018). However, there are currently relatively few studies that validate OOC systems, and the great majority of OOC studies concentrate on demonstrating the micro-anatomy of a certain tissue or organ and describing basic operations. The primary actions and important factors that must be performed to enable OOC technology acceptance into the industrial laboratory are described by Ewart et al. (2017).

Uses of OOC Systems

(i) Mechanistic Investigation

The OOC system will facilitate the quick evaluation of novel drug pharmacology or toxicity pathways. The biology or pathobiology of the targeted site under inquiry would set the technical requirements for use in the mechanistic investigation (Ewart et al., 2017).

(ii) Preclinical Safety Screening

The primary purpose of in vitro models is to offer datasets that can precisely anticipate the negative effects in vivo. Deprioritizing certain compounds with a high-risk profile will allow for the advancement of others with reduced-risk profiles (McKim, 2010).

(iii) Drug Absorption, Distribution, Metabolism, and Excretion (ADME)

In order to understand more about the metabolism and possible medication interactions, ADME assays are essential. With the use of computational methods (physiology-based pharmacokinetics/toxicokinetics), reliable in vivo predictions can be made using well-characterized in vitro models.

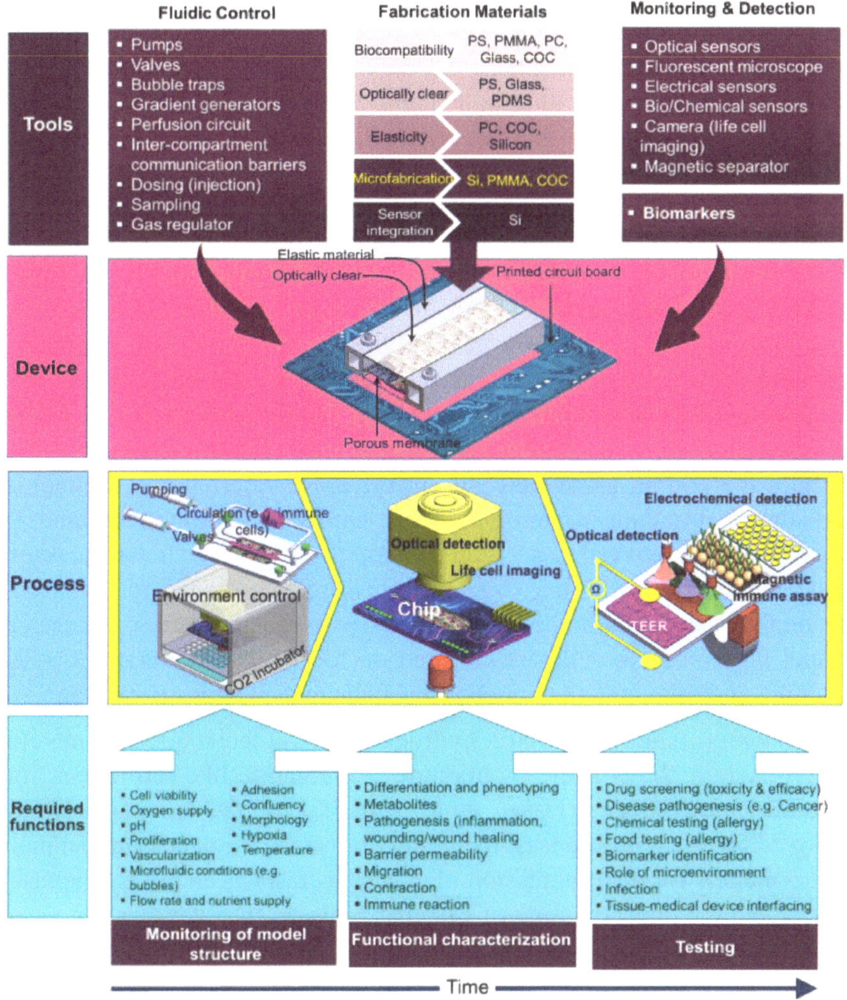

Figure 8 : Comprehensive presentation of the integrated OOC tools, device, process, and necessary functionality through the stages of in vitro model production (model building, functional characterization, and testing) (Ramadan & Zourob, 2020).

Various Organ-On-A-Chip Platforms

Vessel-On-A-Chip

The Blood is transported throughout the body by blood vessels. Arteries transport blood away from the heart. The capillary endothelial barrier is breached by nanoparticles in blood arteries, allowing for the movement of chemicals and water between blood and tissues. Finally, veins return blood from the capillaries to the heart. It is crucial to research how these carriers interact with the vasculature as well as how endothelial cells are affected by fluid shear stress and cyclic strain. Studies have looked into the geometry of the vasculature, including its simple channels (Kim et al., 2014), complicated parameters, and the buildup of micro- and nanoparticles inside its microchannels (Namdee et al., 2013). The fluid shear stress and cyclic stretch on endothelial cells are additional significant factors that need to be taken into account. The microfluidic vessel-on-a-chip is a device designed to resemble blood vessels. In the majority of research, vascular endothelial cells are cultured on membranes using two parallel-flow PDMS chambers that are isolated from one another. Syringe pumps are attached to microfluidic chambers to regulate flow rates. One can investigate the endothelial cell shear stresses caused by the flow rate (Srigunapalan et al., 2011). One of the crucial things to look into is the cyclic stretching of vascular endothelial cells. The elastic membrane that separates the bottom groove layer from the upper microfluidic channel layer exerts cyclic stretching strain on the cells. Under the elastic membrane, there is a groove layer hooked to a vacuum pump that employs suction pressure to extend the elastic membrane. Vacuum pressure is provided to the stretched cells on the elastic membrane while fluid flow is transmitted through the upper channel layer and the cells are cultured on top of the membrane. Cycles of stretching and relaxing are applied to the cells grown on the membrane.

Using photocrosslinkable gelatin hydrogel, a novel technique was developed for producing three-layer biomimetic vasculature-like structures on a microfabricated system (Hasan et al., 2015). It has the potential to be employed for various purposes, including drug screening, the establishment of in vitro models for the progression of cancer metastasis and/or cardiovascular disease, and the understanding of vascular biology. In order to study vascular inflammation and thrombosis in vitro, Mathur et al. (2019) proposed vascular organ-on-chips or "vessel-chips" cultivated with Blood Outgrowth Endothelial Cells (BOECs) from normal or diabetic individuals and built a complete 3D endothelial lumen. It demonstrates that when compared to control BOECs or normal primary endothelial cells, the inactive BOECs of type 1 diabetic pigs exhibit phenotypic behaviour of the disease, including significant vascular dysfunction and thrombogenicity. Brouns et al. (2020)

produced a vessel-on-a-chip model, which consists of a microfluidic device encased with the thrombogenic collagen and tissue factor (TF) and wrapped with patches of human endothelial cells, to evaluate the local inhibitory effects of a discontinuous endothelium. Multicolor fluorescence microscopy was used to track the heterogeneous production of platelet aggregation and fibrin clot formation by circulation of human blood and plasma. It was revealed that a 40% to 60% covering of human umbilical vein endothelial cells on collagen/TF coatings led to a significant overall delay in deposition of platelet and fibrin fiber production under flow. Fibrin production was limited to the spaces between endothelial cells and co-localized with the deposited platelets, suggesting that the clotting process was immediately localized and suppressed.

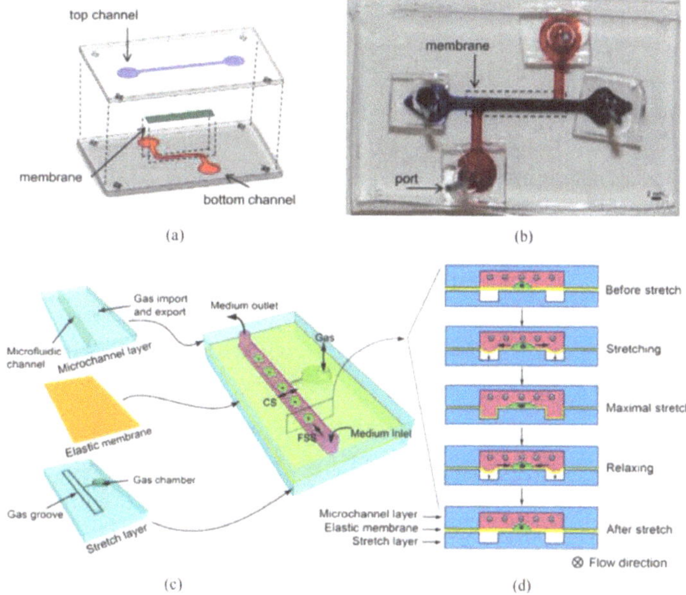

Figure 9 : Vessel-on-a-chip

(a) Schematic of two membrane-connected PDMS chambers (Srigunapalan et al., 2011). **(b)** Implementation of the devices depicted in **(a)**. **(c)** A lengthy microfluidic conduit with alternating flows. **(d)** Stretched and un-stretched flexible membrane consecutively (Zheng et al., 2012).

Liver-On-A-Chip

The liver is an essential organ in the body and is responsible for digestion, detoxification, protein synthesis, and blood-filtering processes. To know and understand more about how the liver works and to investigate drug hepatotoxicity and metabolism, platforms called "liver-on-a-chip" have been created (Lee et al., 2013). Three main sections make up a microfluidic device designed to resemble the porous sinusoidal endothelial barrier between hepatocytes and the liver sinusoid such as a central channel to house the hepatocytes, a microfluidic sinusoid barrier structured with a series of narrow (2 mm wide) microchannels to represent the endothelial barrier, and a microfluidic convection channel encompassing the barrier. The hepatocytes are fed by the flow of nutrients via the microfluidic convection channel. This device uses a designed network of microchannels to imitate the movement among the blood flow and hepatocytes as well as the shear stress placed on hepatocytes. In order to restructure the intricate microenvironment of the liver, 3D microengineered hepatic cell culture platforms have been established (Wang et al., 2015).

The investigation of drug toxicity and metabolism was made possible by the introduction of a perfused multiwell plate. This apparatus consists of an array of 12 bioreactors in a perfused multiwell, comprising liver sinusoidal endothelial cells, stellate and kupffer cells, and hepatocytes, all of which are positioned in scaffold wells. Micropumps are employed to constantly perfuse the scaffold wells. Hepatocytes and hepatic stellate cells (hscs) cell-cell signaling (paracrine signaling) have been studied. The 3D microfluidic liver-on-a-chip used in that investigation had two distinct chambers. Hscs were cultivated on a flat chamber in one chamber (Lee et al., 2013). To prevent direct cell-cell interaction, hepatocytes were cultivated in the other compartment on a concave surface. The flow rate of a nutritious solution was controlled by adjusting the connection between these two various cells, which included a tube and an osmotic pump.

The first liver-based system was created by Kane et al. (2006) which included microfluidic ports wherein the 3T3-J2 fibroblasts and rat liver cells had co-cultured to resemble an airway interface. Rat hepatocytes cultivated in the chip were able to produce albumin and go through metabolism constantly and steadily. A chip created by Lee et al. (2007), in which a culture medium was perfused outside the gap to mimic the interstitial structure of endothelial cells and primary hepatocytes in culture. Hepatocytes were segregated from the exterior sinusoidal region by this porous endothelial gap while yet maintaining effective substance exchange in cord-based structures. Cells were patterned onto circular PDMS chips by Ho et al. (2013) utilizing radial electric field gradients made through electrophoresis. These cutting-edge methods mimicked the anatomy of the hepatic lobules. Ma et al. (2018) designed a biomimetic infrastructure for the in situ perfusion of hepatic spheroids. Systems were developed by Yum et al. (2014) to investigate how hepatocytes influence different cell types. To evaluate the drug toxicity on liver cells, high-throughput tests were established. A unique, multi-throughput Multiorgan-On-a-Chip (MoC) device built on a platform with medium circulation driven by pneumatic pressure was recently developed. The four-organ/tissue (liver, intestine, cancer, and connective tissue) and two-organ (cancer and liver) models could both be sustained by this high-throughput MoC device. The platform, which was created to discover the effects of anticancer medications, could function with just one pressure source (Satoh et al., 2018).

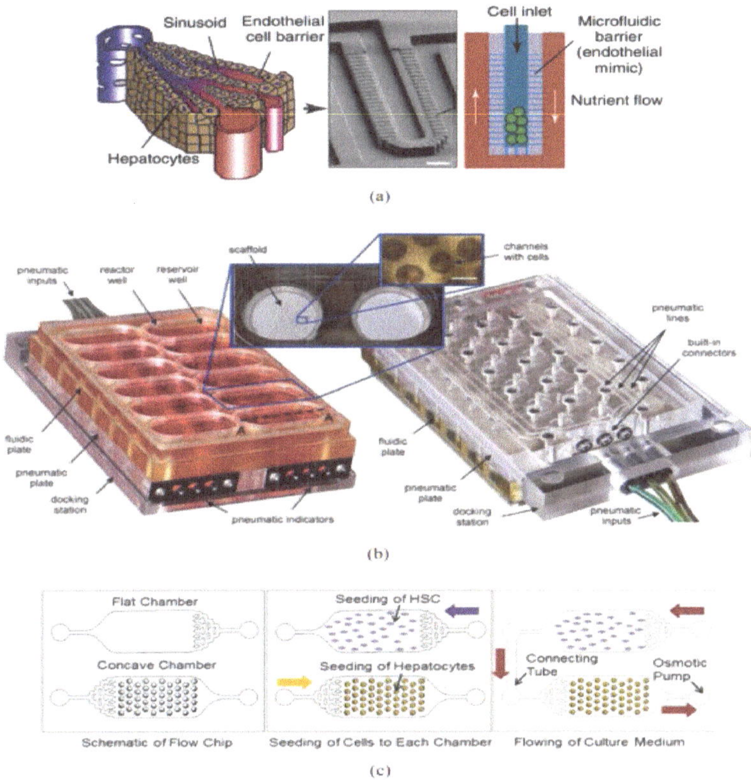

Figure 10 : Liver-on-a-chip

(a) Microfluidic liver on a chip consisting of a central liver cell chamber and surrounding nutrient flow chamber to mimic the structure of the liver. **(b)** Microengineered perfusion of the hepatic culture. **(c)** 3D microfluidic liver on a chip consisting of two different chambers connected with a tube (Lee et al., 2013).

Heart-On-A-Chip

One of the most difficult areas of organ-on-a-chip research is the design of heart-on-a-chip platforms. Simulating complex cardiac tissue microenvironments is difficult. It is challenging to investigate cardiac tissues' electrical responses and contractility at the same time (Grosberg et al., 2011). A few researches have been done using heart-on-a-chip platforms to analyze the contractility and electrophysiological response of cardiac tissue concurrently with the help of cardiac tissue engineering to imitate the microenvironments of cardiac tissue (Shin et al., 2013). In the early stages of the heart-on-a-chip project, a planar electrode array and PDMS microfluidic network were used to monitor the extracellular potential of individual adult cardiomyocytes (CMs). Another microfluidic tool with a variety of electrodes was created to optically and electrically detect cell contractility and the metabolic activity of cardiomyocytes respectively.

A tissue-level heart-on-a-chip was initially developed by Grosberg et al. (2011) to assess the contractility of neonatal cardiac muscle tissue. Eight muscular thin films (mtfs) were constructed on a chip for the design. Spin-coating was used to cover a glass slide with a layer of poly(n-isopropylacrylamide) (pipaam), which dissolved at temperatures below 35 °C. The PIPAAM layer was then covered with a PDMS coating. It is seeded neonatal rat ventricular cardiomyocytes into the PDMS layer. An array of two opposing rows of four rectangular MTF layers was constructed by manually cutting the film layers after the substrate had been seeded with cells and submerged in a bath. When the solution was maintained at a temperature below 35 °C, the mtfs peeled off after the pipaam was dissolved. Finally, electrodes were positioned at the mtfs' top and bottom. The electrical stimulation applied by the field electrode caused the mtfs to begin to curl up. Finally, it was determined how far cells deflected in response to electrical stimulation.

In a PDMS model, hydrogels were used by Zhang et al. (2013) to create self-arrayed myocardial sheets. Differentiated myocardium served as the source of the CMs. The integration of the vascular and myocardial systems was made possible by using 3D printing, which fabricated micro-organ tissue chips. The CMs were placed in the gap in the vascular network formed by the model, which used vascular endothelial cells to build vascular networks. A pharmacological screening platform for cardiovascular (CV)-related drugs was created by the organ chip. In order to evaluate the effectiveness of cardiac drugs, a heart-on-a-chip device was developed that utilized high-speed impedance detection (Zhang et al., 2016). To determine pharmacological effects, the instrument captures the contraction of CMs. A preclinical evaluation of the cardiac drug's effectiveness was represented by the chip. A heart organ system that simulated the mechanical and physiological conditions of CMs was established by Marsano et al. (2016). It was possible to do quantitative analysis and direct visualization, which were not enabled in conventional cell culture or animal models. This system offers standard functional 3D heart models and reflects a breakthrough in the field. This provides the apparatus with a cutting-edge and affordable screening platform to raise the prognostication of in vitro models. Mastikhina et al. (2020) described a cardiac-fibrosis-on-a-chip model where the expression of collagen and brain natriuretic protein was elevated when tissues were subjected to TGF-b, a pro-fibrotic agent, and downregulated when tissues were subsequently treated to an anti-fibrotic medication. The abrupt interruption of blood supply in ischaemia causes the buildup of metabolic wastes while decreasing the tissue's ability to receive oxygen. This has a local impact on how cardiomyocytes contract. Employing perforated electrodes in a heart-on-a-chip device, Liu et al. (2020) were able to mimic the hypoxic milieu and track the action potential changes that occurred in the cell culture. The reliable regulation that can be accomplished with these devices is demonstrated by the combination of environmental cues and the insights acquired from the electrophysiology of the cells.

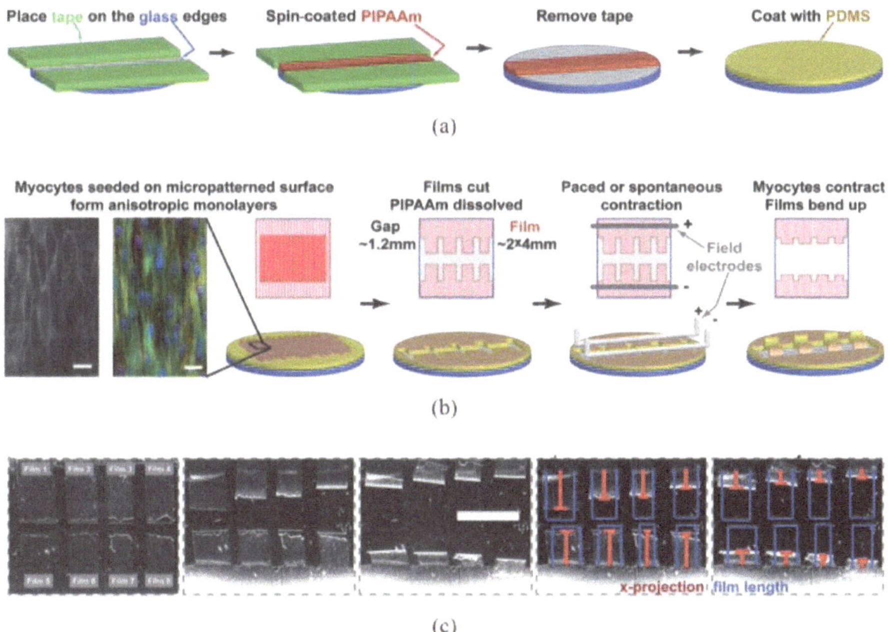

Figure 11: Heart-on-a-chip devices and the image of their beating (Grosberg et al., 2011)

Lung-On-A-Chip

Another essential organ is the lung where blood's carbon dioxide and external oxygen are exchanged. For both physiological studies and medication delivery testing, it is crucial to look into the relationships between cells, blood flow through them, and gas flow through them in the respiratory tract. The purpose of these microfluidic platforms is to cultivate lung cancer cells and test how they react to medications (Zhao et al., 2010) and electrical stimulation (Wang et al., 2015). These platforms, however, do not simulate both the blood-air bi-cellular barriers and the deformation of lung tissue during breathing. Thus, characterizations at the cell level are the limit of this kind of microfluidic technology. Huh et al. (2010) unveiled the first lung-on-a-chip platform. Three channels are implanted in this device such as two side channels, one main channel, and a channel that is separated by a membrane. The primary parts of this system are the membrane and side channels. The main channel can simulate the alveolar-capillary barrier in the lungs after seeding endothelial and epithelial cells on each side of the permeable membrane. This allows interaction among two kinds of cells and two flows in the main channel. For instance, *Escherichia coli* in the flow via the airways can be killed by immune cells from blood flow that can pass through tiny gaps in the membrane barrier. The main channel's membrane stretch, which represents the movement of the lung while breathing, can be altered by adjusting the suction pressure on each side of the chamber. The in vitro microfluidic lung-on-a-chip platform has undergone significant advancements. The lung-on-a-chip offers a practical platform for analyzing dynamic cell behaviours in vitro as well as the thorough testing of medication effects, allowing the examination of complicated chemical, mechanical, and biological impacts in comparison to traditional platforms. Other varieties of lung-on-a-chip systems depend on the fixed features of lung cells to achieve particular functionalities (Huang et al., 2014). The majority of static lung-on-a-chip platforms seek to identify indicators of lung cancer cells or how lung tumour cells react to microenvironmental changes (Hou et al., 2014). The finite element method (FEM) and computational fluid dynamics (CFD) can be used to adjust the device design and flow rate to more closely match in vivo lung functions.

A lung chip that imitated the lung parenchyma was described by Stucki et al. (2015). The system was the first flexible membrane expansion model to simulate breathing, and it comprised an alveolar barrier and 3D cyclic strain to mimic respiration. To determine their applicability as a physiological model, Humayun et al. (2018) cultivated airway epithelium and smooth muscle cells on various sides of a hydrogel membrane. The system was used as a physiological model of chronic lung disease along with microenvironmental cues and toxin exposure. Yang et al. (2018) developed a chip matrix for cell scaffolds using a poly (lactic-co-glycolic acid) (PLGA) electrospinning nanofiber membrane. The system's simplicity makes lung tumour precise treatments and tissue engineering strategies applicable. Organ chips made from lung tissue are beneficial as implantable breathing assistance aids. In order to provide for more gas exchange in the placenta for preterm newborns experiencing respiratory insufficiency, Peng et al. (2018) developed lung assist devices (LAD). In the umbilical arteries and veins, the idea of large-diameter channels was realized, offering the LAD a high extracorporeal blood flow. Since the clinical experiments used to determine the thresholds for umbilical vasodilation were unethical, this has additional value. This experiment was the first to objectively measure the harm caused by catheter extension to the umbilical vessels.

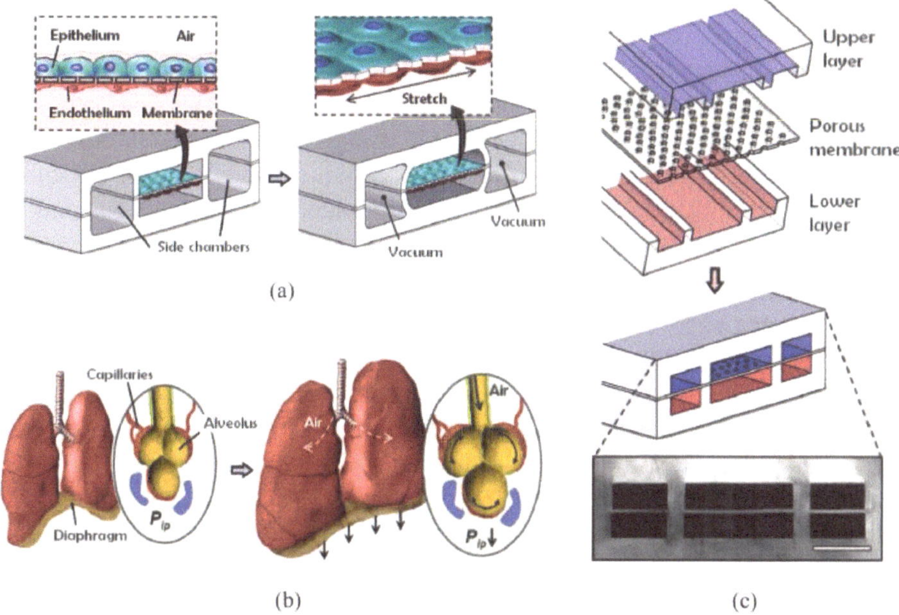

Figure 12 : Lung-on-a-chip system (Wu et al., 2020)

(a) A PDMS-based membrane and vacuum-based deformation controller that simulates the alveolar-capillary barrier and lung breathing respectively. **(b)** Schematic illustrations of variations of lung size during inhalation. **(c)** Bonding and alignment of three-layer PDMS structures to form the devices.

Kidney-On-A-Chip

The control of osmotic pressure drug excretion is the responsibility of the kidney. The need for drug screening systems is highlighted by the loss of the ability of renal filtration caused by kidney toxicity. The nephrons, which are made up of the glomerulus, renal capsule, and renal tubule, are where filtration and reabsorption occur. Microfluidics can mimic the fluid environment that encourages the growth of tubular cells and offers support for permeable membranes necessary to maintain cell polarity (Bhatia and Ingber, 2014).

The first multi-layered microfluidic device with mouse kidney medullary collecting duct cells was created to mimic renal filtration (Jang & Suh., 2010). In response to hormone activation, the device's biomimetic environment promoted cytoskeletal remodeling and molecular transport, which improved the polarity of the inner medullary collecting duct. In 2013, human primary renal epithelial cells were cultured using the same microfluidic apparatus. These were the original primary kidney cell toxicity investigations (Jang et al., 2013). The numerous biological activities of the intact kidney tubule can now be directly observed and quantitatively analyzed by this device in ways that were previously not conceivable using conventional cell culture or animal models. Conventional cell culture techniques have the drawback that functional cell differentiation necessitates prolonged culture durations and an external signal detection device.

In organ culture tools, Musah et al. (2018) developed techniques to generate podocytes obtained from pluripotent stem cells to generate human glomerular chips. These were able to simulate the glomerular capillary wall's structure and function, which was not achievable with earlier techniques. The chip could be used to evaluate nephrotoxicity, design therapeutics, practice regenerative medicine, and study kidney growth and disease. In human proximal tubules and glomeruli, Sakolish et al. (2019) created a reusable microfluidic device that allowed renal epithelial cells to proliferate in varied environments. Nephrotoxicity is caused by shear stress. Stable tubule culture systems were developed by Schutgens et al. (2019) to enable prolonged expansion and examination of human kidney tissue. A system-based multi-purpose model was established for quick and customized molecular and cellular analysis, disease modeling, and drug screening using primary renal epithelial cells. A potent method for producing human islet organoids from human induced pluripotent stem cells was presented by Tao et al. (2019). This approach was useful for a variety of stem cell-based organic engineering and regenerative medicine purposes.

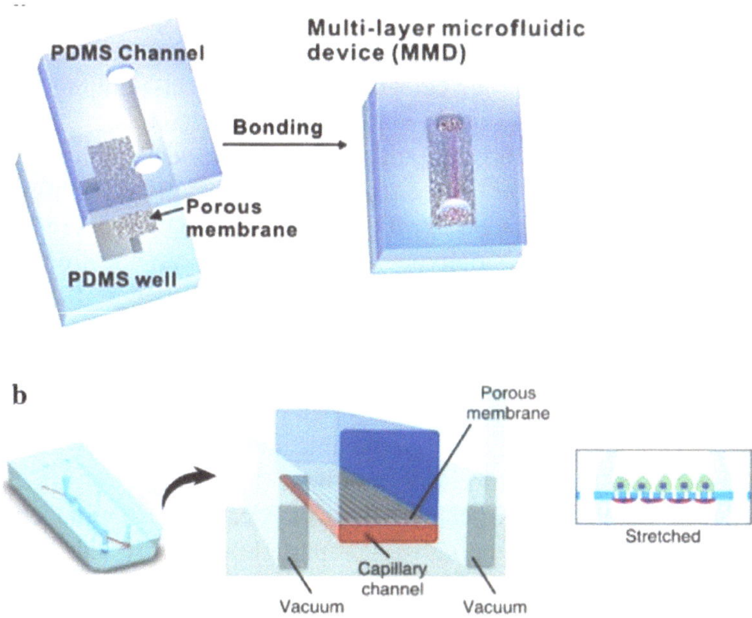

Figure 13 : Kidney-On-A-Chip (Wu et al., 2020)

(a) Kidney tubular chip. The permeable membrane, PDMS reservoir, and PDMS channel are sandwiched together; **(b)** The glomerulus's capillary lumen and urinary cavity can both be mimicked by the channel. The protein laminin can be functionalized to resemble the glomerular basement membrane using the permeable elastic PDMS membrane. Vacuum stretching of the elastic PDMS membrane can result in the creation of cyclical mechanical pressure on the cell layer.

Tumor-on-A-Chip

One of the main areas of focus in organ-on-a-chip research is the tumour. Unknown mechanisms underlie the vascularization and emigration of tumour cells yet. Although therapeutics can be delivered to cancer cells using nanoparticles, it is challenging to understand how nanoparticles work and what effects they have. Three subcategories of tumour-on-a-chip platforms are now available. The first subcategory keeps tabs on tumour cell movement and vascularization, particularly antiangiogenesis (Zervantonakis et al., 2011). The second subcategory tries to examine how drugs included in nanoparticles affect cancer cells (Kwak et al., 2014). The third subcategory is concerned with identifying cancer markers in blood samples to identify cancer in its early stages (Huang et al., 2014). In the beginning, 3D tumour spheroids with adjustable density and diameter were produced using microfluidic devices. Additionally, distinct cancer cells have been sorted using microfluidic tools, and cell density has been balanced for multiple inflows.

Numerous 3D platforms have been designed to co-culture normal and tumour cells in order to study the interactions between normal cells and cancers. Although 3D devices cannot be categorized as organ-on-a-chip technologies, they do allow for a more in-depth analysis of the behaviour of tumour cells when grown in vitro. The most popular cancer marker detection technologies are tumour-on-a-chip devices (Wang et al., 2015). Typically, circulating tumour cells can be collected using microfluidic devices with properly designed channel architectures. Drug delivery and screening systems frequently use nanoparticle screening systems to build microenvironments that resemble tumour tissues and adjacent blood arteries.

The 3D structure mimics a cluster of solid tumour cells and two capillary and lymphatic vessels (Kwak et al., 2014). In the top channel (red channel), a monolayer of endothelial cells on a permeable membrane imitates the capillary with endothelium. The bottom layer has two side channels that resemble the lymphatic system and a core channel that mimics breast tumour cells. This platform allows for the screening of nanoparticle transport and the complex interaction between cancer cells and nanoparticles. During the screening, nanoparticles were moved via the permeable membrane from the top channel to the tumour channel. The consequences of the nanoparticles' interaction with the tumour cells were evident in the side lymphatic channel. The effectiveness of nanoparticles to deliver drugs was affected by factors such as fluid pressure and size of the membrane pore, which was helpful advice for nanoparticle design. To improve nanoparticles, researchers have looked at the movement of nanoparticles in vascular tissues in addition to investigating how tumour cells and nanoparticles interact (Tan et al., 2013).

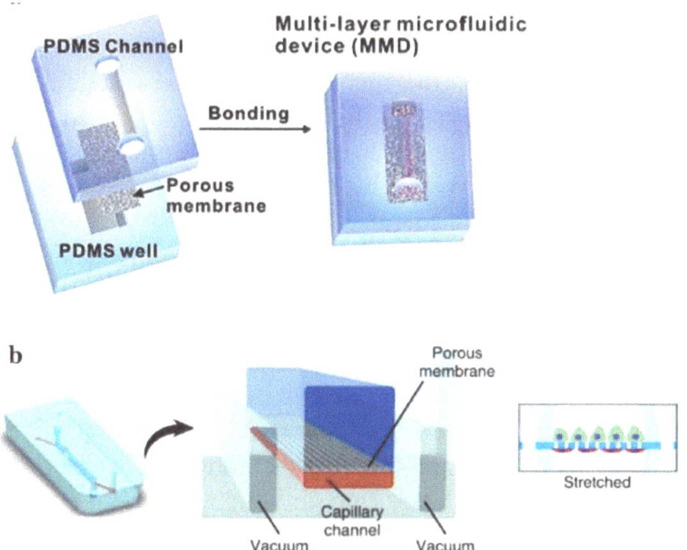

Figure 14 : Tumour-on-a-chip devices (Kwak et al., 2014)

(a) In vivo breast tumour microenvironment formed vascular endothelial, lymphatic endothelial, and tumour cells. **(b)** Conceptual design: An illustration of the tumour-microfluidic-on-chip. **(c)** Constructed prototype and tumour-on-a-chip device's perfusion setup design.

Fish-On-A-Chip

To enhance the ability to comprehend the values of the expanding vertebrate-organism model, high-efficient zebrafish (embryo) handling systems are essential. The limited availability of zebrafish manipulation platforms which primarily use traditional, "static" microtiter plates or glass slides with rigid gel limits the ability to gather dynamic, three-dimensional (3D), tissue- and organ-oriented information from intact larvae with typical developmental dynamics. Additionally, these conventional platforms are unable to accurately manage the development microenvironment by delivering stimuli in a well-defined spatiotemporal manner, making them unsuitable for the high-throughput handling of certain swimming multicellular living creatures at the single-organism level. Utilizing specially designed microscale structures or structured arrays that interact with or connect with functional elements (like imaging systems) to enable quantitative accurate readings of small objects (like zebrafish larvae and embryos) under physiologically normal settings, microfluidics has recently been developed to overcome these technical issues (Yang et al., 2016).

For real-time developmental investigation of transgenic Zebrafish embryos, a 3D microfluidic embryo array has been developed (Akagi et al. 2013).

The PMMA chip made it possible for the embryo to be loaded, docked, and exposed to micro-perfusion therapy automatically. A microfluidic device also designed to investigate the diverse behavioural reactions of Zebrafish larvae to various levels of hypoxia (Erickstad et al., 2015). In order to screen zebrafish embryos, Popova et al. (2017) designed a microchip with arrays of very hydrophilic regions divided by superhydrophobic barriers. They were able to manually disseminate zebrafish embryos across the droplet microarray substrate by utilizing the impact of interrupted dewetting. They looked at the distribution of fluorescently tagged peptoids in various organs and assessed the toxicity of various ZnCl and AgNO concentrations. In 2019, the first fish gut-on-a-chip model is developed (Drieschner et al., 2019). Two intestinal cell lines from rainbow trout epithelial RTgutGC and fibroblastic RTgutF were cultured in an artificial microenvironment to recreate the intestinal barrier for this model. In order to accurately imitate the transition between the intestinal lumen and the organism's interior to investigate how fish epithelial cells responded to shear stress rates of 0.002-0.06 dyne per cm^2 that mimic in-vivo fluid flow in the intestinal lumen and how epithelial and fibroblast cells interacted in ideal flow circumstances.

Chen et al. (2022) developed a low-cost 3D-printed microfluidic System with a blind-hole structure capable of effective automatic rotating and trapping for the injection of zebrafish embryos, with a 99% success rate for trapping and a time for trapping a single embryo as little as 0.2 s. The embryo can be rotated with accuracy of 5° and at a maximum speed of 3.5 r/s. With the aid of computer vision, it can be automatically controlled and carried out. An acoustofluidic rotational tweezing device was demonstrated for high-speed, contactless, 3D multispectral image capturing and digital modelling of zebrafish larvae for quantitative phenotypic characterization. In this system, Zebrafish larvae rotate in a contactless and quick (1 s/rotation) manner due to acoustically induced polarised vortex streaming. This makes it possible to image the internal organs of zebrafish in a multispectral fashion from various vantage points (Chen et al., 2021). Bansal et al. (2021) used microfluidic devices in high-throughput drug screening, high-throughput brain activity mapping, examining the neural and behavioural responses of zebrafish to various external stimuli, microinjection, and neuronal regeneration. Microfluidic platforms have also been designed for multiple zebrafish larvae (one to four larvae) to screen their behavioural responses at a time in different sectors including toxicology and pharmacology applications (Khalili et al., 2021). For the first time, Chi et al. (2022) established an automated microinjection device specifically for the high-throughput injection of zebrafish larvae. A batch of zebrafish larvae can be automatically identified and injected at high efficiency using the suggested custom-built microarray Petriplate in conjunction with machine vision-based motion control. It can be used in a variety of zebrafish microinjection-based biomedical investigations. For the first time, high-throughput intramuscular activities monitoring technique was demonstrated employing microfluidic chip technology to trap numerous 5 dpf of zebrafish larvae. In this mentod, three different chemicals were used to modify the locomotor activity of zebrafish and evaluated the efficacy of therapeutics (Cho et al., 2019). Subendran et al. (2021) designed and developed a novel microfluidic platforms and paired it with a shape memory alloy (SMA) actuator to immobilize zebrafish within the observation area for hydrodynamic quantification of the tail-beating behavioural reactions that may be brought on by excessive exposure to food additives (cochineal red edible food additive). The zebrafish's behaviour throughout its early developmental phases, which resulted in vortex circulation, was extensively explored in this study. To facilitate efficient and rapid zebrafish screening without the support of a fish facility, a completely automated IoT (Internet of Things) microfluidic device was also designed recently. This smart microfabricated device was introduced that allowed for highly precise control of temperature with an accuracy of ±0.1°C, remote zebrafish transport using light patterning, and perfusion and dynamic culturing of zebrafish, all of which could be arranged for on-chip high-quality imaging (Mani and Chen, 2021).

Panuska et al. (2021) demonstrated a novel millifluidic platform design made possible by 3D printing, and we assessed its performance using diluted ethanol as a teratogen in toxicity tests and long-term perfusion culturing of *Danio rerio* embryos. The cultivation chip, which had two inlets and one output, was made up of two independent channels stacked on top of one another and divided by a wall that included cultivation chambers. The chip design now includes an individual embryo removal functionality that may be used during cultivation studies to carefully remove any of the produced embryos off the chip. A newly developed microfluidic system was also suggested for the vascular and phenotype assessment of non-anesthetized zebrafish with typical imaging configurations for the heart, pectoral fn beating, and vasculature to establish the relationship between the cardiac parameters and behavioural endpoints by producing the optomotor activation and hydrodynamic pressure control. In this system, computer-animated moving grids were produced via an in-house control interface that was triggered by the larval optomotor reflex in combination with the pressure suction control in order to provide the visual signals for improved positioning of zebrafish. Additionally, certain drugs like ethanol and caffeine were used to treat zebrafish (Subendran et al., 2021).

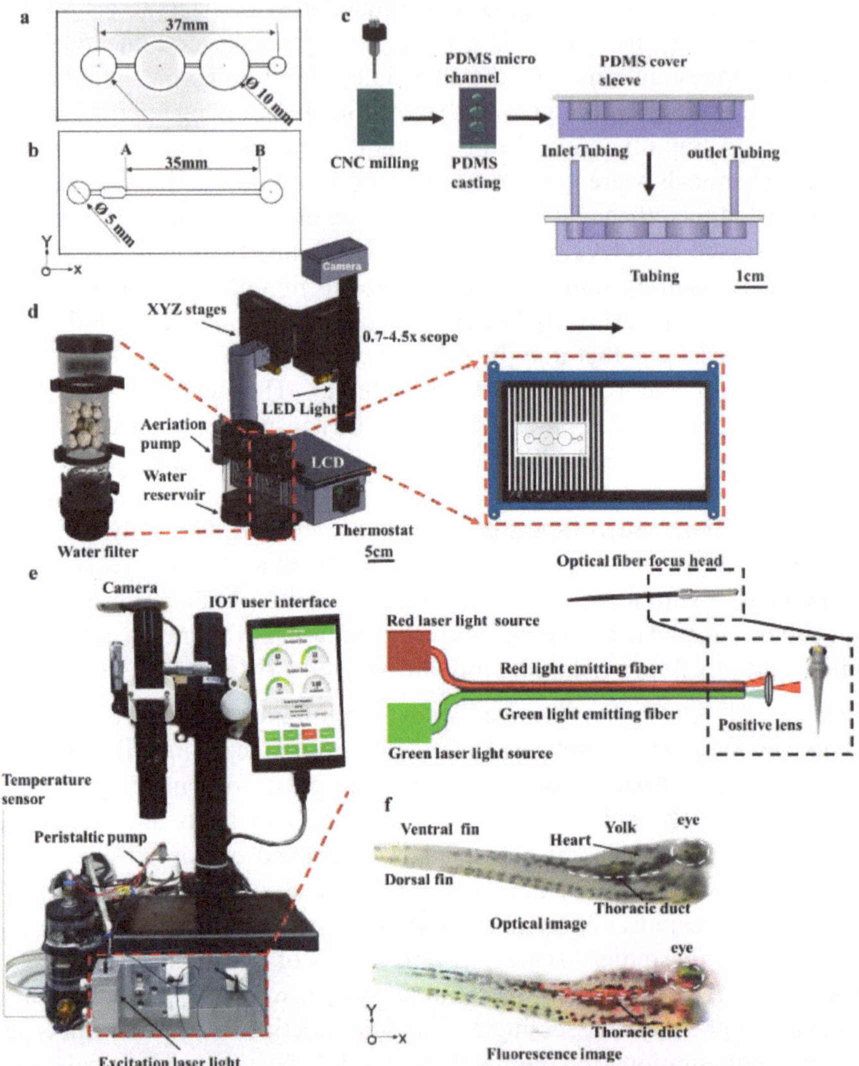

Figure 15 : Schematic illustration of the a fully automated IoT (Internet of Things) microfuidics system for drug screening in Zebrafish (*Mani and Chen*, 2021)

(a) Dimensions of the microchannel used to culture the larvae and **(b)** the microchannel to calculate the time required for migration of the larvae from point A to B. **(c)** The fabrication process of the microchannel. **(d)** The shifting gratings were produced by an LED screen, and the larvae were moved using a microfluidic network. The test zones for measuring fish transit times through light pattern stimulus. **(e)** Image of the system equipped with sensors to provide a well-controlled environment for zebrafsh and the laser excitation light source. **(f)** A fluorescence image of larvae with custom-developed fiber-optic laser excitation light.

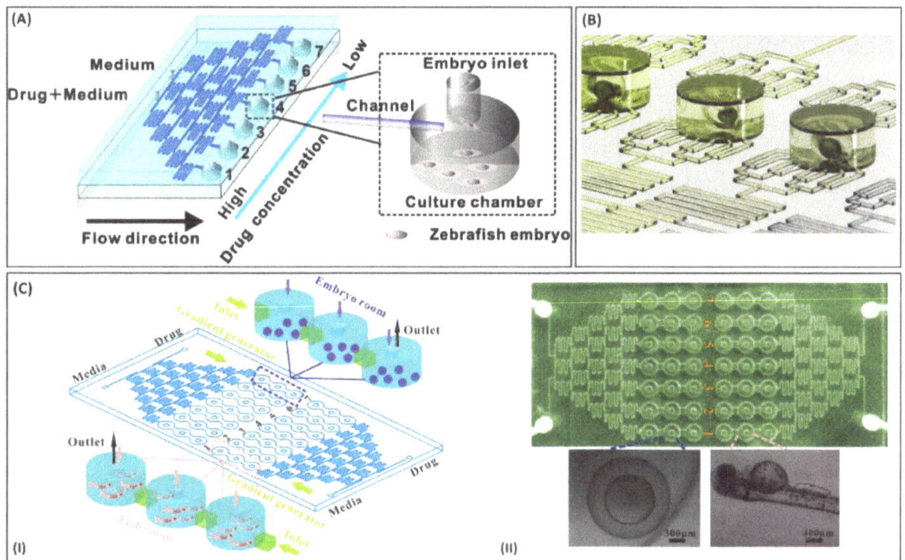

Figure 16 : Microfluidic platforms for zebrafish embryonic experiments (Khalili & Rezai, 2019)

(A) A microfluidic system with two distinct inlets to produce drug concentration gradients in culture compartments 1 to 7 for exposing trapped zebrafish embryos in them. **(B)** A microfluidic perfusion device to monitor zebrafish development. **(C)** (i) Illustration of a microfluidic with two distinct zones for determining the drug toxicity to embryos and larvae. Each zone used to have a gradient generator, a drug inlet, a media inlet, and seven sets of zebrafish compartments; (ii) Images of the microfluidic chip, as well as the larvae and embryos in the chip.

Conclusion and Future Outlook

The tissue/ organ-on-a-chip, also known as in vitro microphysiological models, is an optimistic technique that has the potential to transform many fields of science and business, especially preclinical to clinical translation in the pharmaceutical industry. But before using this method as an alternative to experimental animals and transferring them from the laboratory to the industry, there are still many technological hurdles, as well as standardization and regulatory endorsement that should be resolved. A complex functional tissue structure "micro-anatomy" must be built in order to achieve physiologically realistic in vitro models, and this calls for the establishment of instruments for cells and reagents manipulation as well as for monitoring, sensing, and readout recording. Recent noteworthy efforts that highlight the strong momentum toward the implementation of the "human-on-a-chip" and "fish-on-a-chip" concept for networked multi-organ platforms for drug screening were addressed. To maximize the main physiological robustness and increase their dependability, distinct organ models must be developed. Targeted therapy and the modelling of organs and diseases are both intriguing uses for iPSC research, which is actively being transferred to many other fields, such as OOC technology. To enable bridging the gap between the lab and industry, trans-disciplinary research integrating stem cell biologists, engineers, physiologists, and pharmaceutical industry partners is essential. Additionally, over the past few years, a number of partnerships between pharmaceutical companies, start-ups, and academia have emerged. These partnerships highlight the significance of this technology and its enormous potential to alleviate current business needs in drug discovery while also promising commercial success.

OOC is a technology that is driven by engineering and takes advantage of a variety of well-established engineering disciplines, including microfabrication, microfluidics, microscopy, sensors, etc. The recent integration of 3D (bio) printing with microfluidics, LOC, and OOCs gives the technology and its commercialization even more traction. Incredible progress in our understanding of living tissues and how they respond to stimuli would be possible due to OOCs, which would depict a paradigm shift in physiology research. They would also facilitate constant monitoring of living cells and early identification of the mode of action, offering a plethora of data at a low cost and in a short amount of time. OOC technology, however, goes far beyond the realm of medicine and holds great promise for deepening our comprehension of the intricate interactions between the body and exogenic stimuli. It also offers a more precise assessment of the as-yet-unknown effects of drugs, food, chemicals, pathogens, and environmental toxins.

References

Akagi, J., Khoshmanesh, K., Hall, C.J., Cooper, J.M., Crosier, K.E., Crosier, P.S., Wlodkowic, D., 2013. Fish on chips: Microfluidic living embryo array for accelerated in vivo angiogenesis assays. Sensors and Actuators B: Chemical, 189: 11-20.

Ashammakhi, N., Darabi, M. A., Çelebi-Saltik, B., Tutar, R., Hartel, M. C., Lee, J., Hussein, SM., Goudie, MJ., Cornelius, MB., Dokmeci, MR., and Khademhosseini, A., 2019. Microphysiological Systems: Next Generation Systems for Assessing Toxicity and Therapeutic Effects of Nanomaterials. Small Methods, 1900589: (1-19). doi:10.1002/smtd.201900589

Bansal Pushkar, Abraham Abhinav, Garg Jay, Erica E. Jung, 2021. Neuroscience Research using Small Animals on a Chip: From Nematodes to Zebrafsh Larvae. BioChip Journal, 15:42–51. https://doi.org/10.1007/s13206-021-00012-5

Barnes LA, Marshall CD, Leavitt T, Hu MS, Moore AL, Gonzalez JG, Longaker MT, Gurtner GC., 2018. Mechanical Forces in Cutaneous Wound Healing: Emerging Therapies to Minimize Scar Formation. Adv Wound Care (New Rochelle)., 7(2):47-56. doi: 10.1089/wound.2016.0709

Bavli D, Prill S, Ezra E, Levy G, Cohen M, Vinken M, Vanfleteren J, Jaeger M, Nahmias Y., 2016. Real-time monitoring of metabolic function in liver-on-chip microdevices tracks the dynamics of mitochondrial dysfunction. Proc Natl Acad Sci U S A., 113(16):E2231-40. doi: 10.1073/pnas.1522556113.

Bhatia, S. N., & Ingber, D. E., 2014. Microfluidic organs-on-chips. Nature Biotechnology, 32(8), 760–772. doi:10.1038/nbt.2989

Brouns SLN, Provenzale I, Geffen JV, Meijden PVD, Heemskerk JWM., 2020. Localized endothelial-based control of platelet aggregation and coagulation under flow: a proofof-principle vessel-on-a-chip study. J. Thromb. Haemost.,18(4):931–941.

Calvo, F., & Sahai, E., 2011. Cell communication networks in cancer invasion. Current Opinion in Cell Biology, 23(5), 621–629. doi:10.1016/j.ceb.2011.04.010

Castellone RD, Leffler NR, Dong L, and Yang LV., 2011. Inhibition of tumor cell migration and metastasis by the proton-sensing GPR4 receptor. Cancer Lett., 312(2):197-208. doi: 10.1016/j.canlet.2011.08.013.

Chen Chuyi, Gu Yuyang, Philippe Julien, Peiran Zhang, Hunter Bachman, Jinxin Zhang, John Mai, Joseph Rufo, John F. Rawls, Erica E. Davis, Nicholas Katsanis & Tony Jun Huang, 2021. Acoustofluidic rotational tweezing enables highspeed contactless morphological phenotyping of zebrafish larvae. Nature Communications, 12:1118, Pg.: 1-13. https://doi.org/10.1038/s41467-021-21373-3

Chi, Z., Xu, Q., Ai, N., and Ge, W., 2022. Design and Implementation of an Automatic Batch Microinjection System for Zebrafish Larvae. IEEE Robotics and Automation Letters, vol. 7, no. 2, pp. 1848-1855. doi: 10.1109/LRA.2022.3143286

Cho, S.-J., Kang, Y. J., & Kim, S., 2019. High-throughput Zebrafish Intramuscular Recording Assay. Sensors and Actuators B: Chemical, Vol. 304 : 1-20, 127332. doi:10.1016/j.snb.2019.127332

Drieschner Carolin, Konemann Sarah, Renaud Philippe and Schirmer Kristin, 2019. Fish-gut-on-chip: development of a microfluidic bioreactor to study the role of the fish intestine in vitro. Lab on a Chip, Royal society of chemistry, Pg.: 1-9

Erickstad, M., Hale L.A., Chalasani, S.H., and Groisman, A., 2015. A microfluidic system for studying the behavior of zebrafish larvae under acute hypoxia. Lab on a Chip, 15(3): 857-866.

Esch, M. B., Sung, J. H., Yang, J., Yu, C., Yu, J., March, J. C., & Shuler, M. L., 2012. On chip porous polymer membranes for integration of gastrointestinal tract epithelium with microfluidic "body-on-a-chip" devices. Biomedical Microdevices, 14(5), 895–906. doi:10.1007/s10544-012-9669-0

Ewart L, Fabre K, Chakilam A, Dragan Y, Duignan DB, Eswaraka J, Gan J, Guzzie-Peck P, Otieno M, Jeong CG, Keller DA, de Morais SM, Phillips JA, Proctor W, Sura R, Van Vleet T, Watson D, Will Y, Tagle D, Berridge B., 2017. Navigating tissue chips from development to dissemination: A pharmaceutical industry perspective. Exp Biol Med (Maywood)., 242(16):1579-1585. doi: 10.1177/1535370217715441

Grosberg, A., Alford, P. W., McCain, M. L., & Parker, K. K., 2011. Ensembles of engineered cardiac tissues for physiological and pharmacological study: Heart on a chip. Lab on a Chip, 11(24), 4165. doi:10.1039/c1lc20557a

Guenat OT, and Berthiaume F., 2018. Incorporating mechanical strain in organs-on-a-chip: Lung and skin. Biomicrofluidics., 12(4):042207. doi: 10.1063/1.5024895.

Hasan A, Paul A, Memic A, Khademhosseini A. 2015. A multilayered microfluidic blood vessel-like structure. Biomed Microdevices;17(5):88. doi:10.1007/s10544-015-9993-2.

Ho, C.-T., Lin, R.-Z., Chen, R.-J., Chin, C.-K., Gong, S.-E., Chang, H.-Y., Peng, H-L., Hsu, L., Yew, T-R., Chang, S-F., and Liu, C.-H. (2013). Liver-cell patterning Lab Chip: mimicking the morphology of liver lobule tissue. Lab on a Chip, 13(18), 3578. doi:10.1039/c3lc50402f

Hou, H.-S., Tsai, H.-F., Chiu, H.-T., & Cheng, J.-Y. (2014). Simultaneous chemical and electrical stimulation on lung cancer cells using a multichannel-dual-electric-field chip. Biomicrofluidics, 8(5), 052007. doi:10.1063/1.4896296

Huang, T., Jia, C.-P., Jun-Yang, Sun, W.-J., Wang, W.-T., Zhang, H.-L., Cong H, Jing F-X, Mao H-J, Jin Q-H, Zhang Z, Chen Y-J, Li G, Mao G-X, and Zhao, J.-L. (2014). Highly sensitive enumeration of circulating tumor cells in lung cancer patients using a size-based filtration microfluidic chip. Biosensors and Bioelectronics, 51, 213–218. doi:10.1016/j.bios.2013.07.044

Huh D, Matthews BD, Mammoto A, Montoya-Zavala M, Hsin HY, Ingber DE., 2010. Reconstituting organ-level lung functions on a chip. Science., 328(5986):1662-8. doi: 10.1126/science.1188302.

Humayun, M., Chow, C.-W., & Young, E. W. K. (2018). Microfluidic lung airway-on-a-chip with arrayable suspended gels for studying epithelial and smooth muscle cell interactions. Lab on a Chip, 18(9), 1298–1309. doi:10.1039/c7lc01357d

Ingber DE., 2018. Developmentally inspired human 'organs on chips'. Development., 145(16): 1-4, dev156125. doi: 10.1242/dev.156125.

Jang, K.-J., & Suh, K.-Y. (2010). A multi-layer microfluidic device for efficient culture and analysis of renal tubular cells. Lab Chip, 10(1), 36–42. doi:10.1039/b907515a

Jang, K.-J., Mehr, A. P., Hamilton, G. A., McPartlin, L. A., Chung, S., Suh, K.-Y., & Ingber, D. E. (2013). Human kidney proximal tubule-on-a-chip for drug transport and nephrotoxicity assessment. Integrative Biology, 5(9), 1119-29. doi:10.1039/c3ib40049b

Jiang, B., Zheng, W., Zhang, W., & Jiang, X., 2013. Organs on microfluidic chips: A mini review. Science China Chemistry, 57(3), 356–364. doi:10.1007/s11426-013-4971-0

Kaarj, K & Yoon, J-Y., 2019. Methods of Delivering Mechanical Stimuli to Organ-on-a-Chip. Micromachines, 10(10), 700:1-22. doi:10.3390/mi10100700

Kane, B. J., Zinner, M. J., Yarmush, M. L., & Toner, M., 2006. Liver-Specific Functional Studies in a Microfluidic Array of Primary Mammalian Hepatocytes. Analytical Chemistry, 78(13), 4291–4298. doi:10.1021/ac051856v

Khalili, A., & Rezai, P., 2019. Microfluidic devices for embryonic and larval zebrafish studies. Briefings in Functional Genomics, 00(00), 1–14. doi:10.1093/bfgp/elz006

Khalili, A., Wijngaarden, E., Youssef, K., Zoidl, G. R., & Rezai, P, 2021. Microfluidic devices for behavioral screening of multiple Zebrafish Larvae: Design investigation process. Biotechnology Journal, 2100076, Pg.: 1-32. doi:10.1002/biot.202100076

Kim, D., Finkenstaedt-Quinn, S., Hurley, K. R., Buchman, J. T., & Haynes, C. L., 2014. On-chip evaluation of platelet adhesion and aggregation upon exposure to mesoporous silica nanoparticles. The Analyst, 139(5), 906–913. doi:10.1039/c3an01679j

Kosinska, A., & Andlauer, W., 2013. Modulation of tight junction integrity by food components. Food Research International, 54(1), 951–960. doi:10.1016/j.foodres.2012.12.038

Kwak B, Ozcelikkale A, Shin CS, Park K, and Han B., 2014. Simulation of complex transport of nanoparticles around a tumor using tumor-microenvironment-on-chip. J Control Release., 194:157-67. doi: 10.1016/j.jconrel.2014.08.027.

Lee, P. J., Hung, P. J., & Lee, L. P., 2007. An artificial liver sinusoid with a microfluidic endothelial-like barrier for primary hepatocyte culture. Biotechnology and Bioengineering, 97(5), 1340–1346. doi:10.1002/bit.21360

Lee, S.-A., No, D. Y., Kang, E., Ju, J., Kim, D.-S., & Lee, S.-H., 2013. Spheroid-based three-dimensional liver-on-a-chip to investigate hepatocyte–hepatic stellate cell interactions and flow effects. Lab on a Chip, 13(18), 3529. doi:10.1039/c3lc50197c

Lelievre, S. A., Kwok, T., & Chittiboyina, S., 2017. Architecture in 3D cell culture: An essential feature for in vitro toxicology. Toxicology in Vitro, 45, 287–295. doi:10.1016/j.tiv.2017.03.012

Liu, Y., Kongsuphol, P., Chaim, S. Y., Zhang, Q. X., Gourikutty, S. B. N., Saha, S., Biswas, S. K., and Ramadan, Q., 2018. Adipose-on-a-chip: a dynamic micro-physiological in vitro model of the human adipose for immune-metabolic analysis in type II diabetes. Lab on a Chip. 19, 241–253 doi:10.1039/c8lc00481a

Liu H, Bolonduro OA, Hu N, Ju J, Rao AA, Duffy BM, Huang Z, Black LD, Timko BP, 2020. Heart-on-a-chip model with integrated extra- and intracellular bioelectronics for monitoring cardiac electrophysiology under acute hypoxia. Nano Lett., 20: 2585–2593.

Low, L. A., Mummery, C., Berridge, B. R., Austin, C. P., & Tagle, D. A., 2020. Organs-on-chips: into the next decade. Nature Reviews Drug Discovery. doi:10.1038/s41573-020-0079-3

Ma, L.-D., Wang, Y.-T., Wang, J.-R., Wu, J.-L., Meng, X.-S., Hu, P., Mu, X., Liang, Q., Luo, G.-A. (2018). Design and fabrication of a liver-on-a-chip platform for convenient, highly efficient, and safe in situ perfusion culture of 3D hepatic spheroids. Lab on a Chip. 18, 2547-2562. doi:10.1039/c8lc00333e

Mandenius, C.-F. 2018. Conceptual Design of Micro-Bioreactors and Organ-on-Chips for Studies of Cell Cultures. Bioengineering, 5(3), 56:1-22. doi:10.3390/bioengineering5030056

Mani K., and Chen, C.Y., 2021. A smart microfuidicbased fsh farm for zebrafsh screening. Microfuidics and Nanofuidics, 25:22. https://doi.org/10.1007/s10404-021-02423-0

Marsano A, Conficconi C, Lemme M, Occhetta P, Gaudiello E, Votta E, Cerino G, Redaelli A, Rasponi M., 2016. Beating heart on a chip: a novel microfluidic platform to generate functional 3D cardiac microtissues. Lab on a Chip., 16(3):599-610. doi: 10.1039/c5lc01356a.

Mastikhina O, Moon BU, Williams K, Hatkar R, Gustafson D, Mourad O, Sun X, Koo M, Lam AYL, Sun Y, Fish JE, Young EWK, Nunes SS, 2020. Human cardiac fibrosison-a-chip model recapitulates disease hallmarks and can serve as a platform for drug testing. Biomaterials, 233:119741.

Materne, E.-M., Ramme, A. P., Terrasso, A. P., Serra, M., Alves, P. M., Brito, C., Sakharov, D. A., Tonevitsky, A. G., Lauster, R., and Marx, U., 2015. A multi-organ chip co-culture of neurospheres and liver equivalents for long-term substance testing. Journal of Biotechnology, 205, 36–46. doi:10.1016/j.jbiotec.2015.02.002

Mathur T, Singh KA, Pandian NKR, Tsai S-H, Hein TW, Gaharwar AK, Flanagan JM, Jain A, 2019. Organ-on-chips made of blood: endothelial progenitor cells from blood reconstitute vascular thromboinflammation in vessel-chips. Lab on a Chip, 19(15):2500–2511.

McKim JM Jr., 2010. Building a tiered approach to in vitro predictive toxicity screening: a focus on assays with in vivo relevance. Comb Chem High Throughput Screen., 13(2):188-206. doi: 10.2174/138620710790596736.

Musah S, Dimitrakakis N, Camacho DM, Church GM, Ingber DE., 2018. Directed diferentiation of human induced pluripotent stem cells into mature kidney podocytes and establishment of a glomerulus chip. Nat Protoc., 13:1662–85. https://doi.org/10.1038/s41596-018-0007-8.

Muschler GF, Nakamoto C, and Griffith LG., 2004. Engineering principles of clinical cell-based tissue engineering. J Bone Joint Surg Am., 86(7):1541-58. doi: 10.2106/00004623-200407000-00029.

Namdee, K., Thompson, A. J., Charoenphol, P., & Eniola-Adefeso, O., 2013. Margination Propensity of Vascular-Targeted Spheres from Blood Flow in a Microfluidic Model of Human Microvessels. Langmuir, 29(8), 2530–2535. doi:10.1021/la304746p

Panuska, P., Nejedla, Z., Smejkal, J., Aubrecht, P., Liegertova, M., Stofik, M., Havlica, J., and Maly, J., 2021. A millifluidic chip for cultivation of fish embryos and toxicity testing fabricated by 3D printing technology. RSC Advances, 11(33), 20507–20518. doi:10.1039/d1ra00846c

Parittotokkaporn, S., Dravid, A., Bansal, M., Aqrawe, Z., Svirskis, D., Suresh, V., & O'Carroll, S. J., 2019. Make it simple: long-term stable gradient generation in a microfluidic microdevice. Biomedical Microdevices, 21(3):77. doi:10.1007/s10544-019-0427-4

Peng, J., Rochow, N., Dabaghi, M., Bozanovic, R., Jansen, J., Predescu, D., DeFrance, B., Lee, S-Y, Fusch, G., Selvaganapathy, PR., and Fusch, C. (2018). Postnatal dilatation of umbilical cord vessels and its impact on wall integrity: Prerequisite for the artificial placenta. The International Journal of Artificial Organs, 41(7), 393–399. doi:10.1177/0391398818763663

Popova, A. A., Marcato, D., Peravali, R., Wehl, I., Schepers, U., & Levkin, P. A., 2017. Fish-Microarray: A Miniaturized Platform for Single-Embryo High-Throughput Screenings. Advanced Functional Materials, 28(3), 1703486 (1-12). doi:10.1002/adfm.201703486

Prill S, Jaeger MS, and Duschl C., 2014. Long-term microfluidic glucose and lactate monitoring in hepatic cell culture. Biomicrofluidics., 8(3):034102, 1-9. doi: 10.1063/1.4876639.

Pupovac, A., Senturk, B., Griffoni, C., Maniura-Weber, K., Rottmar, M., & McArthur, S. L., 2018. Toward Immunocompetent 3D Skin Models. Advanced Healthcare Materials, 7(12), 1701405, 1-11. doi:10.1002/adhm.201701405

Ramadan, Q., & Gijs, M. A. M., 2015. In vitro micro-physiological models for translational immunology. Lab on a Chip, 15(3), 614–636. doi:10.1039/c4lc01271b

Ramadan, Q., & Zourob, M. 2020. Organ-on-a-chip engineering: Toward bridging the gap between lab and industry. Biomicrofluidics, 14(4), 041501. doi:10.1063/5.0011583

Ramadan Q, and Jing L., 2016. Characterization of tight junction disruption and immune response modulation in a miniaturized Caco-2/U937 coculture-based in vitro model of the human intestinal barrier. Biomed Microdevices., 18(1):11. doi: 10.1007/s10544-016-0035-5.

Rogal, J., Probst, C., & Loskill, P., 2017. Integration concepts for multi-organ chips: how to maintain flexibility?! Future Science OA, 3(2), FSO180. doi:10.4155/fsoa-2016-0092

Ronaldson-Bouchard, K., & Vunjak-Novakovic, G., 2018. Organs-on-a-Chip: A Fast Track for Engineered Human Tissues in Drug Development. Cell Stem Cell, 22(3), 310–324. doi:10.1016/j.stem.2018.02.011

Sakolish CM, Philip B, Mahler GJ., 2019. A human proximal tubule-on-a-chip to study renal disease and toxicity. Biomicrofuidics., 13:14107. https://doi.org/10.1063/1.5083138.

Salehi-Nik N, Amoabediny G, Pouran B, Tabesh H, Shokrgozar MA, Haghighipour N, Khatibi N, Anisi F, Mottaghy K, Zandieh-Doulabi B., 2013. Engineering parameters in bioreactor's design: a critical aspect in tissue engineering. Biomed Res Int., 2013:762132, 1-15. doi: 10.1155/2013/762132.

Satoh, T., Sugiura, S., Shin, K., Onuki-Nagasaki, R., Ishida, S., Kikuchi, K., Kakiki, M., and Kanamori, T. 2018. A multi-throughput multi-organ-on-a-chip system on a plate formatted pneumatic pressure-driven medium circulation platform. Lab on a Chip, 18(1), 115–125. doi:10.1039/c7lc00952f

Schutgens, F., Rookmaaker, M. B., Margaritis, T., Rios, A., Ammerlaan, C., Jansen, J., Gijzen, L., Vormann, M., Vonk, A., Viveen, M., Yengej, F. Y., Derakhshan, S., de Winter-de Groot, K. M., Artegiani, B., van Boxtel, R., Cuppen, E., Hendrickx, A. P. A., van den Heuvel-Eibrink, M. M., Heitzer, E., Lanz, H., Beekman, J., Murk, J. L., Masereeuw, R., Holstege, F., Drost, J., Verhaar, M. C., and Clevers, H. (2019). Tubuloids derived from human adult kidney and urine for personalized disease modeling. Nature Biotechnology, 37(3), 303–313. doi:10.1038/s41587-019-0048-8

See https://www.kirkstall.com/why-quasi-vivo for more info on the Quasi Vivo® System.

Shin SR, Jung SM, Zalabany M, Kim K, Zorlutuna P, Kim SB, Nikkhah M, Khabiry M, Azize M, Kong J, Wan KT, Palacios T, Dokmeci MR, Bae H, Tang XS, Khademhosseini A., 2013. Carbon-nanotube-embedded hydrogel sheets for engineering cardiac constructs and bioactuators. ACS Nano., 7(3):2369-80. doi: 10.1021/nn305559j.

Soucy JR, Bindas AJ, Koppes AN, and Koppes RA., 2019. Instrumented Microphysiological Systems for Real-Time Measurement and Manipulation of Cellular Electrochemical Processes. iScience., 21:521-548. doi: 10.1016/j.isci.2019.10.052.

Srigunapalan S, Lam C, Wheeler AR, and Simmons CA, 2011. A microfluidic membrane device to mimic critical components of the vascular microenvironment. Biomicrofluidics, 5(1):13409. doi: 10.1063/1.3530598.

Stucki AO, Stucki JD, Hall SR, Felder M, Mermoud Y, Schmid RA, Geiser T, Guenat OT., 2015. A lung-on-a-chip array with an integrated bio-inspired respiration mechanism. Lab on a Chip., 15(5):1302-10. doi: 10.1039/c4lc01252f

Subendran, S., Wang, Y.-C., Lu, Y.-H., & Chen, C.-Y. 2021. The evaluation of zebrafish cardiovascular and behavioral functions through microfluidics. Scientific Reports, 11(1): 13801. doi:10.1038/s41598-021-93078-y

Subendran, S.; Kang, C.-W.; Chen, 2021. C.-Y. Comprehensive Hydrodynamic Investigation of Zebrafish Tail Beats in a Microfluidic Device with a Shape Memory Alloy. Micromachines, 12, 68, Pg.: 1-10. https://doi.org/10.3390/mi12010068

Sun, T., Swindle, E. J., Collins, J. E., Holloway, J. A., Davies, D. E., & Morgan, H., 2010. On-chip epithelial barrier function assays using electrical impedance spectroscopy. Lab on a Chip, 10(12), 1611-1617. doi:10.1039/c000699h

Tan J, Shah S, Thomas A, Ou-Yang HD, and Liu Y., 2013. The influence of size, shape and vessel geometry on nanoparticle distribution. Microfluid Nanofluidics., 14(1-2):77-87. doi: 10.1007/s10404-012-1024-5.

Tao, T., Wang, Y., Chen, W., Li, Z., Su, W., Guo, Y., Deng, P. and Qin, J., 2019. Engineering human islet organoids from iPSCs using an organ-on-chip platform. Lab on a Chip., 19, 948-958. doi:10.1039/c8lc01298a

Ucciferri, N., Sbrana, T., & Ahluwalia, A., 2014. Allometric Scaling and Cell Ratios in Multi-Organ in vitro Models of Human Metabolism. Frontiers in Bioengineering and Biotechnology, 2(74) : 1-10. doi:10.3389/fbioe.2014.00074

Wang Zongjie, Roya Samanipour, Kyo-in Koo1, and Keekyoung Kim, 2015. Organ-on-a-Chip Platforms for Drug Delivery and Cell Characterization: A Review. Sensors and Materials, Vol. 27, No. 6 : 487–506.

Wu, Q., Liu, J., Wang, X., Feng, L., Wu, J., Zhu, X., Wen, W., and Gong, X., 2020. Organ-on-a-chip: recent breakthroughs and future prospects. BioMedical Engineering OnLine, 19(1):9. doi:10.1186/s12938-020-0752-0

Yang, F., C. Gao, P. Wang, G. Zhang and Z. Chen, 2016. Fish-on-a-Chip: Microfluidics for Zebrafish Research. Lab on a Chip, Pg.: 1-44

Yang, X., Li, K., Zhang, X., Liu, C., Guo, B., Wen, W., & Gao, X. (2018). Nanofiber membrane supported lung-on-a-chip microdevice for anti-cancer drug testing. Lab on a Chip, 18(3), 486–495. doi:10.1039/c7lc01224a

Yum K, Hong SG, Healy KE, and Lee LP., 2014. Physiologically relevant organs on chips. Biotechnol J., 9(1):16-27. doi: 10.1002/biot.201300187

Zervantonakis IK, Hughes-Alford SK, Charest JL, Condeelis JS, Gertler FB, and Kamm RD., 2012. Three-dimensional microfluidic model for tumor cell intravasation and endothelial barrier function. Proc Natl Acad Sci U S A., 109(34):13515-20. doi: 10.1073/pnas.1210182109.

Zervantonakis IK, Kothapalli CR, Chung S, Sudo R, and Kamm RD., 2011. Microfluidic devices for studying heterotypic cell-cell interactions and tissue specimen cultures under controlled microenvironments. Biomicrofluidics., 5(1):13406. doi: 10.1063/1.3553237.

Zhang, B., Korolj, A., Lai, B. F. L., & Radisic, M., 2018. Advances in organ-on-a-chip engineering. Nature Reviews Materials, 3(8), 257–278. doi:10.1038/s41578-018-0034-7

Zhang X, Wang T, Wang P, Hu N., 2016. High-Throughput Assessment of Drug Cardiac Safety Using a High-Speed Impedance Detection Technology-Based Heart-on-a-Chip. Micromachines (Basel)., 7(7):122. doi: 10.3390/mi7070122.

Zhang YS, Aleman J, Shin SR, Kilic T, Kim D, Mousavi Shaegh SA, Massa S, Riahi R, Chae S, Hu N, Avci H, Zhang W, Silvestri A, Sanati Nezhad A, Manbohi A, De Ferrari F, Polini A, Calzone G, Shaikh N, Alerasool P, Budina E, Kang J, Bhise N, Ribas J, Pourmand A, Skardal A, Shupe T, Bishop CE, Dokmeci MR, Atala A, Khademhosseini A. 2017.

Multisensor-integrated organs-on-chips platform for automated and continual in situ monitoring of organoid behaviors. Proc Natl Acad Sci U S A., 114(12):E2293-E2302. doi: 10.1073/pnas.1612906114.

Zhao, L., Wang, Z., Fan, S., Meng, Q., Li, B., Shao, S., & Wang, Q. (2010). Chemotherapy resistance research of lung cancer based on micro-fluidic chip system with flow medium. Biomedical Microdevices, 12(2), 325–332. doi:10.1007/s10544-009-9388-3

Zheng, W., Jiang, B., Wang, D., Zhang, W., Wang, Z., & Jiang, X., 2012. A microfluidic flow-stretch chip for investigating blood vessel biomechanics. Lab on a Chip, 12(18), 3441. doi:10.1039/c2lc40173h

Zhuo Chen, Xiaoming Liu, Xiaoqing Tang, Yuyang Li, Dan Liu, Yuke Li, Qiang Huang, Tatsuo Arai, 2022. On-Chip Automatic Trapping and Rotating for Zebrafish Embryo Injection. IEEE Robotics and Automation Letters, 7 (4): 1-7.

21

Bacterial Disease of Shrimp

[1]Manimozhi E and [2]Porkodi M

[1]Department of Aquatic Animal Health Management, Dr MGR Fisheries College and Research Institute, Ponneri, Tamil Nadu

[2]Department of Fish Biotechnology, Fish Genetics and Biotechnology Division ICAR-Central Institute of Fisheries Education (ICAR-CIFE), Mumbai Maharashtra

Abstract

Disease is a major hinderance in shrimp farming industry. Diseases of bacterial origin are of opportunistic nature and most of them belong to gram negative group. Vibrio sp. is the most important among the bacterial pathogen infecting shrimp. The most predominant bacterial diseases in shrimp are Septic Hepatopancreatic Necrosis (SHPN) or "Vibriosis" caused by Vibrio harveyi, V. parahaemolyticus, V. alginolyticus and other species; (NHP)Necrotizing hepatopancreatitis Streptococcosis, (AHPND) Acute hepatopancreatic necrosis disease and spiroplasmosis. Acute hepatopancreatic necrosis disease (AHPND) is an emerging shrimp pathogen infecting shrimp. Prophylactic and management strategies can be followed to prevent bacterial infections in shrimp.

Keywords: Shrimp disease; bacterial disease; Shrimp farming; *Penaeus vannamei*

Introduction

Shrimp farming is a lucrative sector in the food production industry. According to SOFIA (2020), 9.4 million tonnes of crustaceans were produced in 2020 worldwide, with 4.96 million tonnes coming from *Penaeus vannamei.* India is one of the largest global producer and exporter of farmed shrimp. In India, shrimp production increased significantly after introducing Pacific white leg shrimp, *Penaeus vannamei,* in 2009. Among the cultured shrimp species, *Penaeus vannamei* is the leading species under culture in India, with an estimated production of 7.12 lakh tonnes (MPEDA, 2020). Intensification of culture and diversification of species provoke the incidence of diseases making diseases the major constraint in aquaculture, leading to tremendous losses. Shrimp diseases are of viral, bacterial, fungal and parasitic origin. Bacterial diseases of farmed shrimp are widespread, can result in high production losses, and are principally caused by opportunistic Gram-negative bacteria. Environmental factors and confinement are two crucial factors that stimulate the rapid duplication of opportunistic bacteria inhabiting the digestive tract, gills, and cuticle of shrimp and the water, feed, and pond sediment. *Vibrio sp.* is one of the ubiquitous bacteria occupying the topmost position among the bacterial species infecting shrimp. The most predominant bacterial diseases in shrimp are Septic Hepatopancreatic Necrosis (SHPN) or "Vibriosis" caused by *Vibrio harveyi, V. parahaemolyticus, V. alginolyticus* and other species; (NHP)Necrotizing hepatopancreatitis Streptococcosis, (AHPND) Acute hepatopancreatic necrosis disease and spiroplasmosis. Necrotizing hepatopancreatitis(NHP), Acute hepatopancreatic necrosis disease (AHPND) are OIE listed among the bacterial infection causing disease in shrimp.

Vibriosis

Vibriosis also referred to as septic hepatopancreatic necrosis (SHPN), sea gull syndrome is caused by gram negative bacteria belonging to the family Vibrionaceae, a part of the microbiome of the shrimp. The disease has been reported in hatcheries as well as in grow-out shrimp farms. The major species of *Vibrio* causing infections in shrimp are *V. harveyi, V. parahaemolyticus, V. alginolyticus, V. campbellii, Vibrio nigripulchritudo* and *V. penaeicida.* Environmental factors play a major role in precipitating the infection. Chitinoclastic nature of *Vibrio sp.* causes shell diseases and wounds leading to its penetration. Bacterial isolation and biochemical characterization, histology and PCR are the major approaches in diagnosis of infection. The presence of haemocytic infiltration and nodules with bacterial colonies in the heart, gills, lymphoid organ, connective tissue, antennal gland, musculature, and caeca; hepatopancreatic tubule lumen hypertrophy; cell sloughing; and tubular melanization, necrosis, and atrophy (moderate to severe) are the major pathology reported in Vibrio infected shrimp.

Streptococcosis

Streptococcus sp. are the causative agents for Streptococcosis. Juveniles of shrimp are more susceptible to the infection. *Streptococcus* sp. are Gram-positive cocci, principally non-motile, oxidase and catalase negative. There are no obvious gross signs associated with this infection. Typical sign is the presence of empty shelled dead shrimps on pond bottoms with the absence of large portions of internal and tail muscle as if eaten or degraded from the inside outwards. Sever infection reveals bacteremia characterized by numerous free cocci within the hemolymph and aggregates of vacuolated hemocytes with notable intravacuolar cocci in histology. Transmission occurs through ingestion or wound contamination. It can be diagnosed with bacterial isolation and culture, biochemical tests, histopathology, and PCR. Prevention and treatment are difficult; however, proper use of chemotherapeutic agents could prevent the infection to a certain extent.

Luminescent Vibriosis

Luminous bacterial diseases are one of the common causes of mortalities in shrimp hatcheries. The disease is devastating to penaeid shrimp larvae and juveniles, resulting in mortalities up to 100%. The aetiological agent of luminescent Vibriosis is gram negative *V. harveyi*. Spawners, whose midgut bacterial flora contains 16 to 17% luminescent *Vibrio sp.* of its total Vibrio population, are significant sources of luminescent Vibriosis. The most common approach to prevent or control luminescent vibriosis in the hatcheries is through the use of chemicals to reduce the numbers of bacterial pathogens in the rearing units or the incoming water. This approach is considered as the last alternative because most drugs are found ineffective and also reported to induce deformities in treated larvae. The most important diagnostic feature of luminescent Vibriosis in is a greenish luminescence in the dark. Affected shrimp larvae are lethargic and become opaque-white. Its hemocoel and internal tissues are densely packed with bacteria.

There are no significant external signs of the disease except for the darkening of the cephalothorax region. The bacterial infection route is oral, resulting in the formation of plaques on the mouth apparatus of the infected larvae. Infected shrimp juveniles often swim to the pond surface and edges. Diseased shrimp manifest loss of appetite. Histology reveals the severe inflammation in and around the hepatopancreas tubules and melanized lesions in the proximal region of the hepatopancreas. Total necrosis and dysfunction lead to death, while partial dysfunction causes slow growth due to poor digestion, absorption, and assimilation of nutrients. Systemic infections result in mortality reaching up to 100%. The occurrence of mortality is headed by a modification of the bacterial profile in the rearing water, dominated by luminous Vibrio.

Acute Hepatopancreatic Necrosis Disease (AHPND)

Acute hepatopancreatic necrosis disease (AHPND), also known as 'Early Mortality Syndrome' is caused by a particular strain of *Vibrio parahaemolyticus* carrying pirAB toxin (Photorabdus insect related toxin). Recent reports have shown that other species of Vibrio such *as V. campbellii and V. owenssii* also carry the pirAB toxin gene on their plasmids and may cause AHPND. This is known to be one among the important threat to the culture industry. AHPND appears during the first 20 to 30 days of culture, causing upto 100% mortality in severe cases. Moribund shrimp sink to bottom of the ponds, have soft shells and partially full or empty gut. Hepatopancreas (HP) is pale to whitish because of loss of pigment.

In the acute phase, the affected hepatopancreas shows tubular epithelial degeneration, rounding up and detachment of epithelial cells from the basement membrane of the tubular region, and sloughing of tubular epithelial cells into the lumen. In the terminal stage, the hepatopancreas shows extensive intertubular, hemocytic aggregations, causing hepatopancreas dysfunction and severe secondary Vibrio infections. Transmission is reported to be horizontal. Bacterial isolation, species identification using 16S rRNA PCR or toxR-targeted PCR and sequencing are the approaches in diagnosis of the infection. Proper disinfection of pond is the only measure to control or prevent the infection.

Necrotizing Hepatopancreatitis (NHP)

Necrotizing hepatopancreatitis is also referred as Texas Pond Mortality Syndrome. NHP is caused by *Hepatobacter penaei*. It is a pleomorphic, gram-negative, intracytoplasmic bacterium. Infection with *H. penaei* has been exhibited in juveniles, adults, and broodstock of *P. vannamei* and is confined to the hepatopancreas. Horizontal transmission of *H. penaei* happens via cannibalism of infected shrimp as well as faecal contamination from the infected shrimp. In *P. vannamei*, infection with *H. penaei* results in an acute, usually catastrophic disease with mortalities approaching 100%. *H. penaei* is prevalent in cultured shrimp in Belize, Brazil, Colombia, Costa Rica, Ecuador, El Salvador, Guatemala, Honduras, Mexico, Nicaragua, Panama, Peru, United States of America, and Venezuela. The incidence of infection with *H. penaei* in farms is known to increase at long periods of heavy temperatures (>29°C) and increased salinity (20–38 ppt). Gross signs associated with NHP include lethargy, anorexia, atrophied hepatopancreas, anorexia and, empty guts, reduced growth, and low length-weight ratios; soft shells and flaccid bodies; darkened gills; increased surface fouling by epicommensals ulcerative cuticle lesions or melanized appendage erosion; and expanded chromatophores resulting in the appearance of darkened edges in uropods and pleopods.

The acute infection with *H. penaei* is characterized by atrophication in the hepatopancreas with bacterial cells and hemocytes infiltration in the tubules (multifocal encapsulations). The transitional phase is characterized by increased haemocytic inflammation into the intertubular spaces in return to the necrosis, cytolysis, and sloughing tubular epithelialcells of the hepatopancreas. The atrophy of hepatopancreas tubule results in large oedematous (fluid-filled or 'watery') areas in the hepatopancreas. At this phase, hemocyte nodules are observed in the presence of masses of bacteria in the center of the nodule. The chronic phase of NHP is associated with tubular lesions, multifocal encapsulation, and oedematous areas, the decrease in abundance and severity, and is replaced by infiltration and accumulation of hemocytes at the sites of necrosis. Immunohistochemistry (IHC), ISH, and PCR are the major diagnostic techniques used in diagnosing NHP infection. The use of antibiotics can help in controlling the infection at its initial stages.

White Muscle Disease (WMD)

White muscle disease in *Penaeus vannamei* is reported to be caused by *Lactococcus garvieae*, a Gram-positive cocci. At the early stages the signs of infection occurs in the tail region of shrimp. Gross signs of this disease are similar to the White Gut Disease. WMD is known as a secondary infection. The disease originates with the destruction of the abdominal muscular organization of the prawns, especially the striated muscles, finally leading to mortality. The disease progresses with the appearance of whitish muscle from the mid of the back giving it an opaque appearance. Complete necrosis of the white muscle occurs at the final stage leading to mortality in 3-5 days. Under microscopic examination, white muscle necrosis exhibits muscle fibre ailments and stripes. It is difficult to observe the whitening of shrimp muscle from the pond's surface. Further, dead shrimp gets submerged underwater. WMD is reported to cause mortalities from 30% to 100%. Low feed intake and lethargy of the prawns are the first sign of this disease, especially during the first five days of PL settlement resulting in low mortality. The symptoms are focal to extensive necrotic areas, displaying white, tail muscle tissues, opaque appearance.

White Feaces Disease (WFD)

White Feaces Disease is one of the serious issues in shrimp aquaculture. This disease becomes apparent when the digestive system of shrimp malfunctions and feaces turns from normal (brownish color) to pale white color, appearing to be afloat on the water surface. Shrimp hepatopancreas turns into whitish and soft. The shrimps are observed to eat less, tend to be darker in color, bodies losing firmness, becoming soft and limp, and eventually will die. This disease occurs one - two months after stocking and is manifested as reduced feed intake and feed absorption in the shrimp's gut, causing significant losses to farmers and variable sizes during harvest. Infected shrimps manifest dark discoloration of the gills, white and pale appearance of hepatopancreas and gut, slow growth and loose shell. *Vibrio* sp. is isolated from infected shrimps in addition to Gregarine Protozoa. The environmental factors and poor management of the culture facility are the triggering factors for the cause.

Conclusion

Bacterial diseases of shrimp are mostly opportunistic, hence effective health management by maintaining the intricate balance between the host, pathogen, and environment can play a major role in preventing and controlling the infections. Conditions such as high stocking density, poor water quality, and sudden changes in environmental factors precipitate diseases in shrimps. Strategies like the application of prebiotics, probiotics, phytobiotics can be followed as sustainable approaches to modulate the gut microflora in farmed shrimps for preventing gut diseases by favoring the development of beneficial bacteria and inhibiting potentially pathogenic micro-organisms in the gut which inturn can supplement the well being of the animal. Management strategies like molecular screening of broodstock and PL before stocking can help to avoid the entry of pathogens into the aquaculture system. Total disinfection and sanitization of the whole culture system can help in eradicating the pathogens and other diseases causing agents. Use of immunostimulants is been proven to have role in enhancing the immune system and thus prevent the occurrence of diseases in culture animals. Sustainable approaches to prevent and control the occurrence of infections are to be implemented to enhance the production.

Conflict of Interest

The authors have no conflict of interest.

References

Aranguren, L.F., Briñez, B., Aragón, L., Platz, C., Caraballo, X., Suarez, A., Salazar, M., 2006. Necrotizing hepatopancreatitis (NHP) infected Penaeus vannamei female broodstock: Effect on reproductive parameters, nauplii and larvae quality. Aquaculture, 258 (1-4): 337-343.

Cuéllar-Anjel, J., Brock, J., Aranguren, L., Bador, R., Newmark, F., Suárez, J. (1998). A survey of the main diseases and pathogens of penaeid shrimp farmed in Colombia. Paper presented at the Book of abstracts of the World Aquaculture Society Conference "Aquaculture.

Cuéllar-Anjel, J., Pacheco, V., Diez, J., De León, E., Vega, Z., Salazar, H. (2000). Main diseases of farmed penaeid shrimp in Panama. Paper presented at the Book of abstracts of the 4th Latin american Aquaculture Congress and Exhibition. Panamá, República de Panamá.

Dabu, I.M., Lim, J.J., Arabit, P.M.T., Orense, SJAB, Tabardillo Jr, JA, Corre Jr, V.L., Maningas, M.B.B., 2017. The first record of acute hepatopancreatic necrosis disease in the Philippines. Aquaculture research, 48 (3): 792-799. FAO, 2015. Highlights of the First International Technical Seminar/Workshop: "EMS/AHPND: Government, Scientist and Farmer Responses" (June 2015, Panama City,Panama) Under the FAO Project. TCP/INT/3502.

Frelier, P., Sis, R., Bell, T., Lewis, D., 1992. Microscopic and ultrastructural studies of necrotizing hepatopancreatitis in Pacific white shrimp (Penaeus vannamei) cultured in Texas. Veterinary Pathology, 29 (4): 269-277.

Frelier, P.F., Loy, J.K., Kruppenbach, B., 1993. Transmission of necrotizing hepatopancreatitis in Penaeus vannamei. Journal of Invertebrate Pathology, 61 (1): 44-48.

Gauger, E.J., Gómez-Chiarri, M., 2002. 16S ribosomal DNA sequencing confirms the synonymy of Vibrio harveyi and V. carchariae. Diseases of aquatic organisms, 52 (1): 39-46.

Gracia-Valenzuela, M.H., Ávila-Villa, L.A., Yepiz-Plascencia, G., Hernández-López, J., Mendoza-Cano, F., García-Sanchez, G., Gollas-Galván, T., 2011. Assessing the viability of necrotizing hepatopancreatitis bacterium (NHPB) stored at- 20° C for use in forced-feeding infection of Penaeus (Litopenaeus) vannamei. Aquaculture, 311 (1-4): 105-109.

Ha, NTH, Ha, D.T., Thuy, N.T. and Lien, V.T.K., 2010. Enterocytozoon hepatopenaei Parasitizing on tiger shrimp (Penaeus monodon) infected by white feces culture in Vietnam, has been detected. Agriculture and rural development: science and technology (translation from Vietnamese), 12: 45-50.

Han, J.E., Tang, K.F., Aranguren, L.F., Piamsomboon, P., 2017. Characterization and pathogenicity of acute hepatopancreatic necrosis disease natural mutants, pirABvp (-) V.parahaemolyticus, and pirABvp (+) V. campbellii strains. Aquaculture 470, 84–90.

Han, J.E., Tang, K.F., Tran, L.H., Lightner, D.V., 2015. Photorhabdus insect-related (Pir) toxin-like genes in a plasmid of Vibrio parahaemolyticus, the causative agent of acute hepatopancreatic necrosis disease (AHPND) of shrimp. Diseases of aquatic organisms, 113 (1): 33-40.

Hasson, K.W., Wyld, EM, Fan, Y., Lingsweiller, S.W., Weaver, S.J., Cheng, J., Varner, P.W., 2009. Streptococcosis in farmed Litopenaeus vannamei: a new emerging bacterial disease of penaeid shrimp. Diseases of aquatic organisms, 86 (2): 93-106.

http://www.ciba.res.in/wp-content/uploads/2019/06/2019-AAHED-Handouts-AHPND.pdf

https://mpeda.gov.in/?page_id=651

https://www.viv.net/articles/news/prevention-and-treatment-of-white-feces-disease-wfd-in-shrimp

Ibarra-Gámez, J., Galavíz-Silva, L., Molina-Garza, Z., 2007. Distribución de la bacteria causante de la necrosis hepatopancreática (NHPB) en cultivos de camarón blanco, Litopenaeus vannamei, en México. Ciencias marinas, 33 (1): 1-9.

Joshi, J., Srisala, J., Truong, V.H., Chen, I.-T., Nuangsaeng, B., Suthienkul, O., Lo, C.F., et al., 2014. Variation in Vibrio parahaemolyticus isolates from a single Thai shrimp farm experiencing an outbreak of acute hepatopancreatic necrosis disease (AHPND). Aquaculture, 428: 297-302.

Kalaimani, N., Ravisankar, T., Chakravarthy, N., Raja, S., Santiago, T., Ponniah, A., 2013. Economic losses due to disease incidences in shrimp farms of India. Fish Technol, 50: 80-86. Karunasagar, I., Pai, R., Malathi, G., Karunasagar, I., 1994. Mass mortality of Penaeus monodon larvae due to antibiotic-resistant Vibrio harveyi infection. Aquaculture, 128 (3-4): 203-209.

Kim, Y.B., Okuda, J., Matsumoto, C., Takahashi, N., Hashimoto, S., Nishibuchi, M.1999. Identification of Vibrio parahaemolyticus strains at the species level by PCR targeted to the toxR gene. Journal of Clinical Microbiology, 37 (4): 1173-1177.

Kondo, H., Van, P.T., Dang, L.T., Hirono, I., 2015. Draft genome sequence of non-Vibrio parahaemolyticus acute hepatopancreatic necrosis disease strain KC13. 17.5, isolated from diseased shrimp in Vietnam. Genome Announcements, 3 (5)

Lavilla-Pitogo, C.R., Albright, L.J., Paner, M.G., Sunaz, N. (1992). Studies on the sources of luminescent Vibrio harveyi in Penaeus monodon hatcheries. Paper presented at the Diseases in Asian Aquaculture I. Proceedings of the First Symposium on Diseases in Asian Aquaculture, 26-29 November 1990, Bali, Indonesia.

Lavilla-Pitogo, C.R., Baticados, M.C.L., Cruz-Lacierda, E.R., Leobert, D., 1990. Occurrence of luminous bacterial disease of Penaeus monodon larvae in the Philippines. Aquaculture, 91 (1-2): 1-13.

Lavilla-Pitogo, C.R., Leaño, E.M., Paner, M.G., 1998. Mortalities of pond-cultured juvenile shrimp, Penaeus monodon, associated with dominance of luminescent vibrios in the rearing environment. Aquaculture, 164 (1-4): 337-349.

Leaño, E.M., Lavilla-Pitogo, C.R., Paner, M.G., 1998. Bacterial flora in the hepatopancreas of pond-reared Penaeus monodon juveniles with luminous Vibriosis. Aquaculture, 164 (1-4): 367-374.

Lee, C.-T., Chen, I.-T., Yang, Y.-T., Ko, T.-P., Huang, Y.-T., Huang, J.-Y., Huang, M.-F., et al., 2015. The opportunistic marine pathogen Vibrio parahaemolyticus becomes virulent by acquiring a plasmid that expresses a deadly toxin. Proceedings of the National Academy of Sciences, 112 (34): 10798-10803.

Lightner, D., Redman, R., Pantoja, C., Tang, K., Noble, B., Schofield, P., Mohney, L., et al., 2012. Historic emergence, impact and current status of shrimp pathogens in the Americas. Journal of Invertebrate Pathology, 110 (2): 174-183.

Lightner, D.V., 1993. Diseases of cultured penaeid shrimp. CRC handbook of mariculture, 1: 393-486. Lightner, D.V., 1996. A handbook of shrimp pathology and diagnostic procedures for diseases of cultured penaeid shrimp.

Lightner, D.V., Lewis, D.H., 1975. A septicemic bacterial disease syndrome of penaeid shrimp.

Loy, J.K., Frelier, P.F., Varner, P. and Templeton, J.W., 1996. Detection of the etiologic agent of necrotizing hepatopancreatitis in cultured Penaeus vannamei from Texas and Peru by polymerase chain reaction. Diseases of Aquatic Organisms, 25(1-2), pp.117-122.

Mar. Fish. Rev, 37 (5-6): 25-28. Lightner, D.V., Redman, R.M., 1994. An epizootic of necrotizing hepatopancreatitis in cultured penaeid shrimp (Crustacea: Decapoda) in northwestern Peru. Aquaculture, 122 (1): 9-18.

Morales-Covarrubias, M., Cuéllar-Anjel, J., Varela-Mejías, A., Elizondo-Ovares, C., 2018. Shrimp bacterial infections in Latin America: a review. Asian Fisheries Science S, 31: 76-87.

Morales-Covarrubias, M., Ruiz-Luna, A., Moura-Lemus, A.P., Solís, M., Conroy, G., 2011. Prevalence of diseases in cultured Litopenaeus vannamei in eight regions of Latin America. Revista Cientifica, Facultad de Ciencias Veterinarias, Universidad del Zulia, 21 (5): 434-446.

Morales-Covarrubias, M.S., Lozano-Olvera, R., Hernández-Silva, A., 2010. Necrotizing hepatopancreatitis in cultured shrimp caused by extracellular and intracellular bacteria. Tilapia & Camarones, 5: 33-39.

Morales-Covarrubias, M.S., Tlahuel-Vargas, L., Martínez-Rodríguez, I.E., Lozano-Olvera, R., Palacios-Arriaga, J.M., 2012. Necrotising hepatobacterium (NHPB) infection in Penaeus vannamei with florfenicol and oxytetracycline: a comparative experimental study. Revista Científica, 22 (1): 72-80.

Morales-Covarrubias M.S. (2008). Capítulo 3: Enfermedades bacterianas.En: Patología eInmunología de Camarones Editores Vielka Morales y Jorge Cuellar-Angel. Programa CYTED Red II-D vannamei, Panamá, Rep. De Panamá, 120–134.

Morales Covarrubias, M., Osuna-Duarte, A., Garcia-Gasca, A., Lightner, D.V., Mota-Urbina, J., 2006. Prevalence of necrotizing hepatopancreatitis in female broodstock of white shrimp Penaeus vannamei with unilateral eyestalk ablation and hormone injection. Journal of Aquatic Animal Health, 18 (1): 19-25.

Pedersen, K., Verdonck, L., Austin, B., Austin, D.A., Blanch, A.R., Grimont, P.A., Jofre, J., et al., 1998. Taxonomic evidence that Vibrio carchariae Grimes et al. 1985 is a junior synonym of Vibrio harveyi (Johnson and Shunk 1936) Baumann et al. 1981. International Journal of Systematic and Evolutionary Microbiology, 48 (3): 749-758.

Sindermann, C.J., Lightner, D.V., 1988. Disease diagnosis and control in North American marine aquaculture.

Sirikharin, R., Taengchaiyaphum, S., Sanguanrut, P., Chi, T.D., Mavichak, R., Proespraiwong, P., Nuangsaeng, B., et al., 2015. Characterization and PCR detection of binary, Pir-like toxins from Vibrio parahaemolyticus isolates that cause acute hepatopancreatic necrosis disease (AHPND) in shrimp. PloS one, 10 (5): e0126987.

Soto-Rodriguez, S.A., Gomez-Gil, B., Lozano-Olvera, R., Betancourt-Lozano, M., Morales-Covarrubias, M.S., 2015. Field and experimental evidence of Vibrio parahaemolyticus as the causative agent of acute hepatopancreatic necrosis disease of cultured shrimp (Litopenaeus vannamei) in Northwestern Mexico. Applied and Environmental Microbiology, 81 (5): 1689-1699.

Sung, H.-H., Hsu, S.-F., Chen, C.-K., Ting, Y.-Y., Chao, W.-L., 2001. Relationships between disease outbreak in cultured tiger shrimp (Penaeus monodon) and the composition of Vibrio communities in pond water and shrimp hepatopancreas during cultivation. Aquaculture, 192 (2-4): 101-110.

Tran, L., Nunan, L., Redman, R.M., Mohney, L.L., Pantoja, C.R., Fitzsimmons, K. and Lightner, D.V., 2013. Determination of the infectious nature of the agent of acute hepatopancreatic necrosis syndrome affecting penaeid shrimp. Diseases of aquatic organisms, 105(1), pp. 45-55.

Vincent, A.G., Breland, V.M., Lotz, J.M., 2004. Experimental infection of Pacific white shrimp Litopenaeus vannamei with Necrotizing Hepto-pancreatitis (NHP) bacterium by per os exposure. Diseases of Aquatic Organisms, 61 (3): 227-233.

Vincent, A.G., Lotz, J.M., 2005. Time course of necrotizing hepatopancreatitis (NHP) in experimentally infected Litopenaeus vannamei and quantification of NHP-bacterium using real-time PCR. Diseases of Aquatic Organisms, 67 (1-2): 163-169.

Weisburg, W.G., Barns, S.M., Pelletier, D.A., Lane, D.J., 1991. 16S ribosomal DNA amplification for phylogenetic study. Journal of Bacteriology, 173 (2): 697-703.

22

Types of Markers and Their Application in Fisheries

Sanjay Chandravanshi[1], Narshing Kashyap[2] Sudhan Chandran[3] and Hari Prasad Mohale[4]

[1,4]Department of Fisheries Biology and Resource Management Fisheries College and Research Institute, TNJFU, Tamil Nadu, India

[2]Department of Fish Genetics and Breeding, Institute of Fisheries Post Graduate Studies Vaniyanchavadi, TNJFU, Tamil Nadu, India

[3]Department of Fisheries Resource Management, Fisheries Resources Harvest and Post-Harvest Management Division, ICAR-CIFE Mumbai, Maharashtra India

Abstract

A species ability to adapt to a changing environment is improved by genetic variety, which is essential for the survival of the species. Individuals' genetic differences cause diversity at the population, species, and higher cognitive taxonomic levels. The information on genetic variety has several uses, including genetic improvement programmes, conservation and management of natural resources, and research on evolution. A valuable tool for detecting genetic studies of people, groups, or species is the development of molecular genetic markers. These molecular markers, along with recent advances in statistics, have changed the analytical capacity required to investigate genetic variation. Fisheries and aquaculture today use a variety of molecular markers, including protein and DNA (mitochondrial DNA (mt DNA), microsatellites and RAPD). Fisheries and aquaculture today use a variety of molecular markers, proteins, or DNA based markers like mitochondrial DNA markers or some nuclear type DNA markers such as microsatellites, minisatellites, anonymous cDNA, or RAPDs etc. Aquaculture studies of genetic diversity and inbreeding, parentage determination, species and strain identity, and the creation of strong genetic linkage for aquaculture species have all advanced quickly thanks to the application of DNA markers.The methods are causing miscommunications, especially among field researchers and scientists or the intended audience for their research. More significantly, choosing which of these markers to use for a given application is not always easy and frequently depends on the researcher' earlier research.

Keywords: Molecular markers, Genetic markers, Allozyme, RAPD, RFLP, Microsatellites, EST, SNP.

Introduction

A genome marker or a genetic marker is words commonly used to refer to a marker, which is identical to the term "marker." Genes, genetic variants, or random sequences can all be represented by markers. Markers act as milestones or signposts while creating DNA and genomic maps. Markers are DNA sequences that are frequently identified by their position in a genome. The transmission of traits or the risk of sickness within families can be tracked using markers. It is essential to the continued existence of the species and boosts an organism's ability to adjust environmental conditions. In conjunction with other evolutionary forces like genetic drift and selection, genetic diversity inside an individual evolves, leading to divergence at the population/communities, species, as well as higher order taxonomic levels (Askari et al., 2013). Numerous fields, including evolutionary research, resources management and conservation, and genomic improvement programmes, can benefit from the conclusion derived from genetic variety data (Liu and Cordes, 2004).

Genetic variety measures are essential for the comprehension, understanding, and effective management of wild fish species or aquaculture stocks. Genetic diversity has been examined inferentially and indirectly through investigations of genetic selection and performance or by the conventional, systematic measurement of phenotypic characteristics. Blood pigments, immunogenetics, cytogenetics, ecological tagging, physiological or behavioural factors, meristics and immunogenetics are only a few of the many traits and techniques used to analyse the stock structure in fish populations (Okumus and Ciftci, 2003).

Molecular Markers

A molecular marker, also known as a marker gene, is a DNA pattern that is used to "mark" or monitor a specific locus on a certain chromosome. The study of how a trait or gene is inherited is made easier by genes that have a known location, a distinct phenotypic expression that can be identified by analytical techniques, or a recognizable DNA sequence. DNA markers are also being used in population genetics together with protein markers.The complete genome might theoretically be observed and used for genetic research using DNA markers. There are several uses for mitochondrial and genomic DNA. Aquaculture genetics uses a number of widely popular marker types. In the past, mtDNA and allozyme markers were frequently used in studies on aquaculture genetics (Ferguson and Danzmann, 1998).

The 2 types of molecular markers are mostly used: type I markers and type II markers. Type I markers are linked to known-functioning genes, whereas type II markers are linked to unidentified genomic areas. The type I marker is Allozyme, Considering that the protein that encode serves a specific purpose, they fall under this classification. Because RAPD bands are amplified by the PCR from anonymous genomic areas, RAPD is a type II marker. Microsatellite markers are indeed class II markers unless they are connected with recognized function genes. Type I markers are becoming increasingly significant for aquaculture genetics. Type I markers act as a link for the assessment and movement of genomic data from a species with a rich genomic map to a species with a relatively sparse genomic map. Type II markers including RAPDs, microsatellites, and AFLPs are typically thought of as non-coding and so preferentially neutral. In population genetics studies, these markers are frequently used to describe genetic differences between and within populations or species (Askari et al., 2013).

Allozyme

A successful management approach must include genetic diversity measures in aquaculture stocks. Examining allozyme loci continues to be most important common methods for addressing population genetics & stock structure issues in fish (Chauhan et al., 2010). The process of locating genetic diversity at the levels of enzymes that are directly coded by DNA is known as allozyme electrophoresis. Allozymes, which differ a bit in electrical charge, are protein variants created by allelic variants (Okumu and Cifti, 2003). In aquaculture, allozymes are used for stock identification, parentage analysis, and tracking inbreeding. Some allozyme markers showed associations with performance characteristics (Liu and Cordes, 2004).

Allelic variations give rise to protein variants known as allozymes, and they will have a tiny difference in electric charge. Allozymes have codominant markers that exhibit Mendelian expression in a heterozygous person. As a result, allozyme analysis offers information on single loci genetic differences that can address a variety of queries concerning fish and fish population stocks. The initial step in detecting allozyme variation is to isolate allozymes from tissues utilizing particular methods. The variation is then identified using electrophoresis on a gel made of cellulose acetate or acrylamide. Individuals who are homozygous display a single band, while those who are heterozygous display two bands (Kucuktas and Liu, 2007).

Allozyme are among the most researched types of molecular variation because of their ease of use, low cost, and minimal need for specialist equipment. All soluble proteins can be used for allozyme analysis. At one moment, a large number of loci could be tested. Due to the technique's drawbacks, which include the need for a lot of tissue, small organisms cannot be used with this procedure.

It is necessary to sacrifice the fish and freeze the tissue cryogenically because the tissue sampling process is intrusive. Protein electrophoresis might not pick up a point mutation in a nucleotide sequence since it might not even alter an amino acid. Additionally, changes in DNA that affects one amino acid would not affect the protein's total charge and is therefore undetectable. Despite these drawbacks, allozyme has been widely used in fisheries, particularly in analyses of mixed stock fisheries, population structure, and conservation genetics (Kucuktas and Liu, 2007).

Mitochondrial DNA Markers

An increasing number of recent demographic and phylogeny investigations of organisms involve mitochondrial DNA (mtDNA) analyses. An increasing number of recent demographic & phylogenetic assessments of organisms involve mitochondrial DNA (mtDNA) analyses. Investigations of vertebrate species as a whole have demonstrated that mitochondrial DNA sequence deviation increases more quickly than nuclear DNA. Due to the haploid mitochondrial genome's exclusive female inheritance, this has been linked to a greater mutation rate in mtDNA that may be caused by a lack of repair mechanisms during replication and a smaller effective population size. Excluding the roughly 1-kb regulatory region (D-loop), when mtDNA replication and transcription are started, almost the whole molecule is transcribed. The non-coding segments like the D-loop typically have higher levels of diversity than coding sequences like the cytochrome b gene. This is likely because of loosened functional restrictions and less selection pressure (Liu and Cordes, 2004).

Numerous fish species, including eels (Avise et al., 1986), bluefish (Graves et al., 1992), red drum (Gold et al., 1993), snappers (Chow et al., 1993) and sharks, have had their stock structure thoroughly studied using mtDNA marker analyses (Maqsood and Ahmad, 2017). Due in part to how useful they are for identifying broodstocks, mitochondrial markers are extremely popular among aquaculture geneticists. In the initial days of molecular studies, population difference in aquaculture was achieved by taking advantage of high levels of mtDNA variation relative to allozymes (Askari et al., 2013). Numerous techniques can be used to directly or indirectly identify changes in the nucleotide sequence of the DNA molecule in the mitochondria. Numerous population genetic studies have used Restriction Fragment Length Polymorphism analysis of mitochondrial DNA to understand population genetic variation. This can be done by either DNA sequencing of short segments of mtDNA obtained by PCR amplification or by digesting the entire purified mtDNA molecule with restriction endonucleases. Mitochondrial DNA analysis has become increasingly popular because to these procedures with better resolution and maximal information. The mtDNA molecule is regarded as a single locus because of its non-Mendelian pattern of inheritance. Fish stock structure has been widely studied using mtDNA marker analyses in a range of vertebrate species.

Due to its conservation across such a wide range of taxa, the mitochondrial Cytochrome C Oxidase I gene has been selected as the global barcode for species level identification. DNA barcodes, which are roughly 600 base pair fragments of the mitochondrial COI gene, are a quick, low-cost, and effective method for cataloguing biodiversity. Fish biology, taxonomy and population genetics have all benefited from the widespread use of mitochondrial DNA genes over the past two decades (Sukumaran and Gopalkishnan, 2019).

Multiple Arbitrary Primer Markers

This category includes tests that focus on a section of DNA with an unidentified function. Multiple arbitrary amplicon profiling, also known as anonymous nDNA markers, is a general class of PCR-based methods used to find such arbitrary or anonymous sequences. The two main techniques are amplified fragment length polymorphism DNA (AFLP) and Randomly Amplified Polymorphic DNA (RAPD, read rapid).

Randomly Amplified Polymorphic DNA (RAPD)

Random PCR amplification of unidentified loci is known as RAPD. It has a number of benefits and has been used in fisheries studies pretty frequently. The procedure is quick, easy, and affordable; it requires little DNA and doesn't require molecular hybridization; most crucially, it doesn't necessitate any prior knowledge of the genetic makeup of the organism in issue (Okumu and Ciftci, 2004). RAPD markers are amplified by-products of less functioning regions of the genome that don't exhibit a robust phenotypic response to selection. Such DNA areas might develop more nucleotide mutations, which could be used to evaluate the genetic divergence between populations. Using random or arbitrary nucleotide sequence primers and PCR to amplify genomic DNA, RAPD may identify significant amounts of DNA polymorphisms. The method finds repetitive (non-coding) DNA sequences in the genome as well as encoding sequences of DNA, and some of the most relevant polymorphic sequences are those that are generated from these sequences.

Since 90% of the vertebrate nuclear genome is non-coding, it is expected that the majority of amplified sites will be neutral to selection. RAPD loci are evaluated as present or absent and transmitted as dominant Mendelian markers. With the additional benefit that primers are readily accessible on the market and don't require any prior knowledge the target DNA sequence structure, RAPDs have all the benefits of a PCR-based marker. Since RAPD is a dominant marker, it is impossible to distinguish between homozygous and heterozygous states, and these structures are sensitive to even the smallest adjustments to the amplification settings. One of the other advantages of RAPDs is the ease with which several loci and individuals can also be tested concurrently. The presence of paralogous PCR product (various DNA segments with the same lengths that appear to be a single locus), inadequate reproducibility because of the low annealing temperature utilized in the PCR amplification, and other problems have further limited the utility of this marker in fisheries science.

The technique is based on short, arbitrary-sequence oligonucleotide primers for PCR amplification of particular genomic regions. First, RAPD is generated using PCR.Major prerequisites include a modest quantity of genome, one or even more oligonucleotide primers (typically about 10 base pairs in length), free nucleotides, and polymerase with an appropriate reaction buffer. The fundamental problem with RAPDs is that the pattern of bands that is produced is extremely susceptible to changes in the reaction parameters, DNA quality, and PCR temperature profile. The use of RAPDs in fish research has been constrained by these issues.

Restriction Fragment Length Polymorphism (RFLP)

In the past ten years, RFLP has been successfully used to investigate the mitochondrial DNA (mtDNA) population makeup of fishes, often revealing more diverse genetic variability than isozyme analysis. The genomic revolution was believed to have begun with RFLP markers, ushering in a new age in the biological sciences. Bacterial enzymes called restriction endonucleases can identify certain4, 5, 6, or 8 bp (base pair) nucleotide sequences and cut DNA wherever these sequences are encountered, allowing for the gain, loss, or relocation of a restriction site in response to changes in the DNA sequence caused by indels, base substitutions, or rearrangements involving the restriction sites. Individuals, communities, and species can all produce different numbers and sizes of DNA fragments when their DNA is digested by restriction enzymes. Traditionally, fragments were sorted using Southern blot analysis, which involves digesting genomic DNA, running across an agarose gel, transferring it to a membrane, and seeing it by hybridization to specific probes. Previously, fragments were sorted by a process known as Southern blot analysis, which entails processing genomic DNA, passing it through an agarose gel for electrophoresis, depositing it to a membrane, and then seeing the outcome by crossbreeding it with certain probes (Lo Presti et al., 2009).

The time-consuming Southern blot approach has mostly been replaced by PCR-based methods in recent studies (PCR).The RFLP region's portion is amplified by PCR when the flanking sequences for a locus are known. If a very significant deletion or insertion caused the length polymorphism, gel electrophoresis of the PCR findings should reveal the size variation. However, if the length polymorphism is caused by base substitutions at a restriction site, the PCR products should be digested with a restriction enzyme to expose the RFLP. The ability of RFLP markers to reveal genetic variety is fairly restricted when compared to more recently developed markers and approaches described below. Although indels and reorganizations of sections containing restriction sites may be frequent in the genomes of most animals, the possibility of such an occurrence inside a particular locus under study should be minimal.

The main advantage of RFLP markers is that both of an individual's alleles are seen in the study because they are co-dominant markers. Scoring is generally simple because the size disparity is frequently considerable. The comparatively modest amount of polymorphism is the main drawback of RFLP. In addition, either sequence information (for PCR analysis) or probes (for Southern blot analysis) are required, making it difficult and time-consuming to develop markers in species lacking known molecular information.

Amplified Fragment Length Polymorphism (AFLP)

In the same polyacrylamide gel, AFLPs are dominant and numerous markers and alleles are confounded. It overcomes the weaknesses of RAPD and RFLP markers by combining their advantages. The method is PCR-based and does not require a probe or prior sequence knowledge that RFLP does. High stringent PCR ensures its reliability, as opposed to RAPD's issue with poor reproducibility. The primary benefit of AFLPs is their ability to store several polymorphisms in a single polyacrylamide gel without the requirement for any prior development or research. Microsatellite loci appear to be substantially less effective than AFLP at identifying an individual's source among hypothetical populations. Similarly to RAPD, AFLP analysis helps the quick and affordable screening of a large number of additional loci throughout the genome. Despite the fact that AFLP markers' utility for applied population genetics is now constrained by their dominant pattern of inheritance, the method appears ideal for providing genomic evidence for intraspecific mtDNA research.

Microsatellites

A microsatellite is a short DNA sequence that is repeatedly found at different locations throughout an organism's DNA. Despite the significant variability of these repeats, these loci can be utilized as markers. Microsatellites are made up of several copies of 1 to 6 base pair long (SSRs) (simple sequence repeats) that are tandemlyorganized. One microsatellite was discovered in fish for every 1.87 kb of DNA (Mojekwu and Anumudu, 2013). Microsatellites are more helpful in genome mapping and population genetics studies since they occur once every 10 kbp while minisatellite loci only occur once every 1500 kbp in fish (Lakra, 2001). Allozymes and mtDNA don't have nearly as much information as microsatellites do, but they nevertheless provide similar analysis benefits. All methods demand a similar level of technical ability for recognition and scoring/analysis when polymorphic loci have been found.

In the genome, microsatellites are typically distributed equally among all chromosomes and chromosomal regions. Data from the whole genome sequencing has, however, somewhat refuted this claim. They have been discovered in introns, non-gene sequences, and gene encoding areas (Liu et al., 2001). The majority of microsatellite loci are tiny, with a few such a few hundred repeats. Changes in the amount of repeat units can lead to a huge quantity of alleles at every microsatellite loci in a population, regardless of the precise mechanisms involved. As co-dominant markers, microsatellites have indeed been inherited in a Mendelian manner. When compared to other markers, several species' microsatellites were shown to be instructive .Microsatellite marker use, however, necessitates a significant upfront expenditure of time and energy. To create PCR primers, every microsatellite locus must be located and its flanking regions sequenced. Small size changes between alleles of a particular microsatellite locus are possible due to polymerase slippage during replication. Recent years have seen a dramatic increase in the use of microsatellites as markers in a variety of genomic studies (Abdul-Muneer, 2014).

Microsatellite markers are employed in fisheries and aquaculture for phylogeny and phylogeographic analysis, population genetic characteristics, biodiversity preservation, stocking effects, and hybridization. Additionally, it is being utilized more and more for behavioural analysis, kinship determination, genomic mapping, and forensic identification of people.

Single Nucleotide Polymorphism (SNP)

SNPs, or single nucleotide polymorphisms, are polymorphisms brought on by point mutations that result in various alleles with different bases at specific nucleotide positions within a locus. Since SNPs are amenable to automation, represent the most prevalent polymorphism in every organism's genomes and disclose hidden variation not discovered by other markers and approaches, they are increasingly becoming a focus of molecular marker research. Various alleles with different bases at a certain nucleotide location result from these point mutations. The most prevalent polymorphisms in any organism's genome (both coding and non-coding) are SNPs. PCR, microchip arrays, or fluorescence technology can all be used to find these single nucleotide mutations. They are referred to as next-generation markers in fisheries and have applications in disease detection, population genetics, and genomics research.

Expressed Sequence Tags (ESTs)

Single-pass sequences called ESTs were produced by randomly sequencing cDNA clones. ESTs can be used locate genes and do expression analysis on their expression. The genes expression in particular tissue types under particular physiological circumstances or developmental processes can be quickly and accurately analysed. Using cDNA microarrays, genes with differential expression could be systematically found. The most useful use of ESTs is in linkage mapping.

DNA Barcoding

The conservation and management of biodiversity is a fundamental tenet of conservation biology. The identification of lineages particularly deserving of preservation or in need of preservation, as well as the difficulty of generating an assessment of this diversity for prioritizing hotspots of species richness are the two main challenges to such an undertaking. The mitochondrial gene cytochrome oxidase I (COI) has portions of DNA that are around 600 base pairs long. These DNA barcodes have been proposed as a quick, effective, and affordable method to record all biodiversity (Hebert et al., 2003; Hebert et al., 2004). In barcoding, a 600-basepair COI gene fragment is amplified and sequenced using universal PCR primers. Then, a distance-based algorithm is used to compare that segment of the sequence with a database of "known" sequences from specimens that have already been classified by taxonomists. It would appear that using DNA barcodes from a tiny section of the mitochondrial genome is a quick and efficient technique to determine at least a minimal amount of biodiversity. Furthermore, molecular studies using barcode-sized sequences have discovered cryptic DNA lineages for taxa that are already well known, such as birds and mammals, which may be valuable.

Fisheries Use for Molecular Markers

Intra specific and inter specific variations

The introduction of molecular genetic markers as an additional marker system can improve taxonomy research's resolution. The degree of DNA or gene divergence can be used as the basis for species classification because the molecular evolution of taxa varies greatly. We occasionally see a weak association between morphological characteristics and gene divergence because the ecological and morphological characters may diverge more quickly than genetic divergence at neutral loci. When a species that occurs in mixed captures need to be identified and morphological identification is highly challenging, molecular markers are most helpful.These techniques can also be used to identify dead and stranded whales, sharks, and dolphins that are endangered or threatened because morphological identification is frequently not achievable. Molecular type markers also have been employed in the identification of subspecies (Sukumaran and Gopalkishnan, 2019).

Phylogeograhical and Phylogeneticstudies

Populations Structure: Within and Between Populations

Phylogeographic studies evaluate geographic distributions, while phylogenetic studies evaluate the historical processes that have shaped relationships. The introduction of mitochondrial DNA markers in population genetic studies triggered the start of phylogenetic and phylogeographic research. Inferring intraspecific phylogenetic patterns from mitochondrial DNA has proven to be a highly effective method for many animal taxa (Sukumaran and Gopalakrishnan, 2019).The evolutionary history of several fish species may be reconstructed, which will yield important information about historical demography. Phylogenetic studies may also be used to extract information about ecological trends and conservation units.

Genetic marking/tagging

Sometimes it's necessary to mark certain fish in order to track their movement or migration, estimate their population size, or assess their contributions to mixed fisheries. Physical tags cannot be used for generations because they are not heritable. By identifying an infrequent allele in an individual or population and tracking them over generations, genetic marking can be used to gather data on migration patterns from source populations, the effect of hatchery programmes on harvest, and the proportion of stocked individuals who contribute to population growth (Sukumaran and Gopalakrishnan, 2019).

Applications in aquaculture

Molecular markers have wide range of applications in aquaculture mainly; in genetic identification and discrimination of hatchery stocks, finding out inbreeding events, assignment of progeny to parents using genetic tags, finding out quantitative trait loci, marker assisted selection for selective breeding trials and assessment of the effect of polyploidy induction and gynogenesis (Tanya and Kumar, 2010). Genetic variability can also be assessed between and within stocks using molecular markers. They are also useful in determining the contribution of possible parents in mass spawning events. Disease diagnosis is another major area where the benefits of molecular markers can be successfully harnessed. PCR assays for different pathogens have become inexpensive, safe and user friendly in many diagnostic laboratories.

Genetic recognition and discrimination of hatchery stocks, determining inbreeding events, assigning progeny to parents using genetic tags, identifying quantitative trait loci, the marker-assisted choice for selective breeding trials, and evaluation of the impact of polyploidy induction and gynogenesis are just a few of the many applications for molecular markers in aquaculture (Muramulla et al., 2022). Molecular markers can also be used to evaluate genetic variation within and between stocks. They are helpful in determining the role that potential parents might have played in mass spawning occasions. Another important area where molecular markers can be successfully used is in the diagnosis of diseases. In many diagnostic facilities, PCR tests for various infections have become affordable, safe, and eco-friendly.

Conclusion

The wide use of molecular markers in aquaculture and fisheries has had a significant impact on fish genetics. To get the highest possible output quality, each marker type selection should be made carefully. A combination of molecular markers is always appropriate because no one molecular marker is better than any other. Recently, the usage of markers created and screened utilizing next-generation sequencing technology in fish genetics has increased. The need for aquaculture products is rising globally, and contemporary molecular techniques and molecular genetics may be key to enhancing aquaculture's quality and sustainability.

References

Abdul-Muneer, P.M., 2014. Application of microsatellite markers in conservation genetics and fisheries management: recent advances in population structure analysis and conservation strategies. Genetics research international. 11pp.

Askari, G.H., Shabani, A. and KolangiMiandare, H., 2013. Application of molecular markers in fisheries and aquaculture. Scientific Journal of Animal Science, 2(4), pp.82-88.

Avise, J.C., Helfman, G.S., Saunders, N.C., & Hales, H.S. (1986). Mitochondrial DNA differentiation in North Atlantic eels: population genetic consequences of an unusual life history pattern. ProcNatlAcadSci, 83, 4350–4354.

Chow, S., Clarke, M.E., &Walse, P.J. (1993). PCR-RFLP analysis on thirteen western Atlantic snappers (subfamily Lutjaninae): a simple method for species and stock identification. Fish Bull, 91, 619-627.

Ferguson, M.M. and Danzmann, R.G., 1998. Role of genetic markers in fisheries and aquaculture: useful tools or stamp collecting?. Canadian Journal of Fisheries and Aquatic Sciences, 55(7), pp.1553-1563.

Gold, J.R., Richardson, L.R., Furman, C., & King, T.L. (1993a). Mitochondrial DNA differentiation and population structure in red drum (Sciaenopsocellatus) from the Gulf of Mexico and Atlantic Ocean. Mar Biol, 116, 175–185.

Graves, J.E., McDowell. J.R., & Jones, M.L. (1992). A genetic analysis of weakfish Cynoscionregalis stock structure along the mid-Atlantic coast. Fish Bull, 90: 469–475.

Kucuktas, H. and Liu, Z., 2007. Allozyme and Mitochondrial DNA Markers. Aquaculture Genome Technologies, p.73.

Lakra, W.S., 2001. Genetics and Molecular Biology in Aquaculture-Review. Asian-Australasian Journal of Animal Sciences, 14(6), pp.894-898.

Liu, Z.J. and Cordes, J.F., 2004. DNA marker technologies and their applications in aquaculture genetics. Aquaculture, 238(1-4), pp.1-37.

Lo Presti, R., Lisa, C. and Di Stasio, L., 2009. Molecular genetics in aquaculture. Italian Journal of Animal Science, 8(3), pp.299-313.

Magoulas, A., Bartley, D. and Basurco, B., 1998. Application of molecular markers to aquaculture and broodstock management with special emphasis on microsatellite DNA. Cahiers Options Mediterrannes, 34, pp.153-168.

Magoulas, A., Kotoulas, G., Batargias, K. and Zouros, E., 1997. Genetic markers in marine biology and aquaculture research: when to use what. Genet. Aquacult. Afr, pp.67-78.

Maqsood, H.M. and Ahmad, S.M., 2017. Advances in molecular markers and their applications in aquaculture and fisheries. Genetics of Aquatic Organisms, 1(1), pp.27-41.

Mojekwu, T.O. and Anumudu, C.I., 2013. Microsatellite markers in Aquaculture: Application in Fish population genetics.4 (5):43-48.

Muramulla, S.M., Babu, K.G., and Rajeswari, J. 2022. Applications of molecular markers in aquaculture and fisheries-a detailed study. International Journal of Advance Research, Ideas and Innovations in Technology. 8 (1):169-178.

Okumuş, İ. and Çiftci, Y., 2003. Fish population genetics and molecular markers: II-molecular markers and their applications in fisheries and aquaculture. Turkish Journal of Fisheries and Aquatic Sciences, 3(1).51-79.

Sukumaran, S. and Gopalakrishnan, A., 2019. Applications of Molecular Markers in Fisheries and Aquaculture. 190-200pp

Tanya, C. and Kumar, R., 2010. Molecular markers and their applications in fisheries and aquaculture. Advances in Bioscience and Biotechnology, 1:281-291.

23

Importance of Essential Fatty Acid Nutrition in Fish

[1]Shivkumar, [2]Dhanalakshmi M, [3]Mukkeri Krantirekha and [4]Akshaya Vinod Mayekar

[1]Department of Fish Nutrition and Feed Technology, Fish Nutrition Biochemistry and Physiology Division, ICAR - CIFE, Mumbai Maharashtra

[2]Department of Fisheries Resource Management, Fisheries Resources Harvest & Post-Harvest Management Division, ICAR-CIFE Mumbai, Maharashtra

[3,4]Department of Fish Genetics, Fish Genetics & Biotechnology Division ICAR-Central Institute of Fisheries Education (ICAR-CIFE) Mumbai, Maharashtra

Abstract

In a finfish feed, the lipid is the second most preferable macro-nutrient next to protein but still it stands first in apparent digestibility. Lipid in their simplest form as a fatty acids, covers a wide range of functionality, starting from structural to the growth and health of fish. Therefore, food deficient of essential fatty acids may lead to a wide range of impacts related to the imbalance of homeostasis of a fish body. Generally, the quality and quantity of essential fatty acids requirements of fish vary according to size, life stage, and water salinity. Hence, freshwater fishes require Linoleic acid and α-linolenic acid (PUFA), and marine water fishes require Docosahexaenoic acid and Eicosapentaenoic acid (HUFA). Furthermore, this difference is mainly because of the bioconversion capacity of fish with the water salinity. So, different sources of fatty acids available in the markets are used in fish feed preparation according to species requirements from larvae to brooders.

Keywords: Lipid, Essential Fatty acids, Freshwater, Marine water, Larvae and Brooders.

Introduction

Fish oil is the major source of essential Fatty acids (EFA) for humans. Generally, Fish requires three long-chain highly unsaturated Fatty acids (HUFA) for their normal growth and development including reproduction like docosahexaenoic acid (DHA, 22:6n-3), eicosapentaenoic acid (EPA, 20:5n-3) and arachidonic acid (ARA, 20:4n-6) (Sargent et al. 1993a, 1995, 1997). Like the terrestrial mammals, DHA, EPA, and ARA are all involved in maintaining the cell membrane structure and function of fish but majorly EPA and DHA are considered. Moreover, the EPA and DHA are the major currency as fatty acid for the synthesis of phospholipids, Prostaglandins, and energy etc. However, the biosynthesis of EFA in fish changes in accordance with the relative bioconversion capacity of the PUFA (Linoleic acid and α-Linolenic acid) as shown in figure 01. Hence the relative bioconversion capacity of PUFA in fish depends on the fish species, diet and season. Generally, EFA being a major nutrient plays a very important role throughout the life cycle of fish. Therefore, this point should be considered for the preparation of feed for different stages of fish.

EFA requirement for Freshwater, Marine water and Brackish water fish varies. Why?

The Marine water fish fatty acid composition is a better source of ω-3 EFA while freshwater fish are a good source of ω-6 EFA. However, the highest percentage of branched and saturated fatty acids in freshwater. Therefore, the energy source of the freshwater ecosystem is derived mainly from photosynthesis accompanied by algae and higher plants.

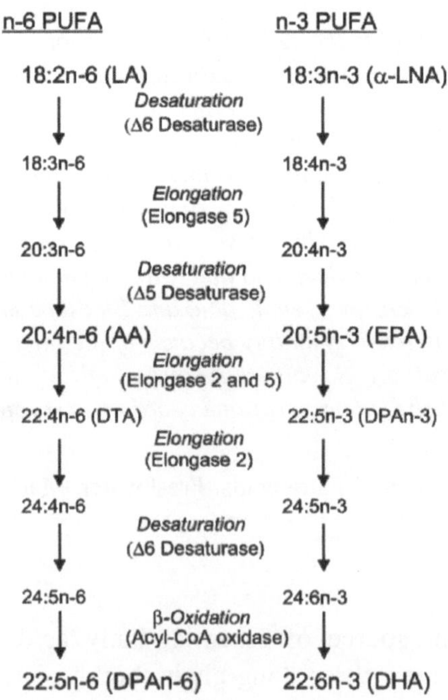

Figure 1: Bioconversion of PUFA to HUFA (Miki et al.,2007)

It was reported that the Fatty acid composition of 10 microalgae reported (Pratoomyot et al.,2005) that the major contribution by the LNA> LA> SFA >MUFA (α-Linolenic acid> Linoleic acid> Saturated fatty acids> Monounsaturated fatty acids). Most of the microalgae showed 0% of EPA and DHA (Pratoomyot et al.,2005). Supportively Choi et al., 1987 reported the microalgae fatty acid composition showed 80% of Total unsaturated fatty acid (16C and 18C) and 38% of EFA out of the total lipid content of algae. So, in freshwater fishes, they received unsaturated Fatty acids are converted into MUFA or PUFA (LA and LNA) by de novo synthesis. Thus, freshwater fishes are having a high level of PUFA in that C18:2 ω-6 is prominent (in the sense of the availability of algae as a food). Now, it is reported that 11 freshwater fish of different families like Mormyridae, Cichlidae, Claridae, Centropomidae and Charidae were investigated for their fatty acid profile and resulted as SFA>MUFA>LA>LNA>ARA>DHA>EPA. Likewise, 10 marine water fish like Anchovy, Horse mackerel, Menhaden, Sardine, Capelin, Herring, Mackerel, Norway pout, Sand eel, and Sprat were investigated for their Fatty acid profile and resulted in EPA>DHA>PUFA>MUFA>SFA (Ugoala et al.,2008). This results that the dietary essential fatty acids requirement for marine fish is ω-3 HUFA higher than freshwater.

The nutritional requirement of marine species for long-chain PUFA n-3 may not be met by C18:3 n-3 due to limited capacity for chain elongation and desaturation (Cowery et al.,1976). Δ6 desaturase expression can be observed in the Brain, Liver, gill, Intestine, and Muscle and this Δ6 desaturase expression is high in the Brain and Liver, weak in Gill too low in the Intestine and Muscle. The Experimental evidence suggests that the dependence of marine fish on dietary HUFA is caused by a deficiency in the activity of one or more key enzymes Δ5, Δ6 desaturase and Elongase (Sargent et al.,2002; Tocher et al., 2003). So, the activity of Δ6 desaturase and Δ5 desaturase is highest in freshwater, once the fish is introduced to seawater both Δ6 desaturase and Δ5 desaturase activity will reduce (Zhang et al.,2005). That means increasing the salinity of the water, decreases the Δ6 desaturase and Δ5 desaturase activity. Additionally, the expression of Δ6 desaturase in *Siganus canaliculatus* fish in 10ppt was about 1.83 times higher than that of 32ppt (Li et al.,2008). However, some exceptions are there like elongase activity is high in seawater compared to freshwater, and the reason behind this is still unknown (Zang et al.,2005). In contrast, Δ5 desaturase has some activity has been seen in Turbot fish but not the elongase activity. Additionally in Seabream fish elongase activity has been seen but not Δ5 desaturase activity. Hence the requirement of EFA varies according to species, life stage, and water salinity as shown in Tables 01, 02, 03, and 04.

Table 1: Essential fatty acids requirement of Freshwater fish species

Freshwater Fish Species		Life stage	EFA	Requirement (% dry wt)	Reference
01	Rainbow trout	Fingerling	18:3n-3	1	March (1993)
02	Grass carp	Subadult	18:2n-6 18:3n-3	1.0 0.5	Takeuchi et al. (1991)
03	Channel catfish	Subadult	18:3n-3	1.0-2.0	Satoh et al. (1989)
04	Tilapia	Juvenile	18:2n-6	0.5	Takeuchi et al. (1983)
05	Ayu	Larvae	18:3n-3 or EPA	1	Kanazawa et al. (1982)
06	Chum salmon	Adult	18:2n-6 18:3n-3	1 1	Takeuchi et al. (1979)
07	Coho salmon	Adult	18:2n-6 18:3n-3	1 1	Yu and Sinnhuber (1979)
08	Common carp	Fingerling	18:2n-6 18:3n-3	1 1	Takeuchi and Watanabe (1977)

Table 2: Essential Fatty acids requirement of Marine water fish species

Marine Fish Species		Life stages	EFA	Requirement (% dry wt)	Reference
01	*Scophthalmus maximus*	Larvae	n-3 HUFA	1.2-3.2	Le Milinaire et al. (1983)
02	Gilthead Seabream	Larvae	EPA and DHA	2.42 1.1	Rodriguez et al. (1994)
03	*Solea solea*	Larvae	n-3 HUFA	0.9	Howell & Tzoumas (1991)
04	*Paralichthys olivaceus*	Larvae	n-3 HUFA	1.8-3.5	lzquierdo et al. (1992)
05	*Seriola quinqueradiata*	Larvae	n-3 HUFA	>3.9	Watanabe (1993)
06	*Oplegniatus fasciatus*	Larvae	n-3 HUFA	3.0	Watanabe (1993)
07	*Pagrus major*	Larvae	n-3 HUFA	3.5	lzquierdo et al. (1989a)
08	Yellow tail	Juvenile	n-3 HUFA	2.0	Deshimaru (1982)
09	Flounder	Juvenile	n-3 HUFA	2.0	Gatesoupe et al. (1977)
10	Striped jack	Juvenile	n-3 HUFA	1.8	Kanazawa (1985)

Marine Fish Species		Life stages	EFA	Requirement (% dry wt)	Reference
11	Turbot	Juvenile	n-3 HUFA	0.8	Wantanabe (1989)
12	Grouper	Juvenile	n-3 -HUFA, DHA > EPA	1.0	Wu et al. (2002)
13	Gilt head seabream	Juvenile	n-3 HUFA n-3 HUFA (DHA:EPA)	0.9 (1:1) 1.9 (1:0.5)	Kalegeropoulos et al. (1992) Ibeas et al. (1994)
14	Japanese flounder	Juvenile	n-3 HUFA	1.4	Takeuchi (1997)

It has been reported that marine larvae Gilthead seabream optimum level of n-3 HUFA requirement EPA/DHA ratio maintained is 1.1 showed better growth (Koven et al., 1990). Additionally, Juvenile Turbot fed with fish oil having n-3/n-6 ratio 5.5, shown high survivability and SGR (Bell et al., 1998). Hence share of n-3 HUFA requirement will change according to species and habitat.

Table 3: Essential Fatty acids requirement of Brackish water fish species

Brackish Fish Species		Life stage	EFA	Requirement (% dry wt)	Reference
01	Red drum	Juvenile (7ppt)	n-3 HUFA	0.5	Lochman and Galtin (1993)
02	Pacific White Shrimp	Juvenile (25ppt)	n-3 HUFA	0.5	Gonzalez-Felix and Galtin (2002)
03	Milkfish	Juvenile (30ppt)	18:2n-6 18:3n-3	0.5 0.5	Bautista and de la Cruz (1988)
04	European Seabass	Larvae	n-3 HUFA	1	Coutteau et al. (1996a)

Table 4: Essential Fatty acids requirement of Shrimp/Prawn

Shrimp/Prawn		Life stage	EFA -	Requirement (% dry wt)	Reference
01	*P. Monodon*	Juvenile (35ppt)	18:2n-6 EPA DHA	1.2 0.9 0.9	Glencross and Smith (1999)
		Post larvae (35ppt)	n-3 HUFA	1.25	Rees et al. (1994)
02	*L. Vannamei*	Juvenile (25ppt)	EPA DHA ARA	0.5 0.5 0.5	Gonzalez-Félix et al. (2003)
		Juvenile (25ppt)	EPA DHA ARA	0.2 0.4 0.17	Jannathulla et al. (2019)

Shrimp/Prawn		Life stage	EFA -	Requirement (% dry wt)	Reference
		Post larvae(25ppt)	EFA	2	Felix et al. (2021)
03	*M. rosenbergii*	Post larvae	ARA DHA	0.08 0.075	D'Abramo and Sheen (1993)

Sources of Essential Fatty Acids used in Fish

1. **Linoleic acid:** Ground nut oil, Soybean oil, Sufflower oil, Sunflower oil, Linseed oil, Rice bran oil, Teel oil, Corn or Maize oil, Cotton seed oil, Olive oil, Canola oil, Cashew oil, Coconut oil, Oat bran, Almond oil, Palm kernel meal (Kaur et al.,2014; Kawashima, 2019)
2. **α-Linolenic acid**: Walnut oil, Linseed oil, Canola oil, Mustard oil, Soybean oil, Soybean meal, Rice bran oil, Sufflower oil, Teel oil, Sunflower oil, Olive oil, Corn oil, Oat bran (Kaur et al.,2014; Kawashima, 2019)
3. **Arachidonic acid**: Liver meal, Meat meal, Fish meal (Kaur et al.,2014; Kawashima, 2019)
4. **EPA and DHA**: Fish meal, Shrimp meal, Various fish oil (Kaur et al.,2014; Kawashima, 2019)

Functions of Essential Fatty Acids in Fish

1. Cell membrane fluidity
2. Digestive tract and renal excretion system
3. Transportation of lipid
4. Transportation of cholesterol
5. Production of prostaglandins and thromboxane
6. Decrease inflammation and increases the phagocytosis
7. Nervous system as messenger
8. Selective permeability
9. Branchial respiration system
10. Stress mitigator
11. Cell multiplication
12. Retinal cell functioning
13. Pigmentation

Fatty acid is a constitute of phospholipid which requires for cell membrane structure and functions to regulate a large number of mechanisms like increasing cell fluidity and improving the ability of selective permeability to avoid toxin entry and phospholipid containing EFA is the major constituent of transport lipoprotein. Generally, EFA act as a precursor for the synthesis of prostaglandins (PGs) especially the PGs produced from Arachidonic acid. Additionally, EFA is the source of eicosanoids which act as a nervous system messenger, development of branchial respiration and digestive tract and renal excretion system and it also helps in the control of the nervous system with increasing retinal rod cells for proper predatory efficiency. Furthermore, EFA improves reproductive ability with the production of spermatozoa. Moreover, it acts as a stress mitigator by regulating the release of cortisol secretion from ACTH. Apart from these, Arachidonic acid involves in the cell multiplication process and by acting as a secondary messenger role in enzyme synthesis. Additionally, EFA helps in the transportation of cholesterol transportation and the prevention of fatty liver syndrome.

Deficiency Symptoms of Essential Fatty Acids in Fish

- Reduce apatite and feed efficiency with increased FCR
- Retardation of growth and development
- Increased mortality especially in a larval stage
- Immuno-suppression thus fish are susceptible to many diseases eg: Caudal fin erosion
- Reddening of fins
- Decrease the reproductive efficiency i.e decrease the fecundity, fertility rate and spawning efficiency
- Reduced hatchability or if hatched deformed larvae and decreased larval survivability
- Reduce vision thus it reduces the predation efficiency
- Fatty liver syndrome with hepatic degeneration and swollen pain liver
- Kidney damage
- Shock syndrome with reduced movement
- If deficiency run for a longer time, degeneration and erosion of gill epithelium can occur
- Lordosis can be seen in some fishes
- If deficiency run for a longer time, decrease the haemoglobin level and RBC count had seen in Trout
- Mitochondrial swelling of liver and brain cell in Trout fish
- Swim bladder inflation has been seen in Gilthead seabream
- Hydrops (a subcutaneous and coelomic deposition of water) formation have seen in Red sea bream fish

- Hypo-melanosis on the ocular side have seen in Marble sole fish
- Poor pigmentation in Turbot

Role of EFA in Larval and Brood-stock Nutrition

Many of the research paper has reported showing the great palatable reason for their nutrition in both larval and brood stock growth and health condition. Like, Larvae of Herring fish eye contain DHA which is correlated with a number of rods in the Photoreceptor population (Bell and Dick, 1993) and helps in growth and development of many fish larvae (Gatesoupe,1985; Izquierdo et al.,1992; Izquierdo et al.,1996). Additionally, the survivability of larvae increased in EFA fed fish of Halibut, Red sea bream, and Striped jack (Holmefjord, 1993; Rodrigur et al.,1998; Salhi et al.,1999) has been reported. Additionally, EFA helps in the pigmentation of fish larvae of Turbot (Rainuzzo et al, 1993) and it proved its vital role in membrane fluidity and as a precursor of eicosanoids resulting in an increase in the immune system (Tocher et al.,2010). Moreover, during larval development, DHA is preferentially incorporated into nervous and retinal tissue (Villalata et al.,2008; Izquiedo et al.,2000). Generally, EPA, ARA, and DHA, through their COX and LOX derivatives, regulate cortisol production by modulating ACTH - stimulated interrenal cells in sea bream (Ganga et al.,2004). It has been reported that the Flounder of 2yr old fed with n-3 HUFA and Gilthead seabream with 1.60% n-3 HUFA increase hatchability, survivability, egg diameter, and egg composition. Cobia fed with 0.4- 0.6% and 1.86% n-3 HUFA and Crescent sweetlip species fed with 2.40% n-HUFA showed increased fecundity, larval survivability, egg composition have been seen (Li, et al.,2005, Fernadez- Palacios 1995).

Conclusion

The essential fatty acid is known for its multi-functional property in fish physiology and the digestive system from the larval stage to brooders. Its requirement is various in accordance with fish species, size, and water salinity. Additionally, the natural fatty acid source also suits to the fish species requirement. Hence, it is necessary to consider fish species, life stage, habitat, lipid source, requirement fatty acid quantity, EPA/DHA ratio, and ω-3/ω-6 fatty acid during feed preparation of fish feed.

References

Bautista, M.N. and De la Cruz, M.C. (1988). Linoleic (ω6) and linolenic (ω3) acids in the diet of fingerling milkfish (Chanos chanos). Aquaculture. 71(4):347-358.

Bell, J.G., Tocher, D.R., Farndale, B.M. and Sargent, J.R. (1998). Growth, mortality, tissue histopathology and Fatty acid compositions, eicosanoid production and response to stress, in juvenile turbot fed diets rich in γ-linolenic acid in combination with eicosapentaenoic acid or docosahexaenoic acid. Prostaglandins, Leukotrienes and Essential Fatty acids. 58(5):353-364.

Bell, M.V. and Dick, J.R. (1993). The appearance of rods in the eyes of herring and increased di-docosahexaenoyl molecular species of phospholipids. Journal of the Marine Biological Association of the United Kingdom. 73(3):679-688.

Cook, H.W. and McMaster, C.R. (2002). Fatty acid desaturation and chain elongation in eukaryotes. In New comprehensive biochemistry. 36:181-204.

Coutteau, P., Van Stappen, G. and Sorgeloos, P. (1996). A standard experimental diet for the study of Fatty acid requirements of weaning and first ongrowing stages of the European sea bass Dicentrarchus labrax L.: comparison of extruded and extruded/coated diets. Archives of animal nutrition. 49:49-59.

Cowey, C.B., Adron, J.W., Owen, J.M. and Roberts, R.J. (1976). The effect of different dietary oils on tissue Fatty acids and tissue pathology in turbot Scophthalmus maximus. Comparative Biochemistry and Physiology Part B: Comparative Biochemistry. 53(3):399-403.

Deshimaru, O., Kuroki, K. and Yone, Y. (1982). Nutritive values of various oils for yellowtail [*Seriola quinqueradiata*]. Bulletin of the Japanese Society of Scientific Fisheries (Japan).

D'Abramo, L.R. and Sheen, S.S. (1993). Polyunsaturated Fatty acid nutrition in juvenile freshwater prawn Macrobrachium rosenbergii. Aquaculture. 115(1-2):63-86.

Felix, S., Menaga, M., Sundari, C.M., Charulatha, M. and Neelakandan, P. (2021). A Study on the Fatty acid Enrichment of Artemia franciscana for the Healthy Rearing of Penaeus vannamei Post-larvae. Indian Journal of Animal Research. 55(3):295-302.

Fernández-Palacios, H., Izquierdo, M.S., Robaina, L., Valencia, A., Salhi, M. and Vergara, J. (1995). Effect of n– 3 HUFA level in broodstock diets on egg quality of gilthead sea bream (*Sparus aurata* L.). Aquaculture. 132(3-4):325-337.

Furuita, H., Tanaka, H., Yamamoto, T., Shiraishi, M. and Takeuchi, T. (2000). Effects of n– 3 HUFA levels in broodstock diet on the reproductive performance and egg and larval quality of the Japanese flounder, Paralichthys olivaceus. Aquaculture. 187(3-4):387-398.

Ganga, R. (2004). Effect of feeding gilthead seabream (" *Sparus aurata*") with vegetable lipid sources on two potential inmunomodulator products: prostanoids and leptins (Master's thesis).

Gatesoupe, F.J. and FJ, G. (1977). Alimentation lipidique du turbot (*Scophthalmus Maximus* l.). II. Influence de la supplementation EN Esters methyliques de l'acide linolenique et de la complementation en acides gras de la serie omega 9 sur la croissance.

Gatesoupe, J. (1985). Dietary value adaptation of live food organisms for covering the nutritional requirements of marine fish larvae. Coll. Fr.-Jp. Oceanogr. Marseille. 8:51-63.

Glencross, B.D. and Smith, D.M. (1999). The dietary linoleic and linolenic Fatty acids requirements of the prawn Penaeus monodon.

González-Félix, M.L., Gatlin III, D.M., Lawrence, A.L. and Perez-Velazquez, M. (2002). Effect of various dietary lipid levels on quantitative essential Fatty acid requirements of juvenile Pacific white shrimp Litopenaeus vannamei. Journal of the World Aquaculture Society. 33(3):330-340.

González-Félix, M.L., Gatlin Iii, D.M., Lawrence, A.L. and Perez-Velazquez, M. (2003). Nutritional evaluation of Fatty acids for the open thelycum shrimp, Litopenaeus vannamei: II. Effect of dietary n-3 and n-6 polyunsaturated and highly unsaturated Fatty acids on juvenile shrimp growth, survival, and Fatty acid composition. Aquaculture Nutrition. 9(2):115-122.

Harel, M., Tandler, A., Kissil, G.W. and Applebaum, S.W. (1994). The kinetics of nutrient incorporation into body tissues of gilthead seabream (*Sparus aurata*) females and the subsequent effects on egg composition and egg quality. British Journal of Nutrition. 72(1):45-58.

Holmefjord, I., Gulbrandsen, J., Lein, I., Refstie, T., Léger, P., Huse, I., Harboe, T., Sorgeloos, P., Olsen, Y., Bolla, S. and Reitan, K.I. (1993). An intensive approach to Atlantic halibut fry production. Journal of the World Aquaculture Society. 24(2):275-284.

Howell, B.R. and Tzoumas, T.S. (1991). The nutritional value of Artemia nauplii for larval sole, Solea solea (L.), with respect to their (n-3) HUFA content. Larvi. 91:63-65.

Ibeas, C., Izquierdo, M.S. and Lorenzo, A. (1994). Effect of different levels of n− 3 highly unsaturated Fatty acids on growth and Fatty acid composition of juvenile gilthead seabream (*Sparus aurata*). Aquaculture. 127(2-3):177-188.

Izquierdo, M.S. (1989). Requirement of larval red seabream Pagrus major for essential Fatty acids. Nippon Suisan Gakk. 55(5):859-867.

Izquierdo, M.S. (1996). Essential Fatty acid requirements of cultured marine fish larvae. Aquaculture Nutrition. 2(4):183-191.

Izquierdo, M.S., Arakawa, T., Takeuchi, T., Haroun, R. and Watanabe, T. (1992). Effect of n-3 HUFA levels in Artemia on growth of larval Japanese flounder (Paralichthys olivaceus). Aquaculture. 105(1):73-82.

Izquierdo, M.S., Socorro, J., Arantzamendi, L. and Hernández-Cruz, C.M. (2000). Recent advances in lipid nutrition in fish larvae. Fish Physiology and Biochemistry. 22(2):97-107.

Jannathulla, R., Chitra, V., Vasanthakumar, D., Nagavel, A., Ambasankar, K., Muralidhar, M. and Dayal, J.S. (2019). Effect of dietary lipid/essential Fatty acid level on Pacific whiteleg shrimp, Litopenaeus vannamei (Boone, 1931) reared at three different water salinities–Emphasis on growth, hemolymph indices and body composition. Aquaculture. 513:734405.

Kalogeropoulos, N., Alexis, M.N. and Henderson, R.J. (1992). Effects of dietary soybean and cod-liver oil levels on growth and body composition of gilthead bream (*Sparus aurata*). Aquaculture. 104(3-4):293-308.

Kanazawa, A. (1985). Essential Fatty acid and lipid requirement of fish. Nutrition and feeding in fish. pp.287-298.

Kanazawa, A., Teshima, S.I. and Sakamoto, M., 1982. Requirements of essential Fatty acids for the larval ayu. Bulletin of Japan Society of Scientific Fisheries. 48:587-590.

Kaur, N., Chugh, V. and Gupta, A.K. (2014). Essential Fatty acids as functional components of foods-a review. Journal of food science and technology. 51(10):2289-2303.

Kawashima, H. (2019). Intake of arachidonic acid-containing lipids in adult humans: dietary surveys and clinical trials. Lipids in Health and Disease. 18(1):1-9.

Koven, W.M., Tandler, A., Kissil, G.W., Sklan, D., Friezlander, O. and Harel, M. (1990). The effect of dietary (n− 3) polyunsaturated Fatty acids on growth, survival and swim bladder development in Sparus aurata larvae. Aquaculture. 91(1-2):131-141.

Le Milinaire, C., Gatesoupe, F.J. and Stephan, G. (1983). Approche du besoin quantitatif en acides gras longs polyinsaturés de la série n-3 chez la larve de turbot (*Scophthalmus maximus*). Comptes rendus des séances de l'Académie des sciences. Série 3, Sciences de la vie. 296(19):917-920.

Li, Y.Y., Chen, W.Z., Sun, Z.W., Chen, J.H. and Wu, K.G. (2005). Effects of n-3 HUFA content in broodstock diet on spawning performance and Fatty acid composition of eggs and larvae in Plectorhynchus cinctus. Aquaculture. 245(1-4):263-272.

Li, Y.Y., Hu, C.B., Zheng, Y.J., Xia, X.A., Xu, W.J., Wang, S.Q., Chen, W.Z., Sun, Z.W. and Huang, J.H. (2008). The effects of dietary Fatty acids on liver Fatty acid composition and Δ6-

desaturase expression differ with ambient salinities in Siganus canaliculatus. Comparative Biochemistry and Physiology Part B: Biochemistry and Molecular Biology. 151(2):183-190.

Lochmann, R.T. and Gatlin, D.M. (1993). Essential Fatty acid requirement of juvenile red drum (*Sciaenops ocellatus*). Fish Physiology and Biochemistry. 12(3):221-235.

March, B.E. (1993). Essential Fatty acids in fish physiology. Canadian Journal of Physiology and Pharmacology. 71(9):684-689.

Miki, I., DeMar, J.C., Kaizong, M., Lisa, C., Bell, J.M. and Rapoport, S.I. (2007). Docosahexaenoic acid synthesis from α-linolenic acid by rat brain is unaffected by dietary n-3 PUFA deprivations. Journal of lipid research. 48(5):1150-1158.

Nguyen, H.Q., Tran, T.M., Reinertsen, H. and Kjørsvik, E. (2010). Effects of dietary essential Fatty acid levels on broodstock spawning performance and egg Fatty acid composition of cobia, Rachycentron canadum. Journal of the World Aquaculture Society. 41(5):687-699.

Pratoomyot, J., Srivilas, P. and Noiraksar, T. (2005). Fatty acids composition of 10 microalgal species. Songklanakarin J Sci Technol. 27(6):1179-1187.

Rainuzzo, J.R. (1993). Fatty acid and lipid composition of fish egg and larvae. Proceedings of the First International Conference on Fish Farming Technology, Trondheim, Norway. pp.43-49.

Rees, J.F., Curé, K., Piyatiratitivorakul, S., Sorgeloos, P. and Menasveta, P. (1994). Highly unsaturated Fatty acid requirements of Penaeus monodon postlarvae: an experimental approach based on Artemia enrichment. Aquaculture. 122(2-3):193-207.

Rodriguez, C., Cejas, J.R., Martin, M.V., Badia, P., Samper, M. and Lorenzo, A. (1998). Influence of n-3 highly unsaturated Fatty acid deficiency on the lipid composition of broodstock gilthead seabream (*Sparus aurata* L.) and on egg quality. Fish Physiology and Biochemistry. 18(2):177-187.

Rodriguez, C., Perez, J.A., Izquierdo, M.S., Mora, J., Lorenzo, A. and FERNANDEZ-PALACIOS, H. (1994). Essential Fatty acid requirements of larval gilthead sea bream, Sparus aurata (L.). Aquaculture Research. 25(3):295-304.

Ruyter, B., Røsjø, C., Einen, O. and Thomassen, M.S. (2000). Essential Fatty acids in Atlantic salmon: time course of changes in Fatty acid composition of liver, blood and carcass induced by a diet deficient in n-3 and n-6 Fatty acids. Aquaculture Nutrition. 6(2):109-117.

Salhi, M., Izquierdo, M.S., Hernández-Cruz, C.M., Bessonart, M. and Fernández-Palacios, H. (1999). Effect of different dietary polar lipid levels and different n-3 HUFA content in polar lipids on the gut and liver histological structure of seabream (Sparus aurata) larvae. Aquaculture. 179:253–264.

Sargent, J.R. (1995). Origins and function of eggs lipids: Nutritional implication. Broodstock management and egg and larval quality. pp.353-372.

Sargent, J.R., Bell, J.G., Bell, M.V., Henderson, R.J. and Tocher, D.R. (1993). The metabolism of phospholipids and polyunsaturated Fatty acids in fish. Coastal and estuarine studies. pp.103-103.

Sargent, J.R., McEvoy, L.A. and Bell, J.G. (1997). Requirements, presentation and sources of polyunsaturated Fatty acids in marine fish larval feeds. Aquaculture. 155(1-4):117-127.

Sargent, J.R., Tocher, D.R. and Bell, J.G. (2003). The lipids. Fish nutrition. pp.181-257.

Satoh, S., Poe, W.E. and Wilson, R.P. (1989). Studies on the essential Fatty acid requirement of channel catfish, Ictalurus punctatus. Aquaculture. 79(1-4):121-128.

Takeuchi, T. (1979). Requirement for essential Fatty acids of chum salmon (Oncorhynchus keta) in freshwater environment. Nippon Suisan Gakkaishi. 45:1319-1323.

Takeuchi, T. (1996). Essential Fatty acid requirements in carp. Archives of Animal Nutrition. 49(1):23-32.

Takeuchi, T. (1997). Essential Fatty acid requirements of aquatic animals with emphasis on fish larvae and fingerlings. Reviews in Fisheries Science. 5(1):1-25.

Takeuchi, T., Satoh, S. and Watanabe, T. (1983). Requirement of Tilapia nilotica for essential Fatty acids. Bulletin of the Japanese Society of Scientific Fisheries (Japan).

Takeuchi, T. and Watanabe, T. (1977). Requirement of carp for essential Fatty acids. Bulletin of the Japanese Society of Scientific Fisheries.

Tocher, D.R. (2003). Metabolism and functions of lipids and Fatty acids in teleost fish. Reviews in fisheries science. 11(2):107-184.

Tocher, D.R. (2010). Fatty acid requirements in ontogeny of marine and freshwater fish. Aquaculture research. 41(5):717-732.

Ugoala, C., Ndukwe, G.I. and Audu, T.O. (2008). Comparison of Fatty acids profile of some freshwater and marine fishes. Internet Journal of Food Safety. 10:9-17.

Villalta, M., Estevez, A., Bransden, M.P. and Bell, J.G. (2008). Effects of dietary eicosapentaenoic acid on growth, survival, pigmentation and Fatty acid composition in Senegal sole (Solea senegalensis) larvae during the Artemia feeding period. Aquaculture Nutrition. 14(3):232-241.

Watanabe, T. (1993). Importance of docosahexaenoic acid in marine larval fish. Journal of the World Aquaculture Society. 24(2):152-161.

Watanabe, T., 1989. Nutritive value of animal and plant lipid sources for fish. In Progress in Fish Nutrition. Proc. Fish Nutrition Symp. 09:151-166.

Wu, F.C., Ting, Y.Y. and Chen, H.Y. (2002). Docosahexaenoic acid is superior to eicosapentaenoic acid as the essential Fatty acid for growth of grouper, Epinephelus malabaricus. The Journal of nutrition. 132(1):72-79.

Yu, T.C. and Sinnhuber, R.O. (1979). Effect of dietary ω3 and ω6 Fatty acids on growth and feed conversion efficiency of coho salmon (Oncorhynchus kisutch). Aquaculture. 16(1):31-38.

Zheng, X., Torstensen, B.E., Tocher, D.R., Dick, J.R., Henderson, R.J. and Bell, J.G. (2005). Environmental and dietary influences on highly unsaturated Fatty acid biosynthesis and expression of fatty acyl desaturase and elongase genes in liver of Atlantic salmon (Salmo salar). Biochimica et Biophysica Acta (BBA)-Molecular and Cell Biology of Lipids. 1734(1):13-24.1

24

Role of Macroalgae in Reducing the Oxidative Stress

P. Chellamanimegalai, Amom Mahendrajit and Geetanjali Deshmukhe

Department of Fisheries Resource Management, Fisheries Resources Harvest and Post-Harvest Management Division, ICAR-CIFE Mumbai Maharashtra

Abstract

Antioxidants are compounds that bind to oxygen free radicals and break the chain of free radical formation. It is an effective mechanism to reduce the rate of oxidation, balance harmful ROS systems, and prevent cell damage. In the nutritional supplement and pharmaceutical industries today, natural antioxidants are gaining popularity. The substance generated from seaweed, a significant marine resource, has the potential to fight free radicals, lower health risks, and improve oxidative stability. They produce different types of antioxidant compounds as secondary metabolites. It is accountable for its bioactive properties such as antioxidant, antimicrobial, anticancer and antiviral agents to promote human health. Taking natural antioxidants has many health benefits with zero side effects.

Keywords: Seaweeds, Antioxidant, DPPH, FRAP, ABTS

Introduction

The ecological stability and food web of intertidal environments are significantly influenced by macroalgae. The majority of marine organisms, including macroalgae, coral reefs, sponges, and the organisms that they are associated with, are well known for producing secondary metabolites and bioactive compounds, which increase their value in a variety of industries, including food, neutraceuticals, pharmaceuticals, and others. Among them, seaweeds play a vital role in promoting health by offering a variety of phytochemicals with antioxidant, antibacterial, anticancer, and antiviral action (Ngo et al. 2012). Worldwide, there have been more than 20,000 macroalgal species identified. In South Asian nations including China, Japan, and Korea, it has been used as sea vegetables. India now has 841 macroalgal species identified, including 434, 191, and 216 species of red algae, brown algae, and green algae, respectively (Oza and Zaidi, 2001). At present, seaweed's antioxidant property is a widely concerned topic due to the emergence of various diseases caused by oxidative stress. The recent awareness about the various side effects caused by synthetic antioxidants such as butylated hydroxytoluene (BHT), tetra butyl hydroquinone (TBHQ), and butylated hydroxyanisole (BHA) and irreplaceable health benefits from seaweeds has created high demand for the natural antioxidants. This article discusses the role of antioxidants in reducing oxidative stress which will be helpful for enhancing the utilization of various seaweeds.

Significance of Antioxidants

Cell destruction, membrane protein damage, and DNA mutations are causing several health problems. The main reason for this is the oxidation process triggered by free radicals or reactive oxygen species (ROS). These reactions indirectly accelerate ageing and cause diseases such as diabetes, liver damage, dermatitis, coronary heart disease, arthritis, atherosclerosis, and cancer in humans (Bhattacharjee and Islam, 2014). Oxidative processes can be stopped by using antioxidant compounds that reaction is given as follows: Free radicals consist of one or more unpaired electrons in the outer layer and must absorb electrons from other substances to achieve neutrality. Overall, another free radical or ROS is formed, leading to a chain reaction (Dontha, 2016). ROS are a series of metabolites made from molecular oxygen (O2). When ROS concentrations exceed that threshold, there is an imbalance between free radical production and reactive metabolites known as 'oxidative stress. This damages cellular structures such as lipids, membranes, proteins and nucleic acids (Balakrishnan et al. 2014). Antioxidants are compounds that bind to oxygen free radicals and break the chain of free radical formation. It is used in effective defence mechanisms that reduce the rate of oxidation, balance harmful ROS systems and prevent cell damage. Cinnamic acid, coumarins, lignans, flavonoids, isoflavonoids, and tannins (phenolic polymers) are some of the effective antioxidants.

Seaweeds and Its Secondary Metabolites

Macro algae are a complex and heterogeneous group. It is characterized as a red, green, and brown alga based on its major photosynthetic pigments. They extend their distribution from temperate to tropical regions by enduring different environmental conditions. Most marine organisms, including algae, secrete secondary metabolites and trigger responses to stressors when the marine environment changes slightly, such as due to seasonal or climate change. Antioxidant compounds are important metabolites and have played a protective role against various stressors such as temperature, UV radiation, dehydration and pollutants (Bhattacharjee and Islam, 2014). These secondary metabolites (phytochemicals) accumulate within cells and contribute to improved adaptations to survive in new environments, provide potential bioactive properties, and protect against competitors and predators. Tropical algae may have more bioactive complexes and antioxidant properties than those reported from higher latitudes (Vasconcelos et al. 2019).

Compounds Responsible for Reducing Free Radicals

Phytochemical compounds are the major group of natural antioxidants in seaweeds that include carotenoids, pigments, polyphenols, and the functional polysaccharides (Morgan et al. 1980; Nakano et al. 1995; Nakamura et al. 1996; Yoshie et al. 2000; Costa et al. 2009). Phenolic compounds such as phlorotannins, fucoxanthins, polyphenols and phyllopheophyllins are the major complexes that determine the efficiency of seaweed antioxidant activity. Many studies have reported a direct relationship between total phenols (TPC) and free radical scavenging activity. Various methods are used to extract phenolic compounds, but the Soxhlet method and methanol solvent are more effective than traditional extraction methods (Foon et al. 2013).

Algal source	**Antioxidant compounds**	
Red and brown algae	Carotenoids	β-carotene, Fucoxanthin, Antheraxanthin, Lutein
Red algae	Phycobilin pigments	Phycoerythrin, Phycocyanin
Red and Brown algae	Phenolic compounds	Stypodiol, Isoeitaondiol, Terpenoids
Green and Brown seaweeds	Polyphenols	Catechin, Flavonoids, Phlorotannins
Brown and Red algae	Sulfated polysaccharides	Fucoidan, Alginic acid, galactans
References: Morgan et al. 1980; Nakano et al. 1995; Nakamura et al. 1996; Yoshie et al. 2000; Costa et al. 2009; Bhattacharjee and Islam, 2014		

Carotenoids have been identified as important components that reduce oxidative stress (Vasconcelos et al. 2019). Polyphenols are a class of phytochemicals found within all plants, including marine flora which are responsible for the antioxidant activity and stabilization of lipid peroxidation (Vijayabaskar and Shiyamala, 2012). Polyphenols contained in green algae are classified into catechin, epicatechin, epigaloctechin, and gallic acid (Yoshie et al. 2000). Phlorotannins are the type of phenolic compounds that consist of eight rings in their structure that exhibits the highest antioxidant activity than terrestrial plants of four rings building. These are abundantly found in brown algal families such as Lacarpaceae, Araliaceae, and Sargassum. The isolation of phlorotannins from *Cystoseira trinodis* has been obtained for the first time, revealing that the identification of particular compounds is essential for effective utilization (Sathya et al. 2017). Phlorotannins have also been implicated in other bioactivities such as antibacterial, chemopreventive, UV protection and antiproliferative effects (Balakrishnan et al. 2014).

Various seaweed-derived polysaccharide compounds with potent antioxidant activity have been reported by Mahendran and Saravanan, (2013). These polysaccharides occur in different forms in each group of macroalgae; carrageenan and agar from red algae, fucoidan, laminarian and alginate from brown algae, galactan, mannan and xylan from green algae (Venkatesan et al., 2019). Carrageenan (kappa-, iota- and lambda-type), fucoidan (homofucan) and fucan (heterofucan) are sulfated polysaccharides obtained from *Gigartina acicularis* (lambda-type), *G. pisilata* (lambda-type), *Euchema cottonii* (kappa-type) and *E. spinosa* (Iota type), *Fucus vesiculosus* (fucoidan) and *Padina gymnospora* (fucan). Among these, fucoidan and lambda-carrageenan have been observed to ensure the highest antioxidant activity (Souza et al. 2007).

Assessment of Antioxidant Activity

Seaweed antioxidant activity can be determined by quantifying the seaweed's ability to reduce free radicals. Several antioxidant assays are used for evaluation. These assays can be mostly classified into two types based on the chemical reactions that occur between antioxidant compounds and free radicals (Dontha, 2016).

1. Assay based on hydrogen atom transfer reaction (HAT assay)

2. Assays based on electron transfer reactions (ET assays)

HAT-based assays measure the ability of antioxidant compounds to donate hydrogen atoms and cleave chains. Here, reactions between synthetic radical generators, oxidizable molecular probes and oxidants are evaluated. HAT-based assays include oxygen radical absorbance capacity (ORAC), ABTS radical scavenger assay, TRAP assay, hydroxyl radical scavenger activity, β-carotene linoleic acid assay and others.

ET-based assays quantify the reducing capacity of antioxidant compounds. This is a simple redox reaction that occurs when free radicals are reduced and themselves oxidized by antioxidant compounds. Here, the endpoint is the colour change of the reagent. The DPPH scavenger assay, superoxide anion scavenger assay, iron reduced antioxidant capacity assay (FRAP), TEAC using ABTS, CUPRAC assay, and Folin-Ciocalteu reagent assay fall into this category of ET-based assays. Among these free radical scavenging methods, the DPPH method is simpler and cheaper than other methods and the ABTS assay can be used for both hydrophilic and lipophilic antioxidants. The most common methods used for determining antioxidant activity are as follows:

1) DPPH Radical Scavenging Activity

The stock solution is made by mixing 24mg DPPH with 100 ml of methanol and stored at -20°C until use. About 10 ml of stock solution is mixed with 45 ml methanol for making the working solution. For the assay, 0.15ml of the seaweed extract is mixed with 2.85ml of the working solution and kept for incubation at room temperature in the dark for 24 hrs. The absorbance of the reaction mixture is measured using a spectrophotometer at 515 nm. Trolox is used as standard (0-1000 μmol/l). The results are expressed as μmol of Trolox per gram of dry seaweed (Thaipong et al. 2006).

2) Ferric Reducing Antioxidant Power Assay

A stock solution preparation includes 300mM acetate buffer (3.1g Sodium acetate trihydrate and 16ml acetic acid) pH 3.6, 10mM TPTZ in 40mM hydrochloric acid and 20mM Ferric chloride hexahydrate solution. A working solution is prepared by mixing 25ml acetate buffer, 2.5ml TPTZ solution and 2.5ml ferric chloride solution. The mixture is warm at 37° C before use. 0.15ml of the extract is mixed with 2.85ml of working solution and kept in dark at room temperature for 30 minutes. This gives a coloured product of the ferrous tripyridyl triazine complex. The absorbance is taken at 593nm in a spectrophotometer. Trolox is used as standard (0-1000 μmol/l). The result is expressed as μmol of Trolox equivalent/gram of dry seaweed (Thaipong et al., 2006).

3) ABTS (2,2'-azino-bis(3-ethylbenzothiazoline-6-sulfonic acid) assay

ABTS stock solution is prepared by mixing 7M ABTS and 60mM in a volume ratio of 1:50 and incubated for 16 hrs. at room temperature. This creates ABTS+ radical. It is used within 24 hr. ABTS stock (2.5ml) is mixed with 97.5ml acetate buffer (0.2M Sodium acetate and 0.2M acetic acids in the ratio of 1:2) of pH 4.3 to prepare the working solution. 0.05ml of seaweed extract is mixed with 4ml of working solution and incubated in the dark for 30 min. The absorbance of the mixture is measured using a spectrophotometer at 734 nm. Trolox is used as the standard (0-1000 μmol/l). The results are expressed as μmol of Trolox equivalent per gram of dry seaweed (Sumczynski et al., 2015).

4) Total Antioxidant Capacity

An aliquot of 0.3ml sample is mixed with 3ml of reagent solution containing 0.6M Sulphuric acids, 28mM Sodium phosphate and 4mM Ammonium molybdate and incubated at 95.8°C for 90 minutes in a water bath. The absorbance of all sample mixtures is measured at 695nm. Ascorbic acid is used as standard (50-1000ppm) (Thanigaivel et al., 2015).

Health Benefits from Antioxidants

Phytochemical compounds contain several abilities such as radical scavenging capacity, lipid peroxidation inhibition, and metal ion chelating capacity, and reducing the power of tissue or drug (Vinayak et al., 2011). Macro-algae have many biological properties, including anticoagulant, antioxidant, antitumor and immune-modulatory effects. Some of them are used as mosquito repellents. These natural antioxidants have no side effects and serve as the best alternatives to synthetic antioxidants such as BHA, PG, PHT, and TBHQ (Mahendran and Saravanan, 2013). Seaweed is used as a dietary supplement to prevent chronic diseases in humans by preventing oxidative damage (Dontha, 2016). Benefits of antioxidants include:

1. Supports kidney function
2. Improvement of reproductive function
3. Maintains healthy teeth
4. Improves nervous system function
5. Anti-aging effect
6. Protects liver from damage
7. Strengthens the immune system
8. Increases body resistance
9. Reduce obesity
10. Enhances digestion
11. Maintains Healthy Eyesight
12. Improves sleep quality
13. Supports the respiratory systemBottom of Form

Conclusion

Algae have many benefits to humans due to their potential ability to reduce free radicals thereby preventing cell damage in oxidative processes. The natural, non-toxic antioxidant chemicals obtained from algae should be commercialised in the neutraceuticals sector (such supplements as FucoXanThin) for effective consumption. We must concentrate on creating functional foods that are high in natural antioxidants in order to promote consumer health. To do this, the local populace and the pharmaceutical sector need to be made more aware of the advantages of seaweeds and the compounds generated from them.

References

Balakrishnan, D., Kandasamy, D. and Nithyanand, P., 2014. A review on antioxidant activity of marine organisms. Int. J. Chem. Tech. Res, 6(7), pp.3431-3436.

Bhattacharjee, S. and Islam, G.M.R., 2014. Seaweed antioxidants as novel ingredients for better health and food quality: bangladesh prospective. Proc Pak Acad Sci, 51, pp.215-233.

Costa, L.S., Fidelis, G.P., Cordeiro, S.L., Oliveira, R.M., Sabry, D.D.A., Câmara, R.B.G., Nobre, L.T.D.B., Costa, M.S.S.P., Almeida-Lima, J., Farias, E.H.C. and Leite, E.L., 2010. Biological activities of sulfated polysaccharides from tropical seaweeds. Biomedicine & Pharmacotherapy, 64(1), pp.21-28.

De Souza, M.C.R., Marques, C.T., Dore, C.M.G., da Silva, F.R.F., Rocha, H.A.O. and Leite, E.L., 2007. Antioxidant activities of sulfated polysaccharides from brown and red seaweeds. Journal of applied phycology, 19(2), pp.153-160.

Dontha, S., 2016. A review on antioxidant methods. Asian J. Pharm. Clin. Res, 9(2), pp.14-32.

Foon, T.S., Ai, L.A., Kuppusamy, P., Yusoff, M.M. and Govindan, N., 2013. Studies on in-vitro antioxidant activity of marine edible seaweeds from the east coastal region of Peninsular Malaysia using different extraction methods. Journal of Coastal Life Medicine, 1(3), pp.193-198.

Ling, A.L.M., Yasir, S., Matanjun, P. and Bakar, M.F.A., 2015. Effect of different drying techniques on the phytochemical content and antioxidant activity of Kappaphycus alvarezii. Journal of Applied Phycology, 27(4), pp.1717-1723.

Mahendran, S. and Saravanan, S., 2013. Purification and in vitro antioxidant activity of polysaccharide isolated from green seaweed Caulerpa racemosa. Int J Pharm Bio Sci, 4(4), pp.1214-1227.

Morgan, K.C., Wright, J.L. and Simpson, F.J., 1980. Review of chemical constituents of the red algaPalmaria palmata (Dulse). Economic Botany, 34(1), pp.27-50.

Nakamura, T., Nagayama, K., Uchida, K. and Tanaka, R., 1996. Antioxidant activity of phlorotannins isolated from the brown alga Eisenia bicyclis. Fisheries science, 62(6), pp.923-926.

Nakano, T., Watanabe, M., Sato, M. and Takeuchi, M., 1995. Characterization of catalase from the seaweed Porphyra yezoensis. Plant Science, 104(2), pp.127-133.

Ngo, D. H., Vo, T. S., Ngo, D. N., Wijesekara, I., and Kim, S. K. (2012). Biological activities and potential health benefits of bioactive peptides derived from marine organisms. Int. J. Biol. Macromol. 51, 378–383. doi: 10.1016/j.ijbiomac.2012.06.001

Oza R.M. & Zaidi S.H., 2001. A Revised checklist of Indian marine algae. Central Salt and Marine Chemicals Research Institute, Bhavnagar, 296 pp.

Sathya, R., Kanaga, N., Sankar, P. and Jeeva, S., 2017. Antioxidant properties of phlorotannins from brown seaweed Cystoseira trinodis (Forsskål) C. Agardh. Arabian Journal of Chemistry, 10, pp.S2608-S2614.

Sumczynski, D., Bubelova, Z., Sneyd, J., Erb-Weber, S. and Mlcek, J., 2015. Total phenolics, flavonoids, antioxidant activity, crude fibre and digestibility in non-traditional wheat flakes and muesli. Food chemistry, 174, pp.319-325.

Suparna Roy, 2020. Screening and Partial Characterization of Natural Antioxidants from Seaweeds Collected From, Rameshwaram Southeast Coast of India, Journal of Marine Science Research and Oceanography. 3(1), pp. 1-12.

Thaipong, K., Boonprakob, U., Crosby, K., Cisneros-Zevallos, L. and Byrne, D.H., 2006. Comparison of ABTS, DPPH, FRAP, and ORAC assays for estimating antioxidant activity from guava fruit extracts. Journal of food composition and analysis, 19(6-7), pp.669-675.

Thanigaivel, S., Hindu, S.V., Vijayakumar, S., Mukherjee, A., Chandrasekaran, N. and Thomas, J., 2015. Differential solvent extraction of two seaweeds and their efficacy in controlling Aeromonas salmonicida infection in Oreochromis mossambicus: a novel therapeutic approach. Aquaculture, 443, pp.56-64.

Vasconcelos, J.B., de Vasconcelos, E.R., Urrea-Victoria, V., Bezerra, P.S., Reis, T.N., Cocentino, A.L., Navarro, D.M., Chow, F., Areces, A.J. and Fujii, M.T., 2019. Antioxidant activity of three seaweeds from tropical reefs of Brazil: potential sources for bioprospecting. Journal of Applied Phycology, 31(2), pp.835-846.

Venkatesan, M., Arumugam, V., Pugalendi, R., Ramachandran, K., Sengodan, K., Vijayan, S.R., Sundaresan, U., Ramachandran, S. and Pugazhendhi, A., 2019. Antioxidant, anticoagulant and mosquitocidal properties of water soluble polysaccharides (WSPs) from Indian seaweeds. Process Biochemistry, 84, pp.196-204.

Vijayabaskar, P. and Shiyamala, V., 2012. Antioxidant properties of seaweed polyphenol from Turbinaria ornata (Turner) J. Agardh, 1848. Asian Pacific Journal of Tropical Biomedicine, 2(1), pp.S90-S98.

Vinayak, R.C., Sabu, A.S. and Chatterji, A., 2011. Bio-prospecting of a few brown seaweeds for their cytotoxic and antioxidant activities. Evidence-based complementary and alternative medicine, 2011.Com

Yoshie, Y., Wang, W.E.I., Petillo, D. and Suzuki, T., 2000. Distribution of catechins in Japanese seaweeds. Fisheries Science, 66(5), pp.998-1000.

25

Herbs in Fish Nutrition As a Boon for Present Fish Feed Consequences

[1]Shivkumar, [2]Dhanalakshmi M., [3]Rinkesh N Wanjari, [4]Harshavarthini M.

[1]Department of Fish Nutrition and Feed Technology, Fish Nutrition Biochemistry and Physiology Division, ICAR - CIFE, Mumbai, Maharashtra

[2]Department of Fisheries Resource Management, Fisheries Resources, Harvest & Post-Harvest Management Division, ICAR-CIFE, Mumbai, Maharashtra

[3]SKUAST-K Division of Fisheries Resource Management, Faculty of Fisheries Rangil, Ganderbal, Jammu & Kashmir

[4]Department of Fish Genetics, Fish Genetics & Biotechnology Division ICAR-Central Institute of Fisheries Education (ICAR-CIFE) Mumbai, Maharashtra

Abstract

In aquaculture, feed is a major source of nutrition and it contributes near about 60% of total investment. This cost of feed is more because of the presence of costly ingredients used in feed preparation like a fish meal which is superior in its nutritional profile and high palatability. Now recently fish meal become a shortage due to increased fish utilization by humans and the use of fish, for its residues and by-products. Hence researchers started to use a variety of plant sources with/without additional amino acid supplementation to replace the fish meal partially or completely. In the notion, some of these plant sources cause a negative impact on feed intake, nutrient utility, and the health status of fish. So, we need a source that is of natural origin and has properties of anti-inflammatory, anti-microbial, anti-carcinogenic, anti-oxidant, and capillary strengthening effects, anti-inflammatory, antihistaminic, and anti-viral agent as well as a flavoring agent. Hence, these kinds of feed consequences can be ameliorated by the use of herbs in fish feed due to the presence of different phytochemicals like alkaloids, tannins, xanthones, stilbenes, lignans, coumarins, quinones, phenolic acids, flavonols, quercetin, catechins, anthocyanins, and proanthocyanins etc. These phytochemicals solitarily or combinatorial function to achieve better growth and health homeostasis of fish.

Keywords: Feed, Fish meal, Plant source, Phytochemicals, Growth and Health

Introduction

Feed is a major source of nutrition in semi-intensive and intensive culture systems. Fish fed with formulated feed is used for their growth, maintenance, and reproduction. Feed Energy utilized for maintenance is mainly for health and survivability. In intensive aquaculture, animals almost exclusively depend on an external supply of high protein feed (generally >20%), usually based on fishmeal (Beveridge and Little, 2002) because of its high protein content, better bioavailability of amino acids, palatability, and unknown growth factors. Hence fish fed with a fish meal-based diet exhibits better growth performance, feed conversion, and fish health than fish fed with other animal and plant-based protein sources (Hardy and Tacon, 2002; Pike and Barlow, 2002; Tacon, 2003; Cook et al., 1996). Still, Global fish meal production is annually decreasing since 1994 due to increased human utilization and increased use of fish residues and by-products, increasingly replacing whole fish for fish meal (FAO, 2020). Hence many scientists were passionate about replacing the fish meal with an alternative feed ingredient without compromising growth. Different ingredients were used as a substitution of fish meal (partially/totally) in fish diets like soybean meal, distillery grains with solubles, caraway seed meal, cowpea, microalgae, Corn gluten meal, soy protein concentrate, Ipil leaf meal, papaya leaf meal, peanut leaf meal, Azolla leaf meal, moringa leaf meal (Webster et al.,1992; Ahmad and Abdel-Tawwab, 2011; Olvera-Novoa et al., 1997; Shah et al.,2018; Kikuchi, 1999; Xie et al.,2016; Amisah et al.,2009; Peñaflorida, 1995; Garduño-Lugo and Olvera-Novoa, 2008; Maity, and Patra, 2008; Richter et al.,2003)as sown in Figure 01. Fish feed is not only concerned with growth performance but also involves immunity. Therefore, different quality feed ingredients and additives were incorporated in fish feed to improve health status. The most widely used feed additives like probiotics, prebiotics, synbiotics, acidifiers, plant extracts, anti-oxidants, nucleotides, and immunostimulants such as β-glucan and lactoferrin (LF) (Misra et al., 2006; Yokoyama et al., 2006; Dawood et al., 2015a,b, 2016c, 2017; Dawood and Koshio, 2016a; Hossain et al., 2016) used for the development of the animal health and performance.

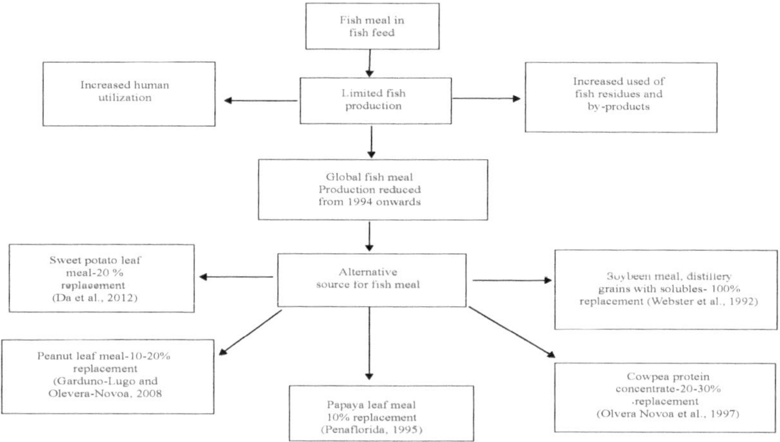

Figure 1 : Graphical representation of Introduction

Some Consequences of Feed/ Feed Ingredients on Fish

After getting the trend of reduced fishmeal annually, researchers have attempted to replace the fishmeal with different plant-based alternative sources to achieve sustainable fish production. Still, some of the alternative sources showed a negative impact on fish health and growth, which are discussed below. Some feed ingredients potentially affect the feed intake, and nutrient utilization capacity of fish like Rainbow trout were fed with commercial soybean meal resulted in decreased amino acid absorption capacity (Dabrowski et al., 1989). Similarly, the Atlantic salmon fed with full-fat soybean meal resulted in abnormal intestinal morphology (Van den ingh et al., 1991). Additionally, Rainbow trout fed with solvent-extracted Soybean meal resulted with lower protein utilization (Rumsey et al., 1993), and Red drum fish and Tilapia fed with solvent-extracted soybean meal and Autoclaved jack bean meal resulted in depressed feed intake (Reigh and Ellis, 1992; Martinez-palacios et al., 1988). Soybean concentrate in fish increases the leucocyte count, an elevation of leukocyte numbers often indicates the commencement of an inflammatory response, onset of a hypersensitivity reaction, or response to an immunogen (Ellis, 1986; Suzuki and Takashi, 1988). Sometimes fish feed act as a source of potential allergic peptides from the fish parasite (Prions); however, the replacement of fish meal with plant source is not a remedial effect because the legumes are also a potential source of these Prions (Faeste et al., 2015). Like *Anisakis simplex* a parasite existing in Fish muscle mainly in Zebrafish and legumes, hence this *A. simplex* parasite from feed may cause allergic reactions like Angioedema, Urticaria, anaphylaxis, and asthma in the host (Faeste et al., 2015) as shown in figure 02. Histamine is a biogenic amine, naturally present in various living organisms and responsible for many physiological and pathophysiological functions of humans via fish. The fish meal made up off scombroids fishes are a potential source of histamine, present at the highly toxic level (>2 g/kg) for more than three years if stored at 4–8 °C, so the fish feed is also a source of histamine (Macan et al., 2006). Additionally, gastric abnormalities have been reported in rainbow trout, and salmon fed histamine rich fish meal (Fairgrieve et al., 1994; Lumsden et al., 2002). Studies performed on shrimps fed biogenic amines (histamine, cadaverine, putrescine, tyramine) rich diet are controversial, showing reduced feed consumption and growth (Ricquemarie et al., 1998; Tapia-Salazar et al., 2001).

Gastrointestinal evacuation reflects the time required for an animal to digest a meal and is usually affected by several parameters such as the fish species and its physiology as well as the physical and chemical composition of the diet (Jobling, 1987). Gastrointestinal evacuation rate has been recognized to be an essential parameter for modeling of daily feed intake (Jobling, 1981) as is moderately responsible for the control of fish appetite (Riche et al., 2004). European seabass (*Dicentrarchus labrax*) fed with a graded level of Legumes (Faba bean, chickpea) resulted in delaying gastric evacuation time (GET), which further reduces feed intake and appetite (Adamidou et al., 2009). African catfish (*Clarias gariepinus*), Nile Tilapia (*Oreochromis niloticus*), and Atlantic salmon (*Salmo salar*) fed with cereal grains as carbohydrate sources cause a delay in GET (Leenhouwers et al., 2006, 2007a, 2007b; Refstie et al., 1999). Tehrany et al. (2018) reported that an increment in GET in common carp (*Cyprinus carpio*) fed with a barley diet. Fish fed with high soya possibly indicating an inflammatory or hypersensitivity response, and this sensitivity generally manifests itself as gastrointestinal hypersensitivity reactions creating major disturbances in the animal digestive processes. Bacterial and viral infections mainly transmitted via trash fish used for fish meal preparation, which includes mainly *Streptococcus spp, Salmonella sp*, Iridovirus, etc. because of improper hygiene and handling (Lunestad et al., 2017; Kim et al., 2007). Aflatoxins are produced by mold, *Aspergillus* species and are highly toxic and carcinogenic. These fungi are commonly found in most soils, and they invade grains and other farm products used for animal feeds production and fish feed during storage (Wagacha et al., 2008). Lipid can undergo peroxidation in a chain reaction with reactive oxygen species (ROS), superoxide anion ($^{\cdot}O_2^{-1}$), peroxide ($^{\cdot}O2^{-2}$), and hydroxyl radical ($^{\cdot}OH^{-1}$) from cells (German, 1999). The dismutation of superoxide (O_2) generates H_2O_2 that can trigger lipid peroxidation in erythrocytes. It is possible that the fish erythrocyte system could be used as a model of lipid oxidation in food and feed ingredients (Li et al., 2016). Lipid oxidation leads to the breakdown of nutritional ingredients, change in taste, scent, and color, development of toxic metabolites, and a decrease in the shelf life of foods and feeds (Błaszczyk et al.,

2013; Smet et al., 2008). Diets with oxidized lipids can result in a decrease in animal health, performance, and quality (Han et al., 2012; Zhang et al., 2011). Hence, there is an imperative feed ingredient required in aquaculture, which acts as anti-inflammatory, anti-microbial, anti-carcinogenic, anti-oxidant, and capillary strengthening effects, anti-inflammatory, antihistaminic, and anti-viral agent as well as a flavoring agent.

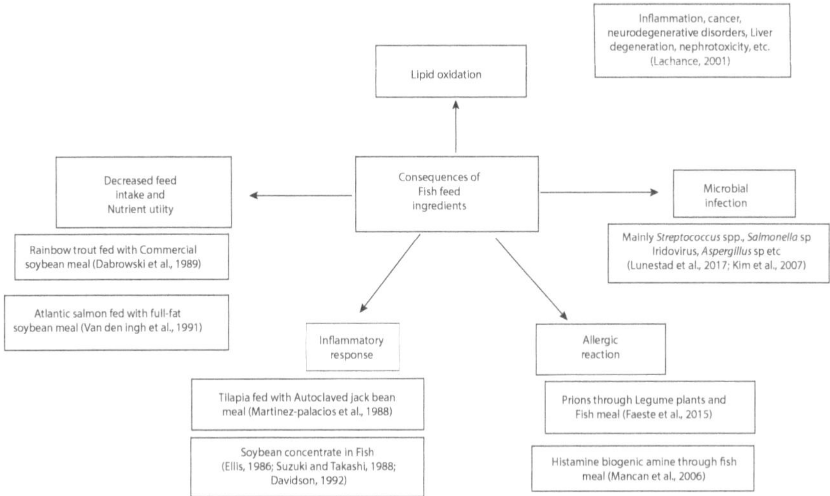

Figure 2 : Consequences of Fish feed ingredients

Herbal feed ingredients - a Remedial Part of Feed Consequences

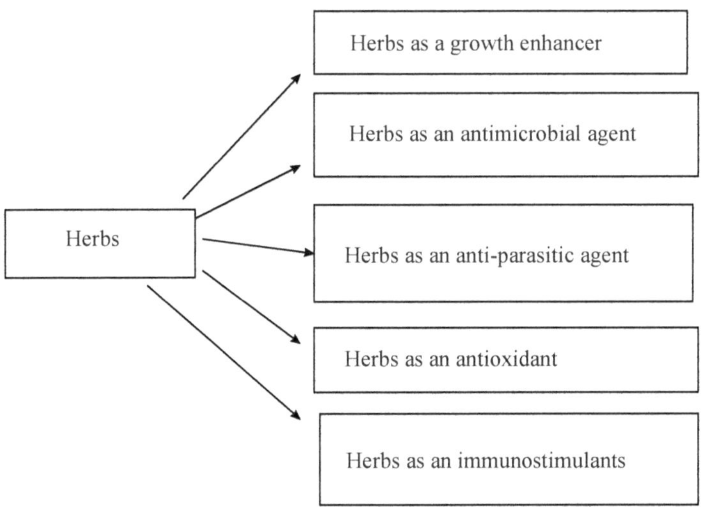

1. Herbs as a Growth Enhancer

Herbs and medicinal plants are the potential to be an essential source of therapeutics in fish culture as a source of treatment and greater accuracy of non-toxicity (Madhuri et al., 2012). Currently, different herbs, plant extracts, and their combination were evaluated as an appetizer, growth promoter, antimicrobial, anti-parasitic, immune-stimulating, and anti-oxidant agents under *in-vitro* and *in-vivo* conditions. Like, tilapia (*Oreochromis niloticus*) fed with a different feed containing different plant based extract like synthesized allicin, culinary herbs, red clover, caraway, and basil helped to increase weight gain and survivability (Zeng,1996; Turan, 2006; ElDakar et al., 2008; Metwally, 2009; Ahmad and Abdel-Tawwab, 2011). Similarly, an improvement in a reproductive performance like early maturation, high fecundity, increased gonadal weight, and reduced intermoult period in *Penaeus monodon* was observed by Mickel Babu (1999) when fed with extracts of different medicinal plants like *ashwagandha* (*Withania somnifera*), monkey tamarind (*Mucuna pruita*), asafetida (*Ferula asafoetida*), and Indian long pepper (*Piper longum*). Since the usage of these herbs promotes sound effects, several commercial herbal additives have been introduced in the aquaculture market. Sangrovit®, a commercial product containing the isoquinoline alkaloid sanguinarine, had shown a positive effect in daily fed Tilapia by increasing appetite, feed intake, and growth.

2. Herbs as An Antimicrobial Agent

Syahidah et al. (2012) reported that Wild *cosmos* (*Cosmos caudatus*), Betle leaf (*Piper betle*), Willow-leaved justicia (*Justicia gendarussa*), and Mango ginger (*Curcuma manga*), had shown potential anti-bacterial property against important aquatic bacteria, *Pseudomonas sp.*, *Aeromonas hydrophilla,* and *Streptococcus agalactiae.* Similarly, cinnamon (*Cinnamomum sp.*) extract exhibited a capable inhibitory action against *Mycobacterium sp., Staphylococcus sp., Enterococcus sp., Pseudomonas sp.,* and *Micrococcus sp.* (Alsaid et al., 2010). Another research proved that Common guava (*Psidium guajava*) was also able to eliminate Vibrio infection in Black tiger shrimp (*P. monodon*) effectively. There are also some reports on the use of herbs in managing the fungal infection. Gormez and Diler (2012) effectively controlled the fungal pathogen (*Saprolegnia parasitica*) by the essential oils of the three Lamiaceae species i.e. oregano (*Origanu monites L.),* black tyme (*Thymbra spicata L.*), and savory (*Satureja tymbra L.*) through in-vitro study. Moreover, Satar (*Zataria multiflora*) essential oil has a promising anti-fungal effect and eliminates *Candida albicans* and *Fusarium solani*. The herbal medication also could fight against the virus, so it secured its utilization part in Aquaculture as an anti-viral compound. About 20 species of Indian traditional medicinal 32 plants such as Bael fruit (*Aegle marmelos*), Bermuda grass (*Cynodon dactylon*), common lantana (*Lantana camara*), Bitter melon (*Momordica charantia*), and Gale of the wind (*Phyllanthus amarus*) have anti-viral activity against White Spot Syndrome Virus (WSSV) (Balasubramanian et al., 2007). Other than that olive tree leaf (*Olea europaea*), *Punica granatum* solvent extracts, *Clinacanthus nutans* ethanol extract has shown successful anti-viral property against Salmonid rhabdovirus, Viral Haemorrhagic Septicaemia virus (VHSV), Lymphocystis Disease virus (LDV) in fish (Direkbusarakom, 2004; Micol et al., 2005; Harikrishnan et al.,2010 a,b).

3. Herbs as An Anti-parasitic Agent

Herbs and Plant material are also reported as anti-parasitic agents in the treatment of many parasitic diseases like myxobolasis, trichodiniasis, gyrodactylosis, argulosis, and scuticociliates in farm fishes (Micol et al., 2005; Harikrishnan et al., 2010 a,b). Garlic extract capable against the Protozoan parasites such as *Entaemoeba histolytica* and *Giardia lamblia* (Ankri and Mirelman, 1999), *Opalina ranarum*, *O. dimidicita*, *Balantidium entozoon*, Trypanosoma, Leishmania, Leptomonas, Crithidia (Reuter et al., 1996). Similarly, the Indian almond (*Terminalia catappa*) extract potentially fight against the all *Trichodina sp.* in Tilapia after two days of treatment (Chitmanat et al., 2003). Additionally, the purified *Commiphora myrrha* extract was used to treat the monogenetic trematodes (gill flukes), infesting the gills of common carp fingerlings when mixed with feed (Abdelhadi, 2007). Moreover, the green tea crude extract has reported being functional against flagellate fish parasites, *Ichthyobodon ecator* in Masu salmon (*Oncorhynchus masou*) and Chum salmon (*Oncorhynchus keta*) (Suzuki et al., 2006).

4. Herbs As An Antioxidant

Different physical, chemical, and biological stressors disturb the homeostasis in an animal, and it liberates different free radicals. These free radicles can be neutralized by a natural anti-oxidative system by utilizing different enzymes and nutrients (Syahidah et al.,2015). Herbs and plant extract exhibiting anti-oxidative properties would counter the action for this oxidation of enzymes and nutrients in an animal. Metwally (2009) observed an increase in activities of glutathione peroxidase, superoxide dismutase (SOD), and catalase (CAT) in Nile tilapia (*Oreochromis niloticus*) when fed with garlic, *Allium sativum* extract. Similarly, elevated activities of SOD and CAT were reported in common carp, *C. carpio* fed with anthraquinone extract of rhubarb, *Rheum officinale* (Xie et al., 2008). Anti-oxidative properties of herbs-derived extracts are associated with tannins, xanthones, stilbenes, lignans, coumarins, quinones, phenolic acids, flavonols, catechins, anthocyanins, and proanthocyanins. These antioxidants act as hydrogen donors or reducing agents for superoxide (Marwah et al., 2007) and peroxide radicles and could enhance immune factors, thus indirectly raise the resistant capacity of fish to various stresses (Chakraborty and Hancz, 2011).

Flavonoids have been labeled as high-level natural anti-oxidants based on their abilities to scavenge free radicals and active oxygen species (Fukumoto and Mazza, 2000; Klahorst, 2002). Flavonoids comprise a large group of polyphenolic compounds that are characterized by a benzo-y-pyrone structure, which is ubiquitous in vegetables and fruits like apples, hops, tea, beer, wine, fruit juice, black tea, citrus fruits, cumins, orange, peppermint, herbs, cereals, parsley, thyme vegetables, onion, cherry, broccoli, kale, tomatoes, berries, tartary buckwheat, lemon, and aurantium, etc. (Yao et at,.2004). In flavonoids, the highest anti-oxidant property has been exhibited by the flavanone *viz.* neoericitrin, hesperetin, hesperidin, eriodictyol (Yao et al., 2004). These flavanones are concentrated at very high levels in citrus fruits.

5. Herbs as An Immunostimulants

From ancient times herbs are using as traditional medicine for the better health of an animals in Egypt, India, South Korea, China, and Thailand, and African countries (Bulfon et al. 2014; Aragona et al. 2018; Van Doan et al. 2018; Garcıa Beltran et al. 2020) these all herbs are potential candidates for aquaculture because of the presence of phytochemicals as mentioned above. Hence some of these herbs are used in fish to enhance the immune function of an animal by directly triggering immune responsible organs like the kidney and spleen as well as increasing antibody production (Goa and Wu. 1994), some of them are discussed below. Rainbow trout (*Oncorhynchus mykiss*) fed with Turmeric (*Curcuma longa*) has potentially increases blood haemoglobin content and lysozyme activity (Yonar et al., 2019). Furthermore, Oregano (*Origanum vulgare*) has increased the cytotoxic activity in Gilthead seabream (*Sparus aurata*) (Garcia Beltran et al.,2018). Additionally, Devil's horsewhip (*Achyranthes aspera*) fed to Rohu has succeed in triggering the activity of lysosome, myeloperoxidase and nitric oxide synthase as reported by Singh et al., 2019. Generally Chinese herbs has vital role in enhancing immunity of fishes like, Common carp (*Cyprinus carpio*) fed with *Astragalus* polysaccharide extracted from *Astragalus membranaceus* root has shown effect on cytokine production and immune-related gene expression in different organs of fish (Yuan et al.,2008). Additionally, the same herb has potentially showed its impact on *Aeromonas hydrophila* challenging study on carp (Yin et al., 2009). Similarly, the other Chinese herbs like *Lonicera japonica*, *Ganoderma lucidium*, *Scutelaria baicalensis* also attain the vital role in shooting out the nonspecific immune response of carp and tilapia (Guo-Jun et al., 2004; Yin et al., 2009; Ardo et al., 2009).

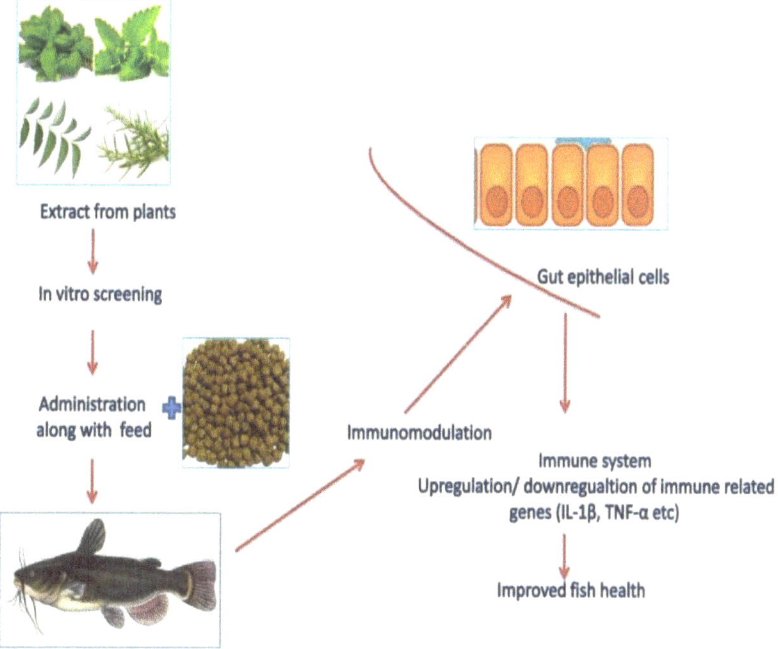

Figure 3: Overview of Possible outcome of action of herbal immunostimulant on fish immunity (Elumalai et al., 2020)

Conclusion

In the present senario fish feed needs a herbs in its natural form or extracted to mitigate the negative impact of plant originated ingredients on fish and to divert the feed gross energy into growth and maintance not for the stress management. One need to be carefull while choosing the herb for fish feed because sometime micro-quantity of antinutrition factors in herb will give beneficial impact on fish health. This will be cost effective feed with favourable effect on growth of a fish.

References

Abdel-Hadi, Y.M. (2007). Prevalence of some parasites infecting the gills of fingerlings of common carp, Cyprinus carpio with trials for treatment. Egyption Journal of Aquaculture Biology and Fish. 11:589-601.

Adamidou, S., Nengas, I., Henry, M., Grigorakis, K., Rigos, G., Nikolopoulou, D., Kotzamanis, Y., Bell, G.J. and Jauncey, K. (2009). Growth, feed utilization, health and organoleptic characteristics of European seabass (Dicentrarchus labrax) fed extruded diets including low and high levels of three different legumes. Aquaculture. 293(3-4):263-271.

Ahmad, M.H. and Abdel-Tawwab, M. (2011). The use of caraway seed meal as a feed additive in fish diets: Growth performance, feed utilization, and whole-body composition of Nile tilapia, Oreochromis niloticus (L.) fingerlings. Aquaculture. 314(1-4):110-114.

Alsaid, M., Daud, H., Bejo, S.K. and Abuseliana, A. (2010). Antimicrobial activities of some culinary spice extracts against Streptococcus agalactiae and its prophylactic uses to prevent streptococcal infection in red hybrid tilapia (Oreochromis sp.). World Journal of Fish and Marine Sciences. 2(6):532-538.

Amisah, S., Oteng, M.A. and Ofori, J.K. (2009). Growth performance of the African catfish, Clarias gariepinus, fed varying inclusion levels of Leucaena leucocephala leaf meal. Journal of Applied Sciences and Environmental Management. 13(1):21-26.

Ankri, S. and Mirelman, D. (1999). Antimicrobial properties of allicin from garlic. Microbes and Infection. 1(2):125-129.

Aragona, M., Lauriano, E.R., Pergolizzi, S. and Faggio, C. (2018). Opuntia ficus-indica (L.) Miller as a source of bioactivity compounds for health and nutrition. Natural Product Research. 32(17):2037–2049.

Ardó, L., Yin, G., Xu, P., Váradi, L., Szigeti, G., Jeney, Z. and Jeney, G. (2008). Chinese herbs (Astragalus membranaceus and Lonicera japonica) and boron enhance the non-specific immune response of Nile tilapia (Oreochromis niloticus) and resistance against Aeromonas Hydrophila. Aquaculture. 275(1-4):26-33.

Balasubramanian, G., Sarathi, M., Kumar, S.R. and Hameed, A.S. (2007). Screening the antiviral activity of Indian medicinal plants against white spot syndrome virus in shrimp. Aquaculture. 263(1-4):15-19.

Beltrán, J.M.G., Espinosa, C., Guardiola, F.A. and Esteban, M.Á. (2018). In vitro effects of Origanum vulgare leaf extracts on gilthead seabream (Sparus aurata L.) leucocytes, cytotoxic, bactericidal and antioxidant activities. Fish & Shellfish Immunology. 79:1-10.

Beltrán, J.M.G., Silvera, D.G., Ruiz, C.E., Campo, V., Chupani, L., Faggio, C. and Esteban, M.Á. (2020). Effects of dietary Origanum vulgare on gilthead seabream (Sparus aurata L.) immune and antioxidant status. Fish & Shellfish Immunology. 99:452-461.

Beveridge, M.C. and Little, D.C., 2002. The history of aquaculture in traditional societies. Ecological Aquaculture. The evolution of the Blue Revolution,. pp.3-29.

Bulfon, C., Volpatti, D. and Galeotti, M. (2014). In vitro antibacterial activity of plant ethanolic extracts against fish pathogens. Journal of the world aquaculture society. 45(5):545-557.

Błaszczyk, A., Augustyniak, A. and Skolimowski, J. (2013). Ethoxyquin: An Antioxidant used in animal feed. International Journal of Food Science 2013. 0:1-12.

Chakraborty, S.B. and Hancz, C. (2011). Application of phytochemicals as immunostimulant, antipathogenic and antistress agents in finfish culture. Reviews in Aquaculture. 3(3):103-119.

Chitmanat, C., Tongdonmuan, K., Khanom, P., Pachontis, P. and Nunsong, W. (2003). Antiparasitic, antibacterial, and antifungal activities derived from a Terminalia catappa solution against some tilapia (Oreochromis niloticus) pathogens. In III WOCMAP Congress on Medicinal and Aromatic Plants-Volume 4: Targeted Screening of Medicinal and Aromatic Plants, Economics. 678:179-182.

Citarasu, T. (2010). Herbal biomedicines: a new opportunity for aquaculture industry. Aquaculture International. 18(3):403-414.

Cook, N.C. and Samman, S. (1996). Flavonoids—chemistry, metabolism, cardioprotective effects, and dietary sources. The Journal of Nutritional biochemistry. 7(2):66-76.

Da, C.T., Lundh, T. and Lindberg, J.E. (2012). Evaluation of local feed resources as alternatives to fish meal in terms of growth performance, feed utilisation and biological indices of striped catfish (Pangasianodon hypophthalmus) fingerlings. Aquaculture. 364:150-156.

Dabrowski, K., Poczyczynski, P., Köck, G. and Berger, B. (1989). Effect of partially or totally replacing fish meal protein by soybean meal protein on growth, food utilization and proteolytic enzyme activities in rainbow trout (Salmo gairdneri). New in vivo test for exocrine pancreatic secretion. Aquaculture. 77(1):29-49.

Direkbusarakom, S. (2004). Application of medicinal herbs to aquaculture in Asia. Walailak Journal of Science and Technology. 1(1):7-14.

El-Dakar, A., Hassanien, G., Gad, S. and Sakr, S. (2008). Use of dried basil leaves as a feeding attractant for hybrid tilapia, Oreochromis niloticus X Oreochromis aureus, Fingerlings. Mediterranean Aquaculture Journal. 1(1):35-44.

Ellis, A.E. (1986). The function of teleost fish lymphocytes in relation to inflammation. International Journal of Tissue Reactions. 8(4):263-270.

Faeste, C.K., Levsen, A., Lin, A.H., Larsen, N., Plassen, C., Moen, A., Van Do, T. and Egaas, E. (2015). Fish feed as source of potentially allergenic peptides from the fish parasite Anisakis simplex (sl). Animal Feed Science and Technology. 202:52-61.

Fairgrieve, W.T., Myers, M.S., Hardy, R.W. and Dong, F.M. (1994). Gastric abnormalities in rainbow trout (Oncorhynchus mykiss) fed amine-supplemented diets or chicken gizzard-erosion-positive fish meal. Aquaculture. 127(2-3):219- 232.

FAO. (2020). The State of World Fisheries and Aquaculture 2020. Sustainability in Action FAO Rome.244p.

Fukumoto, L.R. and Mazza, G. (2000). Assessing antioxidant and prooxidant activities of phenolic compounds. Journal of Agricultural and Food Chemistry. 48(8):3597- 3604.

Gao, X. and Wu, W. (1994). The effect of dang-gui (Angelica sinensis) and ferulic acid on mice immune system. Chin J Biochem Pharm. 15:107-110.

Garduño-Lugo, M. and Olvera-Novoa, M.Á. (2008). Potential of the use of peanut (Arachis hypogaea) leaf meal as a partial replacement for fish meal in diets for Nile tilapia (Oreochromis niloticus L.). Aquaculture Research. 39(12):1299-1306.

German, J.B. (1999). Food processing and lipid oxidation. Impact of processing on food safety. pp.23-50.

Gormez, O. and Diler, O. (2014). In vitro antifungal activity of essential oils from Tymbra, Origanum, Satureja species and some pure compounds on the fish pathogenic fungus, Saprolegnia parasitica. Aquaculture Research. 45(7):1196- 1201.

Guo-Jun, Y., Wiegertjes, G.F., Yue-Ming, L., Schrama, J.W., Verreth, J.A.J., Pao, X.U. and Hong-qi, Z. (2004). Effect of Astralagus radix on proliferation and nitric oxide production of head kidney macrophages in Cyprinus carpio: an in vitro study. Journal of Fisheries of China. 28(6):628-632.

Halver, J.E. (1976). Formulating practical diets for fish. Journal of the Fisheries Board of Canada. 33(4):1032-1039.

Han, Y.Z., Ren, T.J., Jiang, Z.Q., Jiang, B.Q., Gao, J., Koshio, S. and Komilus, C.F. (2012). Effects of palm oil blended with oxidized fish oil on growth performances, hematology, and several immune parameters in juvenile Japanese sea bass, L a t e o l a b r a x japonicas. Fish Physiology and Biochemistry. 38(6):1785-1794.

Hardy, R.W. and Tacon, A.G. (2002). Fish meal: historical uses, production trends and future outlook for sustainable supplies. Responsible marine aquaculture. Pp.311-325.

Harikrishnan, R., Heo, J., Balasundaram, C., Kim, M.C., Kim, J.S., Han, Y.J. and Heo, M.S. (2010). Effect of Punica granatum solvent extracts on immune system and disease resistance in Paralichthys olivaceus against lymphocystis disease virus (LDV). Fish and shellfish Immunology. 29(4):668-673.

Harikrishnan, R., Heo, J., Balasundaram, C., Kim, M.C., Kim, J.S., Han, Y.J. and Heo, M.S., 2010. Effect of traditional Korean medicinal (TKM) triherbal extract on the innate immune system and disease resistance in Paralichthys olivaceus against Uronema marinum. Veterinary Parasitology. 170(1-2):1-7.

Jobling, M. (1981). Dietary digestibility and the influence of food components on gastric evacuation in plaice, Pleuronectes platessa L. Journal of Fish Biology. 19(1):29- 36.

Kikuchi, K. (1999). Partial replacement of fish meal with corn gluten meal in diets for Japanese flounder Paralichthys olivaceus. Journal of the World Aquaculture Society. 30(3):357-363.

Klahorst, S. (2002). Exploring antioxidants. The World of Food Ingredients. pp.54-59.

Kim, J.H., Gomez, D.K., Choresca Jr, C.H. and Park, S.C. (2007). Detection of major bacterial and viral pathogens in trash fish used to feed cultured flounder in Korea. Aquaculture. 272(1-4):105-110.

Leenhouwers, J.I., Adjei-Boateng, D., Verreth, J.A.J. and Schrama, J.W. (2006). Digesta viscosity, nutrient digestibility and organ weights in African catfish (Clarias gariepinus) fed diets supplemented with different levels of a soluble non- starch polysaccharide. Aquaculture Nutrition. 12(2):111-116.

Leenhouwers, J.I., Ortega, R.C., Verreth, J.A. and Schrama, J.W. (2007). Digesta characteristics in relation to nutrient digestibility and mineral absorption in Nile tilapia (Oreochromis niloticus L.) fed cereal grains of increasing viscosity. Aquaculture. 273(4):556-565.

Leenhouwers, J.I., ter Veld, M., Verreth, J.A. and Schrama, J.W. (2007). Digesta characteristiscs and performance of African catfish (Clarias gariepinus) fed cereal grains that differ in viscosity. Aquaculture. 264(1-4):330-341.

Li, H., Zhou, X., Gao, P., Li, Q., Li, H., Huang, R. and Wu, M. (2016). Inhibition of lipid oxidation in foods and feeds and hydroxyl radical-treated fish erythrocytes: a comparative study of Ginkgo biloba leaves extracts and synthetic antioxidants. Animal Nutrition. 2(3):234-241.

Lumsden, J.S., Clark, P., Hawthorn, S., Minamikawa, M., Fenwick, S.G., Haycock, M. and Wybourne, B. (2002). Gastric dilation and air sacculitis in farmed chinook salmon, Oncorhynchus tshawytscha (Walbaum). Journal of Fish Diseases. 25(3):155-163.

Lunestad, B.T., Nesse, L., Lassen, J., Svihus, B., Nesbakken, T., Fossum, K., Rosnes, J.T., Kruse, H. and Yazdankhah, S. (2007). Salmonella in fish feed; occurrence and implications for fish and human health in Norway. Aquaculture. 265(1-4):1-8.

Macan, J., Turk, R., Vukušić, J., Kipčić, D. and Milković-Kraus, S. (2006). Long-term follow-up of histamine levels in a stored fish meal sample. Animal Feed Science and Technology. 127(1-2):169-174.

Madhuri, S., Mandloi, A.K., Govind, P. and Sahni, Y.P. (2012). Antimicrobial activity of some medicinal plants against fish pathogens. International Research Journal of Pharmacy. 3(4):28-30.

Maity, J. and Patra, B.C. (2008). Effect of replacement of fishmeal by Azolla leaf mael on growth, food utilization, pancreatic protease activity and RNA/DNA ratio in the fingerlings of Labeo rohita (Ham.). Canadian Journal of Pure and Applied Science. 2(2):323-333.

Martinez-Palacios, C.A., Cruz, R.G., Novoa, M.A.O. and Chávez-Martinez, C. (1988). The use of jack bean (Canavalia ensiformis Leguminosae) meal as a partial substitute for fish meal in diets for tilapia (Oreochromis mossambicus Cichlidae). Aquaculture. 68(2):165-175.

Marwah, R.G., Fatope, M.O., Al Mahrooqi, R., Varma, G.B., Al Abadi, H. and Al- Burtamani, S.K.S. (2007). Antioxidant capacity of some edible and wound healing plants in Oman. Food Chemistry. 101(2):465-470.

Metwally, M.A.A. (2009). Effects of garlic (Allium sativum) on some antioxidant activities in tilapia nilotica (Oreochromis niloticus). World Journal of Fish and Marine Sciences. 1(1):56-64.

Michael Babu, M. (1999). Developing bio-encapsulated ayurvedic products for maturation and quality larval production in Penaeus monodon (Doctoral dissertation, Manonmaniam Sundaranar University).

Micol, V., Caturla, N., Pérez-Fons, L., Más, V., Pérez, L. and Estepa, A. (2005). The olive leaf extract exhibits antiviral activity against viral haemorrhagic septicaemia rhabdovirus (VHSV). Antiviral Research. 66(2-3):129-136.

Olvera-Novoa, M.A., Pereira-Pacheco, F., Olivera-Castillo, L., Pérez-Flores, V., Navarro, L. and Sámano, J. C. (1997). Cowpea (Vigna unguiculata) protein concentrate as replacement for fish meal in diets for tilapia (Oreochromis niloticus) fry. Aquaculture. 158(1-2):107-116.

Penaflorida, V.D. (1995). Growth and survival of juvenile tiger shrimp fed diet where fish meal is partially replaced with papaya (Carica papaya L.) or camote (Ipomea batatas Lam.) leaf meal. The Israeli Journal of Aquaculture-Bamidgeh. 47(1):25- 33.

Pike, I.H. and Barlow, S.M., 2002. Impacts of fish farming on fish, Bordeaux Aquaculture and Environment Symposium. Bordeaux, France. pp.6-7.

Refstie, S., Svihus, B., Shearer, K.D. and Storebakken, T. (1999). Nutrient digestibility in Atlantic salmon and broiler chickens related to viscosity and non-starch polysaccharide content in different soyabean products. Animal Feed Science and Technology. 79(4):331-345.

Reigh, R.C. and Ellis, S.C. (1992). Effects of dietary soybean and fish-protein ratios on growth and body composition of red drum (Sciaenops ocellatus) fed isonitrogenous diets. Aquaculture. 104(3-4):279-292.

Reuter, H.D. (1996). Therapeutic effects and applications of garlic and its preparations. Garlic.

Riche, M., Haley, D.I., Oetker, M., Garbrecht, S. and Garling, D.L. (2004). Effect of feeding frequency on gastric evacuation and the return of appetite in tilapia Oreochromis niloticus (L.). Aquaculture. 234(1-4):657-673.

Richter, N., Siddhuraju, P. and Becker, K. (2003). Evaluation of nutritional quality of moringa (Moringa oleifera Lam.) leaves as an alternative protein source for Nile tilapia (Oreochromis niloticus L.). Aquaculture. 217(1-4):599-611.

Ricque-Marie, D., Abdo-de La Parra, M.I., Cruz-Suárez, L.E., Cuzon, G., Cousin, M. and Pike, I.H. (1998). Raw material freshness, a quality criterion for fish meal fed to shrimp. Aquaculture. 165(1-2):95-109.

Rumsey, G.L., Hughes, S.G. and Winfree, R.A. (1993). Chemical and nutritional evaluation of soya protein preparations as primary nitrogen sources for rainbow trout (Oncorhynchus mykiss). Animal Feed Science and Technology. 40(2- 3):135-151.

Shah, M.R., Lutzu, G.A., Alam, A., Sarker, P., Chowdhury, M.K., Parsaeimehr, A., Liang, Y. and Daroch, M. (2018). Microalgae in aquafeeds for a sustainable aquaculture industry. Journal of Applied Phycology. 30(1):197-213.

Singh A, Sharma J, Paichha M, Chakrabarti R. (2019). Achyranthes aspera (Prickly chaff flower) leaves- and seeds-supplemented diets regulate growth, innate immunity, and oxidative stress in Aeromonas hydrophila-challenged Labeo rohita. Journal of Applied Aquaculture. 31:1–18.

Smet, K., Raes, K., Huyghebaert, G., Haak, L., Arnouts, S. and De Smet, S. (2008). Lipid and protein oxidation of broiler meat as influenced by dietary natural antioxidant supplementation. Poultry Science. 87(8):1682-1688.

Suzuki, K., Misaka, N. and Sakai, D.K. (2006). Efficacy of green tea extract on removal of the ectoparasitic flagellate Ichthyobodo necator from chum salmon, Oncorhynchus keta, and masu salmon, O. masou. Aquaculture. 259(1-4):17-27.

Suzuki, Y. and Hibiya, T. (1988). Dynamics of leucocytic inflammatory responses in carp. Fish Pathology. 23(3):179-184.

Syahidah, A., Saad, C.R., Daud, H.M. and Abdelhadi, Y.M. (2015). Status and potential of herbal applications in aquaculture: A review. Iranian Journal of Fisheries Sciences. 14(1):27-44.

Syahidah, A., Saad, C.R. and Daud, H.M. (2012). Potential antibacterial activity of local herb extracts on fish pathogenic bacteria. In 31st Symposium of the Malaysian Society for Microbiology. Microbiology Research in the Omics Era. 63:13-15.

Tacon, A.J. (2003). Aquaculture production trends analysis. Review of the state of world aquaculture. FAO fisheries circular. 886(2):5-29.

Tapia-Salazar, M., Smith, T.K., Harris, A., Ricque-Marie, D. and Cruz-Suarez, L.E. (2001). Effect of dietary histamine supplementation on growth and tissue amine concentrations in blue shrimp Litopenaeus stylirostris. Aquaculture. 193(3-4):281- 289.

Tehrany, Z.M., Amirkolaie, A.K. and Oraji, H. (2018). The effects of different grain sources on gut evacuation rate and nutrient digestibility in common carp, Cyprinus carpio. International Journal of Aquatic Biology. 6(2):104-113.

Turan, F. (2006). Improvement of growth performance in tilapia (Orepochrimis aureus Linnaeus) by supplementation of red clover Trifolium pratense in diets. The Israeli Journal of Aquaculture – Bamidgeh. 58(1): 34-38.

Van den Ingh, T.S.G.A.M., Krogdahl, Å., Olli, J.J., Hendriks, H.G.C.J.M. and Koninkx, J.G.J.F. (1991). Effects of soybean-containing diets on the proximal and distal intestine in Atlantic salmon (Salmo salar): a morphological study. Aquaculture. 94(4):297-305.

Webster, C.D., Tidwell, J.H., Goodgame, L.S., Yancey, D.H. and Mackey, L. (1992). Use of soybean meal and distillers grains with solubles as partial or total replacement of fish meal in diets for channel catfish, Ictalurus punctatus. Aquaculture. 106(3-4):301-309.

Xie, J., Liu, B., Zhou, Q., Su, Y., He, Y., Pan, L., Ge, X. and Xu, P. (2008). Effects of anthraquinone extract from rhubarb Rheum officinale Bail on the crowding stress response and growth of common carp Cyprinus carpio var. Jian. Aquaculture. 281(1-4):5-11.

Xie, S.W., Liu, Y.J., Zeng, S., Niu, J. and Tian, L.X. (2016). Partial replacement of fish-meal by soy protein concentrate and soybean meal based protein blend for juvenile Pacific white shrimp, Litopenaeus vannamei. Aquaculture. 464:296-302.

Yao, L.H., Jiang, Y.M., Shi, J., Tomas-Barberan, F.A., Datta, N., Singanusong, R. and Chen, S.S. (2004). Flavonoids in food and their health benefits. Plant Foods for Human Nutrition. 59(3):113-122.

Yin, G., Ardó, L.Á.S.Z.L.Ó., Thompson, K.D., Adams, A., Jeney, Z. and Jeney, G. (2009). Chinese herbs (Astragalus radix and Ganoderma lucidum) enhance immune response of carp, Cyprinus carpio, and protection against Aeromonas hydrophila. Fish & Shellfish Immunology. 26(1):140-145.

Yonar, M.E., Yonar, S.M., İspir, Ü. and Ural, M.Ş. (2019). Effects of curcumin on haematological values, immunity, antioxidant status and resistance of rainbow trout (Oncorhynchus mykiss) against Aeromonas salmonicida subsp. achromogenes. Fish & shellfish immunology. 89:83-90.

Yuan, C., Pan, X., Gong, Y., Xia, A., Wu, G., Tang, J. and Han, X. (2008). Effects of Astragalus polysaccharides (APS) on the expression of immune response genes in head kidney, gill and spleen of the common carp, Cyprinus carpio L. International Immunopharmacology. 8(1):51-58.

Zhang, W., Xiao, S., Lee, E.J. and Ahn, D.U. (2011). Consumption of oxidized oil increases oxidative stress in broilers and affects the quality of breast meat. Journal of Agricultural and Food Chemistry. 59(3):969.

Zeng, H., Ren, Z.L. and Guo, Q. (1996). Application of allicin in tilapia feed. China Feed. 21:29-30.

26

Impacts of Xenobiotics in Fish Reproduction

[1]Sanjay Chandravanshi, [2]Sudhan Chandran, [3]Sahil Rai, Narsale [4]Swapnil Ananda and [5]H.S. Mogalekar

[1]Department of Fisheries Biology and Resource Management, Fisheries Resources, Fisheries College and Research Institute, TNJFU, Tamil Nadu

[2]Department of Fisheries Resource Management, Fisheries Resources, Harvest and Post-Harvest Management Division, ICAR-CIFE, Mumbai, Maharashtra

[3]Department of Fisheries Resource Management, Fisheries Resources College of Fisheries, Guru Angad Dev Veterinary and Animal Sciences University Ludhiana, Punjab

[4]Department of Fish Pathology and Health Management, Fisheries College and Research Institute,TNJFU, Tamil Nadu

[5]Department of Aquatic Environment Management, College of Fisheries, Dholi Muzaffarpur, RPCAU, Bihar

Introduction

"Xenobiotics" are "foreign compounds," which are alien to the organism and removed from their natural environment. Xenobiotics are substances that are either alien to an organism or do not form a natural part of its diet. Compounds including pharmaceuticals, food additives, & environmental toxins are examples of xenobiotics. Normally, these substances are removed from the body by being broken down into compounds and then being expelled through the kidney, bile and lungs. For the pharmaceutical business, enzymes that metabolize xenobiotics are crucial because they are in charge of breaking down medicines. Similarly, xenobiotic transporters influence how long medications remain in the body.

Xenobiotics include substances such as pesticides insecticides, herbicides, metals and their derivatives, pharmaceuticals such antibiotics, and many more. Numerous studies have demonstrated that both short-period and long-period subjection to these xenobiotics can permanently affect living beings. When xenobiotics enter a living system, they go through four stages:

absorption, distribution, metabolism, and elimination. The typical metabolism of xenobiotics entails continuous biotransformations such as oxidation, reduction of the major molecule to produce reactive groups (-NH2, -COOH-OH), followed by conjugation of hydrophilic molecules (glutathione, sulphate, glucuronic acid) to increase the hydrophilicity of xenobiotics, which leads to intestinal excretion.

Xenobiotics and Toxicity

Numerous synthetic chemicals used in various industrial and domestic processes have been proven to interfere with normal endocrinology and physiology of living things. Three basic categories—neuro-physiological, reproductive, and behavioral—are used to categorize how xenobiotics affect an organism as a whole. These impacts are frequently interconnected; for example, neurological changes can influence behavior, and behavioral changes can influence reproduction. A compound's impact on a target organism or group isn't usually immediate. It is always based on the compound's concentration and the length of exposure. In the end, these impacts may be either acute or persistent. Acute toxicity happens quickly, is well-defined, frequently lethal, and infrequently reversible. Long-term exposure to low levels or long after exposure results in chronic consequences that can eventually result in mortality. When a xenobiotic directly causes death or has the potential to do so, it is said to be deadly. And a toxin is sub deadly when its potency is below that at which it directly results in death. The organism's physiological or behavioral functions subsequently regress, which lowers its overall fitness. Mostly in the case of radioactive pollution, the ecosystem is likely to suffer permanent damage. The impact of pollution on freshwater species is manifested in the extinction of certain species, with possible economic benefits for others. There is typically a decline in species variety, albeit not always in the total number of species, as well as a shift in the relative importance of various processes including predation, competition, and material cycling. Due to the intricacy of pollution, the characteristics of the pollutant also influence how it affects aquatic life. When two or more poisons are mixed in an effluent, they may have an additive, antagonistic, or synergistic effect on an organism. The combined toxicity of cadmium and zinc to fish serves as an illustration of an additive interaction. Lead, zinc, and aluminum are all hostile to calcium. In addition to being a complement to zinc, chlorine, cadmium, mercury and copper also lessens the toxicity of cyanide.

Additionally, it has been discovered that xenobiotics can cause gene mutations that cause cancer. The influence of xenobiotic pollution on aquatic ecosystems has been extensively studied, and the three main categories of their effects on fish and other aquatic creatures are behavioral, neurophysiological, and reproductive. The aforementioned consequences are frequently related, as neurological alterations impact fish behavioural patterns and reproductive system, respectively. In this study we have talked about the current situation involving xenobiotics and marine fish interaction and also to give a literature overview of the biological changes that have been noticed in several marine fish species as a result of external and internal exposure to xenobiotics.

Effects of Xenobiotics on Fish Reproduction

1. Reproductive Dysfunction in Invertebrates

Numerous chemicals that have been discharged into the aquatic system have merely undergone the barest amount of toxicological assessment. Most of the components in industrial and household wastes have not been tested at all, although some of these chemicals, like pesticides, have only undergone rudimentary toxicity tests that establish the "safe" level for people most at risk of exposure. Recent worries about a potential decline in sperm counts, an increase in male genital abnormalities in humans, alligators with abnormal male genitalia, and male fish in rivers that exhibit female characteristics have brought attention to the possibility that several of these chemicals can be harmful at concentrations well below those that are lethal.

The majority of these well-documented worries have been concentrated on substances that imitate the hormone estradiol. These include estradiol derived from both human and animal sources as well as organochlorine pesticides like DDT, PCBs and breakdown products of alkylphenoxy detergents. Since sperm, which are continuously generated in large amounts, are considerably more susceptible to quantification than eggs, which mammals produce in much less quantities during their estrus or monthly cycles, it is clear that a big portion of this research is focused on males. The effects of estrogens are also immediately observable in fish since they are the sole known stimulator of the yolk protein vitellogenin. As a result, the production of this protein by immature fish offers a ready method for determining the presence of environmental pollutants with estrogenic activity. It is crucial to understand the motivations behind the focus on estrogens and men as well as the likelihood that environmental estrogens will also have an impact on female fertility. In addition, there are a variety of additional substances that do not act as estrogens but can still have an impact on fertility by affecting the gametes or endocrine system.

Since all chemicals, whether first discharged on land, into the atmosphere, or directly into rivers, will eventually find their way into the rivers and seas as the final repository, the aquatic ecosystem is most likely to get polluted from pollutants of all the habitats. Pollutants from sewage effluents, industrial wastes, agricultural chemicals, herbicides, pesticides, sea dumping, road runoff water, fossil fuel burning, etc. will all be present in such systems. Drugs used to treat illness and parasites, as well as hormones used to manipulate growth, reproductive cycles, or sexual differentiation, may have an impact on wild fish in specific locations near intense fish culture (Kime, 1999).

2. Fish as an Experimental Model

The reproductive function of vertebrates can be studied using fish as a particularly helpful model for investigating the effects of environmental pollutants. They are the species facing maximum risk to be effected by pollution, they can smoothly be exposed in the lab to environmentally realistic doses, both sexes produce large numbers of gametes that can easily be counted or examined for malformations, fertilization and hatch rates can easily be determined, the offspring can easily be monitored for developmental abnormalities or fertility of the second and third generation offspring of exposed parents, and the endocrine coenzymes can easily be affected. The comparison of samples exposed in the lab under more controlled circumstances with wild fish from polluted habitats is similarly simple. The effects observed in fish would be predicted to have analogs in mammals, despite the fact that the endocrine system of fish differs in detail of its structure and hormones, reflecting the huge variation of reproductive behaviors.

Effects on the Endocrine System

1. The Basics of Fish Reproductive Endocrinology

Similar to mammals, fish have an endocrine system that involves a complicated interplay between external stimuli, pituitary and hypothalamus hormones, gonadal hormones and the liver's deactivation of hormones. Temperature, photoperiod, rainfall, and other environmental factors all act through the hypothalamus to release gonadotrophin releasing hormone (GnRH), which in turn triggers the production of gonadotrophin (GtH) from the anterior pituitary, regulating gonadal recrudescence. Despite the fact that some fish contain at least two gonadotrophins (GtH-I and GtH-II), which are comparable to LH and FSH in mammals, this article will not distinguish between the two and just refer to them as GtH. Gonadotrophins stimulate gonadal development and production of steroid hormones. Fish differ from mammals in that the major testicular androgen is 11-ketotestosterone rather than testosterone, and while estradiol is the ovarian estrogen it acts very differently to mammals. Its major target tissue is the liver which is stimulated to produce the vitellogenin which is the yolk protein which is incorporated into the oocyte under the influence of the gonadotrophin GtH-I.Several momths are required for th Gonadal recrudescence in both male and female till the production of viable gametes. After that, the biosynthetic route switches to producing a progestogen, which causes the final maturity of the oocytes' germinal vesicle breakdown, or GVBD, or the production of sperm. Estradiol and 11-ketotestosterone are subsequently no longer secreted. This progestogen is 17, 20 β-dihydroxy-4-pregnen-3-one (17, 20 βP) in many fish, although 11-deoxycortisol or other progestogens may be present in other fish. The liver functions as the hormone's target tissue for estradiol because it catabolizes the hormone and causes the synthesis of vitellogenin there. Through the hormones they secrete as well as other messengers like inhibins, the gonads can communicate their status to the hypothalamus and pituitary.

2. Effects at the Pituitary and Hypothalamus

Numerous pollutants have been observed to lower the gonadosomatic index, however it is often uncertain whether this is due to the primary dysfunction occurring at the gonad itself or to a lack of pituitary hormone release. While a histological examination of the pituitary may be able to clarify this, this requires much more skill than merely looking at the gonadosomatic index (GSI), as the pituitary is a small & inaccessible gland in most fish. Therefore, it is often better to look at this indirectly via measurement of the pituitary or gonadal hormones rather than physically examining the pituitary to track the xenobiotics impact on its structure. Alterations in the basal secretion of GnRH and GtH as well as changes in plasma GtH in response to GnRH challenge can be used to identify pituitary or hypothalamic dysfunction. These peptides, however, have a high level of species specificity, therefore before a test can be undertaken, hormones from the specific species must be isolated, purified, and produced, along with antisera. However, these tests are accessible for the salmonids and cyprinids that make up the OECD test species, including rainbow trout (*Oncorhynchus mykiss*), carp (*Cyprinus carpio*), goldfish (*Carassius auratus*), zebrafish (*Danio rerio*). The gonads send feedback signals to the pituitary and hypothalamus. Some of this might be due to androgen aromatization to estrogens in the brain or estrogen receptors. Hormonally active pollutants that function as estrogens, anti-estrogens, androgens, and anti-androgens may also have an impact on these receptors, altering GtH secretion as a result.

2. Effects on Gonadal Hormone Production

A precise temporal pattern of hormone production is necessary for the formation of the gametes, and this pattern is established by signaling from the gonad to the pituitary. When the gamete has matured to the point where the synthetic pathway can flip from 11-ketotestosterone/estradiol to progestogens, such signaling may be very crucial. Estradiol synthesis in the ovary needs to be high enough to cause the liver to make vitellogenin, which can then be incorporated into the oocyte. Lower levels of estradiol may cause fewer eggs or fewer eggs that are smaller than usual. These two variables may have an impact on the number of potentially viable offspring. Measuring the gonadosomatic index (gonad weight/body weight) in both control and treated fish is the simplest way to diagnose gonadal dysfunction. Reduced GSI is a sign of decreased hypothalamic, pituitary, or gonadal activity. Ideally, this measurement should be used in conjunction with a histological examination to identify any inhibition of a specific maturation stage, such as the induction of vitellogenesis, the development of spermatogonia, or the growth and maturation of oocytes, which may reveal some information about the pollutant's mode of action.

However in larger species it might be possible to collect a gonadal biopsy of a relatively mature fish, GSI determination typically necessitates the sacrifice of the fish. This is not possible in the early phases of development or when gonadal recrudescence is severely inhibited. Measuring plasma steroid concentrations is a further sign of gonadal state, and it has the benefit, at least in species larger than 100 g, that sacrifice may not be required. For bigger species, serial sampling is also an option during exposure. In particular, the Enzyme Linked Immunosorbant Assay (ELISA), which doesn't require a lot of complicated equipment, makes measuring steroid hormones in fish quite easy.Although Radioimmunoassay is still utilized in several laboratories for hormone assays, radioimmunoassay has the drawback of requiring expensive counting apparatus. The assay of certain steroids, like 11-ketotestosterone and 17,20 βP of fish, is currently constrained by the absence of both commercial antisera and labelled steroid, despite the fact that for the mammalian hormones testosterone and estradiol commercial kits are present.

Radiolabelled steroids can only be produced by chemically derivatizing commercial drugs, such as cortisol or 17-hydroxyprogesterone, and this takes some basic chemical knowledge and experience, even though these antisera are frequently made available through the goodwill of academic colleagues. The preparation of enzyme-labelled steroids for ELISA is potentially considerably simpler. This kind of research could help determine if a lack of steroid production is the result of the primary gonadal steroidogenic failure or secondary failure brought on by pituitary/hypothalamic malfunction.

Endocrine Effects Mediated by the Liver

The rate of gonadal secretion and the rate of hepatic deactivation and excretion together determine the final plasma steroid concentrations. Low plasma steroid levels could be caused by increased hepatic catabolism or inhibited production. Hepatic mixed function oxidase can be stimulated by a variety of contaminants. Steroid catabolism is aided by MFO activation. If the enzyme was accessible for steroid catabolism, such increased activity might lead to more rapid oxidation, but it might also result in a decline in metabolism if the pollutants were preferred as a substrate. Even though benzopyrene in the sediments enhanced MFO activity as assessed by EROD activity, flounder exposed to sediment from the Rhine estuary exhibited higher estradiol, which was linked to decreased steroid catabolism. The fact that changes in MFO activity in the gonad and liver must necessarily result in disruption of the steroidogenic balance of the endocrine system is frequently overlooked when using MFO activity as an indicator of exposure to pollution. Because low estradiol secretion will result in an abnormally low plasma vitellogenin titre, vitellogenin production in the liver can also be utilized as a marker of ovarian activity. It is simpler to test plasma vitellogenin in small fish as only a few microliters of plasma can be recovered from them, and plasma vitellogenin levels are many orders of magnitude greater than those of estradiol. For such measurements, radioimmunoassay and ELISA techniques are both available, however the former suffers from the need for radioiodine. Although vitellogenin has some species specificity, there is a significant level of cross-reactivity within genera, making it possible to employ 'cyprinid' or '*Oncorhynchus*' vitellogenin for a wide range of widespread species.

Since the livers of boys and young children also have receptors for estradiol can react to it by synthesizing vitellogenin, they can also be used to detect environmental pollutants having estrogenic activity. It has been used to identify estrogens in sewage effluent and, in the lab, can create a straightforward in vivo test for contaminants that are estrogenic by exposing male or immature fish, in which vitellogenin is not often detectable. Today, we perform vitellogenin assays using an ELISA, which has the primary benefit of not requiring radioactive iodine. Vitellogenin is a definite sign of exposure to substances with estrogenic activity when it is found in plasma. Conversely, a drop in vitellogenin in female plasma may indicate that an anti-estrogen is having an effect on the liver, but it must be demonstrated that this action is not merely being caused by a reduction in pituitary or gonadal hormone release. The aquatic system is negatively impacted by xenobiotics, which also disrupt fish reproductive at all stages.

3. Direct Effects on the Gametes

Gamete Development in Fish

Pollutants can have an indirect impact on sperm and egg development by disrupting the ideal hormonal environment, as was mentioned above, but if they themselves have hormonal activity, they can also have a direct impact on the gamete's local hormonal environment. Numerous substances that are said to have estrogenic or other hormonal action may also, like non-hormonal substances, have toxic effects, which can interfere with gamete development either directly or indirectly by harming the gametes or the endocrine system. The net impact on the fish's fertility is what matters in these situations, not whether the activity is classified as "endocrine disruptive" or not. The important thing to remember about the impact of any environmental toxin is that it can interfere with reproduction at concentrations below those that have an impact on the fish's general health. Aside from its reduced fertility, however, the fish can still live a normal life. Over focusing on endocrine disruptors can cause us to overlook other impacts that happen at remarkably low concentrations. Zinc is an important element for the growth of both sperm & eggs, and metallothioneins, whose synthesis is controlled by variations in gonadal steroids during the breeding cycle, keep the mineral's natural balance. Therefore, heavy metals that cause metallothionein synthesis in the gonad or liver may interfere with gamete development by upsetting the usual balance of zinc in the body. Hormonally active pollutants may also cause variations in the synthesis of metallothionein, which are typically linked to seasonal cycles. Similar to MFO, induction of metallothionein indicates subjection to heavy metals from environmental contamination as well as the potential impairment of normal reproductive function.

Effects on Sperm

The sperm of teleost fish is different from that of mammals in a number of ways. It is immotile upon ejaculation, only becomes motile upon contact with water, remains motile for only a short period of time, and penetrates the egg via the micropyle as opposed to an acrosome response. Therefore, the initial minute following motility induction is important to its ability to successfully fertilize an egg. Even if it is put on the egg surface, it must swim quickly enough to locate the micropyle. Since fish sperm's ability to quickly move sufficient during first minute of activity may be crucial to its ability to fertilize, any pollutant that delays this mobility is likely to have an impact. Even with videotaping, it is very challenging to determine velocity by direct observation due to the extremely short period of motility. The majority of research has employed a highly subjective scoring system to determine the sperm's travel speed or duration. However, recent advancements in Computer Assisted Sperm Analysis (CASA) and its application to fish sperm have now made it possible to quantify such measures and perform an analysis in real time with some tools. Using an "extender" to which a pollutant has been added, it is possible to assess the effects of exposure for this length of time because sperm may be kept immobile in a "extender" for up to 24 hours. Using this method, it was demonstrated that cadmium at 50 mg/l drastically reduced sperm motility. The permissible level for mercury in drinking water is 1 ppb, yet this mercury had a significant impact on sperm motility, and methyl mercury may be lethal at a concentration of just one hundredth of this. Fish may bioaccumulate mercury at tissue levels of 0.5 ppm, hence the data imply that environmentally plausible mercury levels may be affecting fish reproductive in the wild.The morphological of sperm hasn't received much attention, although research on mammals suggests that mercury, even in little doses, may harm the flagella. We've seen a similar effect on catfish sperm. Therefore, sperm morphology may offer a useful quantitative assessment of these developmental effects, and it has been used in preliminary experiments in our lab to offer such quantitative information on the effects of pollutants on the morphology of fish sperm.

Recent studies make it abundantly evident that figuring out the lowest sperm-to-egg ratio is crucial before investigating a pollutant's impacts because too many sperm can obscure those effects. Pollutants have the ability to kill 99% of the sperm in the presence of a 100 fold excess while yet leaving enough to fertilize the egg. However, has hypothesized that the majority of male fish only release the smallest amount of sperm necessary for fertilization, meaning that any decline in sperm quantity or quality will result in a decreased fertilization rate. It is much simpler to achieve natural fertilization with small fish, such as *danios*, although it may be more challenging to distinguish the effects of pollution on the eggs and sperm.

Effects on Oocytes

The easy access to large quantities of eggs allows it to quantify the effects of pollutants acting through the liver or ovary on egg size, quantity, and rate of fertilization. This is one of the main benefits of employing fish to monitor pollutant effects. Such studies can demonstrate the direct impact of pollutants on the oocyte as well as the cumulative impact of pollutants acting on the reproductive endocrine system components. For instance, mercury directly affects the micropyle and blocks sperm from entering.When huge numbers of batches of eggs need to be collected, measured, and analyzed for fertilization rates, experiments utilizing small fish, such as *danios*, that can be allowed to spawn naturally in small tanks, can be highly helpful but take a lot of time.

We are now looking into how image analysis might be used for this kind of measurement. It has been demonstrated through experiments with *danios* that even parental exposure to estradiol can result in a stop of development in the eggs that they generate. This is likely connected to the known cytostatic activity of estradiol.

Effects on Sexual Differentiation

In mammals, testicular secretions during the embryonic and neonatal stages are necessary for sexual differentiation. Aquaculture has made extensive use of the ability to reverse sex in fish by administering testosterone or estradiol to eggs or larvae.Therefore, it is reasonable to anticipate that exposure to environmental hormone mimics will have similar effects on eggs or larvae. There is proof that estrogenic xenobiotics such the detergents derivative nonylphenol and the organochlorine -BHC can induce intersex in medaka.

An Integrated Approach

Numerous environmental pollutants have been found to have particular impacts on fish reproductive system components. The pituitary and gonads, as well as the hormones they produce, the liver's breakdown of steroids and the production of sperm motility,vitellogenin, larval survival, fertilization rate,and abnormality rate are a few examples of such elements. Such investigations are frequently conducted in isolation and utilized as markers of pollution exposure without any clear knowledge of which components are impacted at what amounts. The most sensitive location for a specific contaminant throughout the entire reproductive system must be identified. If a species is exposed to this type of pollutant, we need to know which components are impacted at any given concentration so that progressively lower dosages will only damage a single location. This would allow us to determine the initial mechanism of action—whether it would be direct impact on the gametes, reduced pituitary or gonadal secretion, or liver function—as well as how this corresponds to ecologically realistic exposure. Although it is obvious that a comprehensive test of this kind is not feasible for all environmental pollutants, if used for one chemical from each major class (e.g., organochlorine and organophosphate pesticides, PCBs, estradiol, etc.), it would serve as the foundation for a more in-depth examination at this location for other similar chemicals. At present it is often assumed that the embryonic stage is the most sensitive, but this is based on limited data of standard toxicity tests and may not be valid at the lower levels characterized by specific endocrine and reproductive disruptors.

References

Arya, P., & Haq, S. A. (2019). Effects of xenobiotics and their biodegradation in marine life. In Smart Bioremediation Technologies 63-81.

Chandana, G. L., Kote, N. V., & SP, S. C. (2020). Recent Scenario of Impact of Xenobiotics on Marine Fish: An Overview. Pharmacognosy Journal, 12(6): 1797-1800.

Da Rocha, C.A.M., De Lima, P.D.L., Santos, R.A.D. and Burbano, R.M.R., 2009. Evaluation of genotoxic effects of xenobiotics in fishes using comet assay—a review. Reviews in Fisheries Science, 17(2), pp.170-173.

Di Paola, D., Capparucci, F., Lanteri, G., Cordaro, M., Crupi, R., Siracusa, R., & Peritore, A. F. (2021). Combined toxicity of xenobiotics Bisphenol A and heavy metals on zebrafish embryos (Danio rerio). Toxics, 9(12), 344.

Kime, D.E., 1999. A strategy for assessing the effects of xenobiotics on fish reproduction. Science of the total environment, 225(1-2), pp.3-11.

Kleinow, K.M., Melancon, M.J. and Lech, J.J., 1987. Biotransformation and induction: implications for toxicity, bioaccumulation and monitoring of environmental xenobiotics in fish. Environmental Health Perspectives, 71, pp.105-119.

Liu, H., Cui, H., Huang, Y., Gao, S., Tao, S., Hu, J., & Wan, Y. (2021). Xenobiotics targeting cardiolipin metabolism to promote thrombosis in zebrafish. Environmental Science & Technology, 55(6), 3855-3866.

Li, X., Richter, W. and Skinner, L.C., 2014. Xenobiotics in Fish from Lake Erie, the Niagara River, Cayuga Creek and Lake Ontario, New York. Bureau of Habitat.

Malakhova, L. V., Skuratovskaya, E. N., Malakhova, T. V., & Viktorovna, L. V. (2021). Relation of integral biochemical index and organochlorine xenobiotics content in the liver of the black scorpion fish Scorpaena porcus LINNAEUS, 1758 from Sevastopol bays and coastal areas. Biology, 13(4), 387-409.

Mesnage, R. (2021). Analytical strategies to measure toxicity and endocrine-disrupting effects of herbicides: Predictive toxicology. In Herbicides (pp. 273-289). Elsevier.

Musatadi, M., González-Gaya, B., Irazola, M., Prieto, A., Etxebarria, N., Olivares, M., & Zuloaga, O. (2020). Focused ultrasound-based extraction for target analysis and suspect screening of organic xenobiotics in fish muscle. Science of the Total Environment, 740, 139894.

Paul, N., Joseph, K., Ruth, W., & Jackson, K. (2021). Assessment of occurrence and concentrations of xenobiotics in selected fish species from lake Naivasha, Kenya.

Piwowarska, D., & Kiedrzyńska, E. (2021). Xenobiotics as a contemporary threat to surface waters. Ecohydrology & Hydrobiology.

Segner, H., Bailey, C., Tafalla, C. and Bo, J., 2021. Immunotoxicity of Xenobiotics in Fish: A Role for the Aryl Hydrocarbon Receptor (AhR)?. International Journal of Molecular Sciences, 22(17), p.9460.

Silvestre, F. (2020). Signaling pathways of oxidative stress in aquatic organisms exposed to xenobiotics. Journal of Experimental Zoology Part A: Ecological and Integrative Physiology, 333(6), 436-448.

Strobel, A., Lille-Langøy, R., Segner, H., Burkhardt-Holm, P., Goksøyr, A., & Karlsen, O. A. (2022). Xenobiotic metabolism and its physiological consequences in high-Antarctic Notothenioid fishes. Polar biology, 45(2), 345-358.

Teixidó, E., Leuthold, D., de Crozé, N., Léonard, M., & Scholz, S. (2020). Comparative assessment of the sensitivity of fish early-life stage, daphnia, and algae tests to the chronic ecotoxicity of xenobiotics: perspectives for alternatives to animal testing. Environmental toxicology and chemistry, 39(1), 30-41.

Tonelli, F. C. P., & Tonelli, F. M. P. (2020). Concerns and threats of xenobiotics on aquatic ecosystems. In Bioremediation and Biotechnology, Vol 3 (pp. 15-23). Springer, Cham.

Witczak, A., Harada, D., Aftyka, A., & Cybulski, J. (2021). Endocrine-disrupting organochlorine xenobiotics in fish products imported from Asia—an assessment of human health risk. Environmental Monitoring and Assessment, 193(3), 1-15.

Zaki, M.S., Hammam, A.M. and Nagwa, S., 2014. Ata. Xenobiotics as stressful factors in aquatic system (in fish). Life Sci J, 11(4), pp.188-197.